FAN HANDBOOK

Other McGraw-Hill Handbooks of Interest

FAN HANDBOOK

Selection, Application, and Design

Frank P. Bleier, P.E.

Consulting Engineer for Fan Design

McGRAW-HILL

New York   San Francisco   Washington, D.C.   Auckland   Bogotá
Caracas   Lisbon   London   Madrid   Mexico City   Milan
Montreal   New Delhi   San Juan   Singapore
Sydney   Tokyo   Toronto

Library of Congress Cataloging-in-Publication Data

Bleier, Frank P.
    Fan handbook : selection, application, and design / Frank P.
Bleier.
        p.   cm.
    Includes index.
    ISBN 0-07-005933-0 (alk. paper)
    1. Fans (Machinery)—Handbooks, manuals, etc.   I. Title.
TJ960.B58   1997
621.6'1—dc21   97-23697
                                                           CIP

McGraw-Hill

A Division of The McGraw·Hill Companies

1 2 3 4 5 6 7 8 9 0   DOC/DOC   9 0 2 1 0 9 8 7

ISBN 0-07-005933-0

*The sponsoring editor for this book was Harold B. Crawford, the editing
supervisor was Paul R. Sobel, and the production supervisor was Tina
Cameron. It was set in Times Roman by North Market Street Graphics.*

Printed and bound by R. R. Donnelley & Sons Company

McGraw-Hill books are available at special quantity discounts to use as
premiums and sales promotions, or for use in corporate training programs.
For more information, please write to the Director of Special Sales,
McGraw-Hill, 11 West 19th Street, New York, NY 10011. Or contact your
local bookstore.

CONTENTS

Chapter 5. Fan Laws   5.1

Chapter 6. System Resistance   6.1

Chapter 7. Centrifugal Fans   7.1

Chapter 8. Fan Selection, Specific Speed, Examples   8.1

Chapter 23. AMCA Standards

Chapter 24. Mechanical Strength

Chapter 25. Trouble Shooting and Problem Solving

Chapter 26. Installation, Safety, and Maintenance

FOREWORD

After receiving a degree in applied physics, Mr. Bleier worked for three fan manufacturers, first as a draftsman in Lyons, France, next as a test engineer in East Moline, Illinois, and then as a director of research in Chicago, Illinois. Since then, he has worked as a consulting engineer for 137 fan manufacturers, most of them in the United States, some in Canada, and one in Germany.

Over the years, Mr. Bleier has designed and tested close to 800 fans, among them most of the units pictured in this book. His designs ranged in size from a 4-in-diameter vaneaxial fan, used to ventilate a copy machine, to a 1000-hp, four-stage turbo blower, producing more than 300 in of static pressure, for pneumatic conveying of grain from cargo ships. Some of his other interesting assignments have included the design of low-noise exhaust fans for invasion ships used by the Navy during World War II and the development of a pressure blower used to pressurize the flotation bags of an Army tank to make it amphibious.

Mr. Bleier has written seven articles for technical magazines and for two engineering handbooks in simple, easy-to-understand language. He also has held twelve seminars at universities and for industrial groups. He is listed in *Who's Who in Engineering* and holds several patents on mixed-flow fans.

By sharing his broad experience with others, Mr. Bleier will help engineering students and people engaged in the design, manufacture, selection, application, and operation of fans. If you are active in one of these fields, you will benefit from reading this book.

Jerome R. Reich, Ph.D.
Chicago, Illinois

PREFACE

Let me say a few words about Robert Andrews Millikan. He was the American physicist who performed the so-called oil-drop experiment to determine the electric charge of an electron. Millikan also was a good teacher and was proud of it. He once made the statement: "I can explain anything to anybody." That's quite a statement. It impressed me. In writing this book, I have kept this statement in mind and have tried to produce an understandable text and to present some effortless reading material.

The story of fans is about airflow considerations, such as velocities, pressures, and turbulence losses. This book will give explanations of these concepts and present sample calculations to enable engineers and nonengineers to design fans and systems, to select and apply fans for systems, and to meet requirements for air volume, static pressure, brake horsepower, and efficiency.

If the reader is familiar with high school mathematics, he or she will be able to understand and apply the principles, graphs, and formulae presented here. Calculus and differential equations are not used in this book. Instead, a "feel" for aerodynamics will be developed gradually, a judgment of what an air stream will or will not do. The early chapters present the basics that will be needed to understand the principles discussed in later, more advanced chapters.

In grateful memory to

Mr. Archibald H. Davis

my former boss and teacher.

Frank P. Bleier
Chicago, Illinois

LIST OF SYMBOLS

Symbol	Meaning	Unit
AR	Air ratio	1
A	Area	ft²
A_a	Annular area	ft²
AR	Aspect ratio	1
α	Angle of attack	°
δ	Air angle past blade	°
ahp	Air horsepower	hp
BP	Barometric pressure	inHg
β	Relative air angle	°
β + α	Blade angle	°
l	Blade width	in
bhp	Brake horsepower	hp
cfm or Q	Rate of flow	ft³/min
D	Wheel diameter	in
d	Hub diameter	in
DB	Dry-bulb temperature	°F
WB	Wet-bulb temperature	°F
t	Temperature	°F
T	Absolute temperature	K
dB	Decibel sound level	dB
DR	Diffuser ratio	1
D_s	Specific diameter	in
N_s	Specific speed	min⁻¹
f	Frequency, musical note	s⁻¹
V	Volts	V
A	Amps	A
W	Watts	W
kW	Kilowatts	kW
ME	Mechanical or total efficiency	%
SE	Static efficiency	%
r	Radius	in
ρ	Air density	lb/ft³
Re	Reynolds number	
rpm	Revolutions per minute	min⁻¹
SP	Static pressure, positive	inWC, psi
SP	Static pressure, negative	inWC, inHg
TP	Total pressure	inWC
V	Vacuum	inHg
V	Volume	ft³
V	Velocity	fpm
VP	Velocity pressure	inWC
φ	Flow coefficient	
ψ	Pressure coefficient	

CONVERSION FACTORS

LENGTH

1 ft = 12 in   1 in = 0.0833 ft   1 yd = 3 ft = 36 in
1 mile = 1760 yd = 5280 ft

AREA

$1 \text{ ft}^2 = 144 \text{ in}^2$ $1 \text{ in}^2 = 0.00694 \text{ ft}^2$ $1 \text{ yd}^2 = 9 \text{ ft}^2 = 1296 \text{ in}^2$

VOLUME

$1 \text{ ft}^3 = 1728 \text{ in}^3$

VELOCITY

1 fpm (foot per minute) = 0.011364 mph (miles per hour)
1 mph = 88 fpm
Acceleration due to gravity $g = 32.17 \text{ ft/s}^2$

PRESSURE

1 inHg = 13.595 inWC = 0.4912 psi (lb/in^2) (used for high vacuums)
1 psi = 2.036 inHg = 27.68 inWC (used for high pressures)
1 atm (atmosphere) = 29.92 inHg = 406.8 inWC = 14.7 psi
1 inWC = 0.0736 inHg = 0.0361 psi

POWER

1 hp = 0.746 kW = 746 W = 42.42 Btu/min
1 kW = 1000 W = 1.341 hp = 56.89 Btu/min

TEMPERATURE

$1°F = (\frac{5}{9})°C$, but $32°F = 0°C$

Thus (example) $80°F = (80 - 32) \times \frac{5}{9} = 26.7°C$

Absolute temperature $T = °F + 459.7$

Example: $80°F = 539.7$ K (absolute temperature)

FAN HANDBOOK

CHAPTER 1

BASICS OF STATIONARY AND MOVING AIR

ATMOSPHERIC PRESSURE

Our planet earth has an average diameter of about 7914 mi or a radius of 3957 mi. It is surrounded by a comparatively thin layer of air. The air pressure is highest close to earth, due to compression by the weight of the air above. At higher altitudes, as the height of the air column above becomes less, the air pressure decreases, as shown in Fig. 1.1. At sea level, the atmospheric or barometric pressure is 29.92 inHg. At an altitude of 15 mi or 79,200 ft, which is only 0.4 percent of the earth's radius, the atmospheric pressure is only 1.00 inHg (3 percent of the sea level pressure). However, some rarified air extends about 500 mi up, which still is only 13 percent of the earth's radius.

The air consists mainly of nitrogen (about 78 percent by volume) and oxygen (about 21 percent by volume) plus less than 1 percent of other gases. Air is a physi-

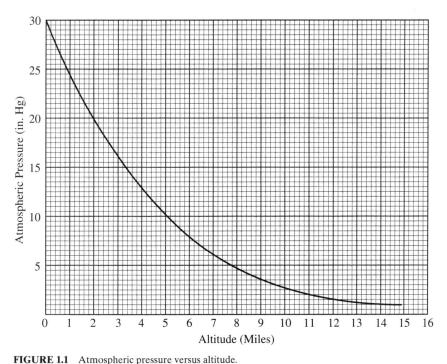

FIGURE 1.1 Atmospheric pressure versus altitude.

cal mixture (not a chemical compound) of these gases. Normally, air also contains some water vapor. This reduces the air density, as will be discussed in Chap. 18, on testing.

According to the National Advisory Committee for Aeronautics (NACA), later succeeded by the National Aeronautics and Space Administration (NASA), temperature, atmospheric pressure, and air density at various altitudes are as shown in Table 1.1.

TABLE 1.1 Temperature, Pressure, and Density versus Altitude

Altitude, ft	Temperature, °F	Atmospheric pressure, inHg	Air density, lbm/ft³
0	59.0	29.92	0.0765
1,000	55.4	28.86	0.0743
2,000	51.8	27.82	0.0721
3,000	48.4	26.81	0.0700
4,000	44.8	25.84	0.0679
5,000	42.1	24.89	0.0659
6,000	37.6	23.98	0.0639
7,000	34.0	23.09	0.0620
8,000	30.6	22.22	0.0601
9,000	27.0	21.38	0.0583
10,000	23.4	20.58	0.0565
11,000	19.8	19.79	0.0547
12,000	16.2	19.03	0.0530
13,000	12.6	18.29	0.0513
14,000	9.2	17.57	0.0497
15,000	5.5	16.88	0.0481
20,000	−12.3	13.75	0.0407
25,000	−30.1	11.10	0.0343
30,000	−48.1	8.88	0.0286
35,000	−65.8	7.04	0.0237
40,000	−69.7	5.54	0.0188
45,000	−69.7	4.35	0.0148
50,000	−69.7	3.42	0.0116
55,000	−69.7	2.69	0.0092
60,000	−69.7	2.12	0.0072
65,000	−69.7	1.67	0.0057

Source: Robert Jorgensen, *Fan Engineering Co.,* Buffalo, N.Y., Buffalo Forge.

The standards used in fan engineering are slightly different: Here the density used for standard air is 0.075 lbm/ft³. This is the density of dry air at an atmospheric pressure of 29.92 inHg or at 29.92 × 25.4 = 760 mmHg.

We are not discussing fans yet, but in order to get a comparative idea of fan pressure versus atmospheric pressure, let us anticipate for a moment and pretend that we know already what static pressure is and how much static pressure can be produced by certain types of fans.

Since 1 inHg equals 13.6 inWC (inches of water column), 29.92 inHg equals 13.6 × 29.92 = 406.8 inWC. In other words, the standard barometric pressure of 29.92 inHg also can be expressed as 406.8 inWC. This means that a fan producing a static pressure of 3 inWC (a good average value) will increase the absolute air pressure by less than 1 percent.

On the other hand, 1 inHg equals 0.491 lb/in² (psi). Therefore, 29.92 inHg equals $0.491 \times 29.92 = 14.7$ lb/in². In other words, the standard barometric pressure of 29.92 inHg also can be expressed as 14.7 lb/in². For high-pressure centrifugal units (either units running at very high speeds or multistage units), static pressure is usually measured in pounds per square inch (psi). Such units often produce as much as 7 lb/in². They will increase the absolute pressure by a significant 48 percent.

STATIC PRESSURE

Figure 1.2 shows a cylinder with a piston that can be moved up or down. It also shows a U-tube manometer indicating zero pressure. This means that the pressure below the piston is the same as the barometric pressure in the surrounding air. As the piston is moved down, the air volume below the piston is compressed, and the manometer will register a positive static pressure relative to the atmospheric pressure, which is considered zero pressure. This compressed air then has potential energy, i.e., the potential to expand to its original volume. If, on the other hand, the piston is raised, the air volume below the piston is expanded, and the manometer will register a negative static pressure relative to atmospheric pressure. This expanded air also has potential energy, i.e., the potential to contract to its original volume. This explains the concept of positive and negative static pressure in stationary air.

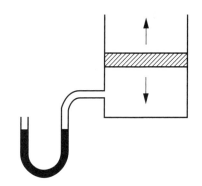

FIGURE 1.2 Cylinder with piston and manometer. As the piston moves, the static pressure below the piston will become either positive or negative.

Positive and negative static pressure exists in moving air as well as in stationary air. A fan blowing into a system (including such resistances as ducts, elbows, filters, dampers, and heating or cooling coils) produces positive static pressure, which is used to overcome the various resistances. A fan exhausting from a duct system produces negative static pressure, which again is used to overcome the resistance of the system.*

AIRFLOW THROUGH A ROUND DUCT OF CONSTANT DIAMETER, VELOCITY PRESSURE

Air flowing through a straight, round duct of constant diameter has a velocity distribution, as shown in Fig. 1.3, with the maximum air velocity near the center and with zero velocity at the duct wall. For small duct diameters of 6 to 10 in, and for air veloc-

---

* Some of the material in this section was taken from Bleier, F. P., Fans, in *Handbook of Energy Systems Engineering,* copyright © 1985 by John Wiley and Sons, New York.

ities of 1000 to 3000 ft/min (fpm), the average velocity V is approximately equal to 91 percent of the maximum velocity at the center. To find the average velocity in larger ducts and for larger air velocities, a so-called Pitot tube traverse across the duct is taken (Fig. 1.4). From the average velocity V (in feet per minute) and the cross-sectional area of the duct (in square feet), we can calculate the volume of air Q (in cubic feet per minute, or cfm) as

$$Q = A \times V \tag{1.1}$$

Furthermore, from the average velocity V (in feet per minute) and the air density d (in pounds mass per cubic foot), we can calculate the velocity pressure VP (in inches of water column, inWC) as

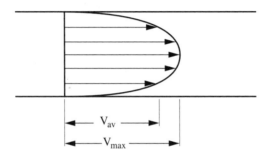

FIGURE 1.3 Velocity distribution airflow through a round duct.

FIGURE 1.4 A 27-in-diameter test duct with two supports for Pitot tube traverses and with a throttling device at the end of the duct.

$$VP = d\left(\frac{V}{1096.2}\right)^2 \tag{1.2}$$

or for standard air density of $d = 0.075$ lbm/ft³,

$$VP = \left(\frac{V}{4005}\right)^2 \tag{1.3}$$

Velocity pressure is the pressure we can feel when we hold our hand in the air stream. It represents kinetic energy.

For a straight, round duct with constant diameter and smooth walls, the friction loss f is

$$f = 0.0195 \frac{L}{D} VP \tag{1.4}$$

where f = pressure loss, in inches of water column
  L = length of the duct, in feet
  D = duct inside diameter, in feet
  VP = average velocity pressure, in inches of water column

Example: Let's consider a 100-ft-long duct, 2 ft = 24 in i.d., so that the duct area A = 3.14 ft². If the airflow through this duct is $Q = 8855$ cfm, the average air velocity will be $V = 8855/3.14 = 2819$ fpm = 47.0 fps, and the corresponding velocity pressure will be $VP = (2819/4005)^2 = 0.50$ inWC. The friction loss then will be $f = 0.0195$ (100/2)0.50 = 0.49 inWC.

Figure 1.5 shows a chart for determining friction loss in straight, round ducts. For our example, proceed as follows: On the horizontal abscissa on top, find the point for 8855 cfm (just slightly below 9000 cfm). From this point, move straight down until you reach the inclined line for a 24-in-diameter duct (pipe). This point will give you two results: (1) Another inclined line near this point indicates that the duct velocity will be slightly more than 2800 fpm. (2) Moving from that point straight across to the vertical ordinate at the right indicates a friction of 0.49 inWC for 100 ft of duct, the same as we obtained above from Eq. (1.4).

Reynolds' Number

If this airflow through the duct were laminar (smooth, streamline, free of eddies), the friction loss would be smaller than that just computed. Unfortunately, laminar airflow is seldom found in fan engineering. In most ventilating systems, the airflow is turbulent. Let's see whether this air flow through the 24-in i.d. duct is really turbulent. We can check this by calculating the Reynolds' number for this example. The English physicist Osborne Reynolds studied experimentally the flow of liquids and gases and arrived at a dimensionless parameter that is characteristic for certain flow conditions. The formula for this Reynolds' number is

$$\text{Re} = \frac{\rho V R}{\mu} \tag{1.5}$$

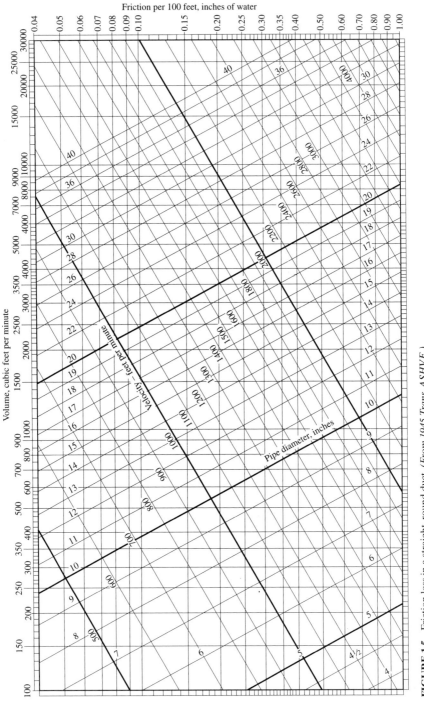

FIGURE 1.5 Friction loss in a straight, round duct. *(From 1945 Trans. ASHVE.)*

where ρ = gas density, in slugs per cubic foot
V = average air velocity, in feet per second
R = one-half the duct inside diameter, in feet
μ = coefficient of viscosity, in pounds second per square foot

For standard air, $\rho = 2.378 \times 10^{-3}$ and $\mu = 3.73 \times 10^{-7}$, resulting in

$$\text{Re} = 6375VR \qquad (1.5a)$$

For the preceding example, $V = 47.0$ and $R = 1$, resulting in Re = 300,000. Reynolds found that whenever Re is smaller than 1000 to 2000, the airflow will be laminar, but above 2000, it starts to become turbulent. Our example is far above this critical range. The transition from laminar to turbulent airflow is gradual. There are various degrees of turbulence, such as slightly turbulent, very turbulent, and extremely turbulent. In most fan applications, the air flow is slightly or very turbulent, and this has to be accepted. But extreme turbulence, as is found in a vaneaxial fan without an inlet duct and without a venturi inlet (see the section on venturi inlet), should be avoided.*

Total Pressure, Air Horsepower, Brake Horsepower, Efficiencies

Total pressure TP is defined as the sum of static pressure *SP* and velocity pressure *VP*:

$$TP = SP + VP \qquad (1.6)$$

In this equation, *VP* is always positive. *SP* and *TP* may be positive or negative. Here are three examples, illustrating this:

$SP = +2.2 \text{ in WC}$ $\qquad$ $SP = -0.5 \text{ in WC}$ $\qquad$ $SP = -1.4 \text{ in WC}$

$VP = 0.8 \text{ in WC}$ $\qquad$ $VP = 0.8 \text{ in WC}$ $\qquad$ $VP = 0.8 \text{ in WC}$

$TP = +3.0 \text{ in WC}$ $\qquad$ $TP = +0.3 \text{ in WC}$ $\qquad$ $TP = -0.6 \text{ in WC}$

Let us now consider another example: a fan having an outlet area $OA = 4.00 \text{ ft}^2$, blowing 16,000 cfm into a system, and producing 3 in *SP* in order to overcome the resistance of the system. The fan will have an average outlet velocity of

$$V = \frac{\text{cfm}}{OA} = \frac{16.000}{4.00} = 4000 \text{ fpm}$$

and a velocity pressure of

$$VP = \left(\frac{4000}{4005}\right)^2 = 1.00 \text{ in WC}$$

The total pressure will be $TP = 3.00 + 1.00 = 4.00$ in WC, and the power output of the fan (called *air horsepower,* ahp) will be

---

* Some of the material in this section was taken from Bleier, F. P., Fans, in *Handbook of Energy Systems Engineering,* copyright © 1985 by John Wiley and Sons, New York.

$$\text{ahp} = \frac{\text{cfm} \times TP}{6356} = 10.07 \text{ hp} \tag{1.7}$$

If the motor output (= fan input) is 15 brake horsepower (bhp), the fan efficiency at this point of operation will be the *mechanical efficiency* (also called *total efficiency*)

$$ME = TE = \frac{\text{ahp}}{\text{bhp}} = \frac{10.07}{15.0} = 0.67 = 67 \text{ percent} \tag{1.8}$$

Another efficiency that is sometimes used is called the *static efficiency*. It can be calculated as follows: First, we calculate the so-called static air horsepower (which, however, is not the real power output of the fan):

$$\text{ahp}_s = \frac{\text{cfm} \times SP}{6356} = \frac{16,000 \times 3}{6356} = 7.55 \text{ hp} \tag{1.7a}$$

And then we calculate the static efficiency:

$$SE = \frac{\text{ahp}_s}{\text{bhp}} = \frac{7.55}{15.0} = 0.50 = 50 \text{ percent} \tag{1.8a}$$

As you can see, the static efficiency is easier to calculate than the total efficiency because we do not have to calculate the velocity pressure first. For this reason, the static efficiency is sometimes used, even though it does not represent the real fan efficiency. The total or mechanical efficiency is the real fan efficiency.

Coming back to our example, to demonstrate the various types of efficiencies, let's assume that the motor input at this point of operation is 12.7 kW. The *motor efficiency* (or *electrical efficiency*) then will be

$$EE = \frac{0.746 \times \text{bhp}}{\text{kW}} = 0.88 = 88 \text{ percent} \tag{1.9}$$

This equation is used more often for calculating the brake horsepower when input (in kilowatts) and electrical efficiency are known:

$$\text{bhp} = \frac{\text{kW} \times EE}{0.746} = \frac{12.7 \times 0.88}{0.746} = 15.0 \tag{1.9a}$$

Finally, the efficiency of the set (fan plus motor), called the *set efficiency*, is

$$\text{Set eff.} = ME \times EE = 0.67 \times 0.88 = 0.59 = 59 \text{ percent} \tag{1.10}$$

In selecting a fan for a certain application, the fan efficiency is of great importance, because with a higher efficiency, we can obtain the same air horsepower with less power input. This not only will reduce the operating cost but also will save energy at the same time. High-efficiency fans, on the other hand, normally are more expensive, as shown in Fig. 1.6. It should be attempted, therefore, to find a balance between first cost and operating cost, taking into consideration that the first cost of the fan unit itself often is only a small portion of the system's total cost.*

---

* Some of the material in this section was taken from Bleier, F. P., Fans, in *Handbook of Energy Systems Engineering,* copyright © 1985 by John Wiley and Sons, New York.

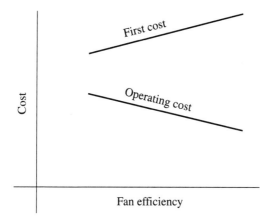

FIGURE 1.6 Cost versus fan efficiency. Selecting a fan of higher efficiency normally results in higher first cost, but in lower operating cost.

Scalars and Vectors

The physical quantities used to describe airflow phenomena can be divided into two groups: scalars and vectors. *Scalars* are quantities such as time, temperature, volume, and mass. They have only magnitude and can be added simply. *Vectors* are quantities such as force, velocity, and acceleration or deceleration. They have magnitude and direction and can be added only by way of vector diagrams, such as the velocity diagrams that will be discussed in later chapters.

AIRFLOW THROUGH A CONVERGING CONE

When cars on a crowded highway reach a point where the highway narrows, one of two things will happen: (1) the cars upstream will have to slow down, or (2) the cars past the point of constriction will have to speed up. Possibly both (1) and (2) will happen. In any case, the cars downstream will travel faster than the cars upstream, but obviously, the number of cars will remain the same.

A similar condition exists when air flows through a converging cone, as shown in Fig. 1.7. The air volume Q is

$$Q = A \times V \tag{1.1}$$

where Q = air volume, in cfm (ft^3/min)
A = duct area, in square feet (ft^2)
V = average air velocity, in fpm (ft/min)

As the airflow passes through the converging cone, the air volume (cfm) obviously will remain the same ahead and past the converging cone. This can be expressed as

$$Q_1 = Q_2 \quad \text{or} \quad A_1 \times V_1 = A_2 \times V_2 \tag{1.1b}$$

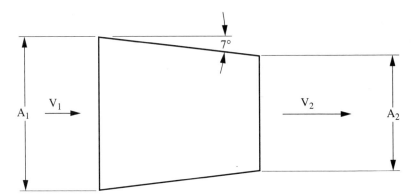

FIGURE 1.7   Air flowing through a converging cone: It accelerates.

Bernoulli's Principle of Continuity

This simple and obvious equation is called the *principle of continuity* because the air volume (cfm) continues to be the same before and after the point of constriction. As the air passes through the converging cone, it will accelerate from V_1 to a larger air velocity V_2 because the area A_2 is smaller than the area A_1.

Example: Let's again assume that the upstream duct i.d. is 24 in, so $A_1 = 3.14$ ft². Next let's assume that the downstream duct i.d. is 18 in, so $A_2 = 1.767$ ft². If $Q = 8855$ cfm, we get $V_1 = 8855/3.14 = 2819$ fpm and $VP_1 = 0.50$ inWC and $V_2 = 8855/1.767 = 5011$ fpm and $VP_2 = 1.57$ inWC.

This substantial increase in velocity pressure from 0.50 to 1.57 inWC, of course, will result in an increased kinetic energy, which will be obtained at the expense of a decreased static pressure. Basically, this is Bernoulli's theorem, which in its simplest form says: When the air velocity increases, the static pressure will decrease. This is easy to understand, but Bernoulli's theorem also says: When the air velocity decreases (as in a diverging cone), the static pressure will increase. This increase is called *static regain*. This is more difficult to understand. It will be discussed in more detail in the section on the diverging cone.

A converging cone past a fan is often used to increase the air penetration for such applications as snow blowing or comfort cooling.

A converging cone past the scroll housing of a centrifugal fan usually works without any problem. Care has to be taken, however, on a converging cone past an axial-flow fan because there often is an air spin past an axial-flow fan, even if it is a vaneaxial fan with guide vanes that are supposed to remove the air spin. If a little air spin remains past the fan, it is multiplied manyfold as the air travels to a smaller duct diameter because it tends to retain its circumferential component. As a result, at the smaller diameter, the revolutions per minute of the air spin becomes considerably larger, just like a watch chain spun around a finger turns faster and faster as the chain becomes shorter.

Here is an example illustrating the phenomenon that the air spin increases as the converging cone becomes smaller. Back in 1949, I designed and tested a 14-in vaneaxial fan with a 12-in hub diameter, resulting in an unusually large hub-tip ratio of 86 percent. This was done because the requirements were for a small airflow (cfm) and a high static pressure. A centrifugal fan would have been a better selection, but

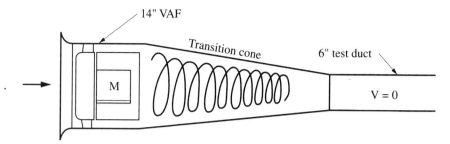

FIGURE 1.8 A 14-in vaneaxial fan with a 12-in hub diameter blowing into a converging transition cone and a 6-in test duct.

the customer insisted on a vaneaxial fan. Since the annular area was so small, I used a 6-in test duct past the unit (in order to get a good air velocity through the duct) plus a transition cone from 14 in down to 6 in, as shown in Fig. 1.8. This test setup, of course, was not in accordance with the Air Movement and Control Association (AMCA) test code, which requires that the test duct area be within 5 percent of the fan outlet area. A better test would have been on a nozzle chamber instead of on a test duct, but in 1949 very few companies had a nozzle chamber. To my surprise, I found zero air velocity in the 6-in test duct. The reason was that the remaining air spin past the fan became so strong in the transition cone that the friction path became excessive and used up all the static pressure available. After I put two longitudinal cross sheets into the transition cone to prevent the air spin, the proper airflow was restored and a fairly normal duct test could be run.

AIRFLOW THROUGH A DIVERGING CONE

As discussed earlier, a converging cone, as shown in Fig. 1.7, will produce an increased air velocity past the cone, resulting in increased kinetic energy, which is obtained at the expense of a decreased static pressure. By the same token, a diverging cone, as shown in Fig. 1.9, will produce a decreased air velocity past the cone, resulting in decreased kinetic energy. Will this difference in kinetic energy be lost?

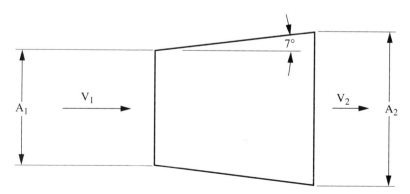

FIGURE 1.9 Air flowing through a diverging cone: It decelerates.

About half of it will be lost, mainly due to turbulence. The other half, believe it or not, will be regained by an increase in static pressure, as stated by Bernoulli's theorem, provided that the cone angle is small. About 7° or less on one side is recommended. While air normally flows from higher static pressure to lower static pressure, here is a case where the opposite takes place: Air is flowing from lower static pressure to higher static pressure.

Using the same dimensions and the same airflow as in Fig. 1.7, the velocity pressure now will decrease from 1.57 to 0.50 inWC, for a reduction of 1.07 inWC. One-half of this, or about 0.53 inWC, can be expected as a static regain. Such a regain is sometimes obtained by the use of a diffuser past a fan.

AERODYNAMIC PARADOX

Normally, as we go along with the airflow through a duct system, the static pressure is highest upstream and gradually decreases from there. This is the reason why the air flows. The high static pressure upstream forces the air through the duct, filters, etc. Therefore, it seems hard to believe that the static pressure will increase as the air passes through a diverging cone. It seems contrary to common sense. It seems paradoxical.

Let us discuss a device that might convince you of the truth of the preceding statements. It is called an *aerodynamic paradox* and is shown in Fig. 1.10*a*. It consists of a circular plate *A* with a pipe *B* on top. Another thin, lightweight disk *C* is suspended about ½ in below *A* in such a way that it can easily move up. If we blow into the pipe on top, we would expect that the air stream will blow the lower disk down. Actually, the lower disk will move up. Let me explain why.

The air stream leaving the pipe will turn 90° and move outward, since it has no other way to go. As it moves outward, the cross-sectional area becomes larger (as in

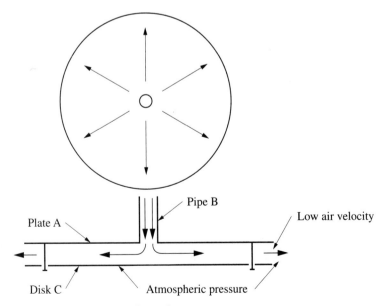

FIGURE 1.10*a*   Aerodynamic paradox.

a diverging cone), so the air velocity becomes smaller. As the air stream reaches the outside of the disks, the air velocity will be quite small, and the static pressure at that point will be close to the atmospheric pressure of the surrounding environment. Further inside, however, where the cross-sectional area is much smaller, the velocity of the air flowing outward is larger, and (per Bernoulli) the static pressure, therefore, is lower than the atmospheric pressure that pushes against the underside of the lower disk. As a result, the lower disk is lifted up against the upper plate.

The moment the lower disk touches the upper plate, the air stream is stopped, and the lower disk will drop again. The phenomenon then will repeat itself.

Conclusion

As the airflow passes through a system of ducts, converging and diverging cones, etc., the velocity pressure (kinetic energy) may increase or decrease and the static pressure (potential energy) also may increase or decrease. These two pressures are mutually convertible. However, the total pressure (total energy), being the sum of velocity pressure and static pressure, will always decrease, since it is gradually used up by friction and turbulence.

TENNIS BALL WITH TOP SPIN

Another example illustrating Bernoulli's principle is a tennis ball moving through the air with top spin, say, from right to left. To analyze the flow conditions, let's examine an equivalent configuration: The ball is spinning in place, and an air stream is moving from left to right, relative to the ball, as shown in Fig. 1.10*b*. On top of the ball, the rotation is opposite to the air velocity. This will slow down the velocity of the air passing over the top of the ball. In accordance with Bernoulli's principle, the slower air velocity will produce a higher pressure in this region. On the other hand, below the ball, the rotation is in the same direction as the air velocity. This will accelerate the velocity of the air passing over the underside of the ball and will produce a lower pressure in this region. As a result, the ball is pushed and pulled down. It therefore will drop faster than it would if it were only pulled down by gravity.

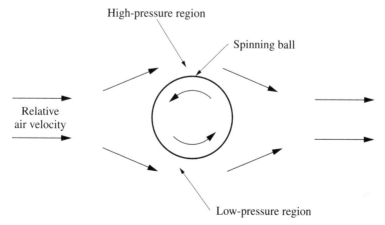

FIGURE 1.10*b*   A tennis ball with top spin will drop faster.

AIRFLOW THROUGH A SHARP ORIFICE, VENA CONTRACTA

When an air stream passing through a round duct of diameter D hits a sharp orifice with a hole of diameter d, a flow pattern as shown in Fig. 1.11 will develop because the upstream airflow will approach the edge of the opening at an inward angle rather than in an axial direction. Obviously, this angular velocity will continue past the orifice. This jet past the orifice will have a minimum diameter of about $0.6d$, and this minimum diameter will occur at a distance of about $0.5d$ past the orifice. After this point, the airflow will gradually spread out again, but only after a distance of $3d$ past the orifice will the airflow fill the duct "evenly," as shown in Fig. 1.3. This contraction of the air stream, shown in Fig. 1.12, is called *vena contracta* (contracted vein).

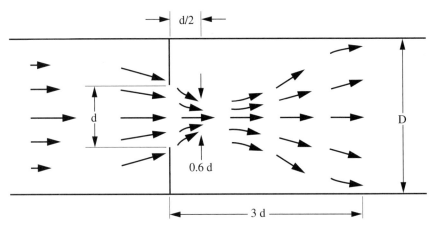

FIGURE 1.11   Airflow through a sharp orifice.

VENTURI INLET

A similar condition (although somewhat less extreme) exists when an airflow enters a round duct without a venturi inlet, as shown in Fig. 1.12. The reason why it is less extreme than in Fig. 1.12 is the upstream flow pattern. In Fig. 1.11, the approaching air is moving; in Fig. 1.12 it is hardly moving. Nevertheless, even in Fig. 1.12, a vena contracta exists, even though it is less pronounced.

Figure 1.13 shows the improved flow pattern obtained when the duct entrance is equipped with an inlet bell, also called a *venturi inlet*. This will reduce the duct resistance and increase the flow (cfm). For best results, the radius should be $r = 0.14D$ or more. If due to crowded conditions the radius has to be made smaller, the benefit will be reduced, but it will still be better than no venturi at all.

A venturi inlet is of particular importance at the entrance to an axial fan (as shown in Fig. 1.14) because without the venturi inlet the blade tips would be starved for air. In a vaneaxial fan, where the blades are short (due to a large hub), we can expect a flow increase of about 15 percent if a venturi inlet (or an inlet duct) is used.

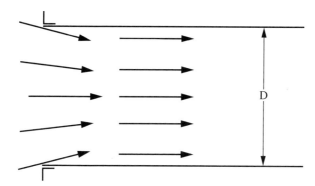

FIGURE 1.12   Airflow entering a round duct.

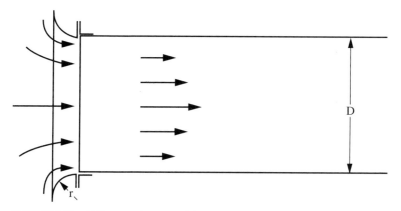

FIGURE 1.13   Airflow entering a round duct with a venturi inlet at the duct entrance.

In a propeller fan, where the blades are longer (since there is no hub or only a small hub), a flow increase of about 12 percent can be expected. Furthermore, the lack of a venturi inlet (when no inlet duct is used) will result in an increased noise level because the blade tips will operate in extremely turbulent air.

In centrifugal fans without an inlet duct, a venturi inlet will boost the flow by about 6 percent. The improvement here is somewhat less, for the following three reasons:

1. The turbulent airflow here will hit the leading edges (not the blade tips), which are moving at lower velocities.

2. Centrifugal fans normally run at lower speeds (rpms) than axial fans.

3. The flow pattern is different in centrifugal fans. The airflow makes a 90° turn before it hits the leading edges of the blades. The airflow ahead of the blades, therefore, contains some turbulence to begin with, and some additional turbulence, due to the lack of a venturi inlet, therefore, is less harmful.

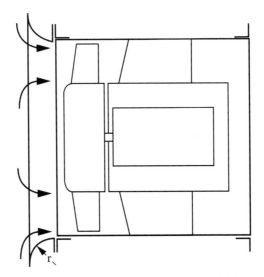

FIGURE 1.14   Venturi inlet ahead of a vaneaxial fan.

AIRFLOW ALONG A SURFACE

Let's look at the flow pattern of air passing through an elbow of rectangular cross section, as shown in Fig. 1.15. It is easy to understand that the airflow will tend to crowd on the inside of the outer wall. Call it inertia or centrifugal force, if you will. Obviously, it is mainly the outer wall of the elbow that keeps the air from flowing straight, as it would like to do, due to inertia.

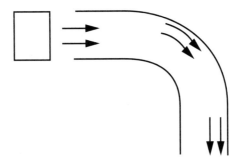

FIGURE 1.15   Airflow through a rectangular elbow.

Now let's look at the flow pattern of the air when the inner wall of the elbow has been removed, as shown in Fig. 1.16. The outer wall still keeps the air from flowing straight, and the flow pattern is quite similar to that of Fig. 1.15. Possibly, the air will crowd a little more near the outer wall.

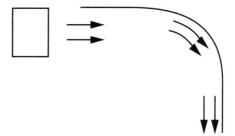

FIGURE 1.16   Airflow along the outer wall only.

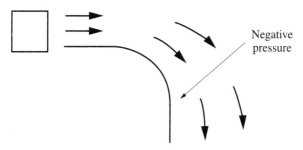

Negative
pressure

FIGURE 1.17   Airflow along the inner wall only.

Now let's go one step further and look at Fig. 1.17, which shows the flow pattern when the outer wall has been removed and only the inner wall has been retained. The air still will not flow straight to the right, as one might expect. It will still attempt to adhere to the inner wall. It may not be 100 percent successful in this attempt, but let's say it will be 70 percent successful. Why does the airflow turn at all, you ask? The reason is that a negative pressure will develop just outside the inner wall, and this negative pressure will tend to keep the airflow fairly close to the inner wall. Thus we can make the following statement: A curved wall will try to keep an airflow on its outside attached to itself. This is an important statement. We will come back to it in Chap. 2, on airfoils.

CHAPTER 2
AIRFOILS AND SINGLE-THICKNESS SHEET METAL PROFILES

DESCRIPTION AND FUNCTION OF AN AIRFOIL

An airfoil is a streamline shape, such as shown in Fig. 2.1. Its main application is as the cross section of an airplane wing. Another application is as the cross section of a fan blade. This is the application we will discuss now. There are symmetric and asymmetric airfoils. The airfoils used in fan blades are asymmetric. Figures 2.1 and 2.2 show an asymmetric airfoil that has been developed by the National Advisory Committee for Aeronautics (NACA). It is the NACA airfoil no. 6512. Table 2.1 shows the dimensions (upper and lower cambers) as percentages of the airfoil chord c. Let us make a list of the features shown in Figs. 2.1 and 2.2:

1. The airfoil has a blunt leading edge and a pointed trailing edge. The distance from leading edge to trailing edge is called the *airfoil chord c.*

2. The airfoil has a convex upper surface, with a maximum upper camber of 13.3 percent of c, occurring at about 36 percent of the chord C from the leading edge.

3. The airfoil has a concave lower surface, with a maximum lower camber of 2.4 percent of c, occurring at about 64 percent of the distance from the leading edge. In some airfoils used in fan blades, the lower surface is flat rather than concave.

4. The airfoil has a baseline, from which the upper and lower cambers are measured. The cambers are not profile thicknesses.

5. The angle of attack α is measured between the baseline and the relative air velocity.

6. As the airfoil moves through the air (whether it is an airplane wing or a fan blade), it normally produces positive pressures on the lower surface of the airfoil and negative pressures (suction) on the upper surface, similar to the phenomenon discussed with Fig. 1.17. While one might expect that the positive pressures do most of the work, deflecting the air stream, this is not the case. The suction pressures on the top surface are about twice as large as the positive pressures on the lower surface, but all these positive and negative pressures push and pull in approximately the same direction and reinforce each other.

The combination of these positive and negative pressures results in a force $\mathbf{F}$, as shown in Fig. 2.1. This force $\mathbf{F}$ can be resolved into two components: a lift force L (perpendicular to the relative air velocity) and a drag force D (parallel to the relative air velocity). The lift force L is the useful component. In the case of an airplane wing, L acts upward and supports the weight of the airplane. In the case of an axial fan blade, L (by reaction) deflects the air stream and produces the static pressure of

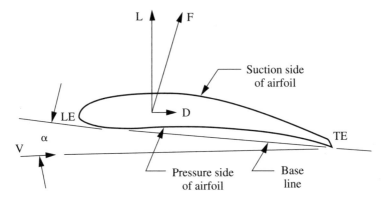

V Relative air velocity L Lift   D Drag   F Resultant force
α Angle of attack   LE Leading edge TE Trailing edge

FIGURE 2.1 Shape of typical airfoil (NACA no. 6512).

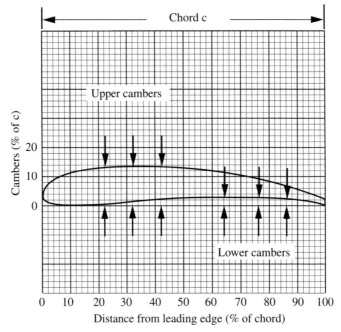

FIGURE 2.2 Dimensions of NACA airfoil no. 6512.

the fan. The drag D is the resistance to the forward motion of the airfoil. It is the undesirable, power-consuming component. We therefore would like to use airfoil shapes that have not only a high lift L but also a good lift-drag ratio L/D. As the angle of attack changes, lift, drag, and lift-drag ratio all change considerably, as will be seen in Fig. 2.4

TABLE 2.1 Dimensions of NACA Airfoil No. 6512

Distance from leading edge (% of chord)	Upper camber (% of chord)	Lower camber (% of chord)
0	2.71	2.71
1.25	5.25	1.34
2.50	6.21	0.83
5	7.60	0.32
7.5	8.65	0.09
10	9.51	0.00
15	10.89	0.05
20	11.88	0.28
30	13.00	0.95
40	13.23	1.61
50	12.71	2.13
60	11.51	2.37
70	9.66	2.29
80	7.18	1.88
90	4.06	1.10
95	2.25	0.58
100	0.24	0

INFLUENCE OF SHAPE ON AIRFOIL PERFORMANCE

The National Advisory Committee for Aeronautics (NACA) and the Göttingen Aerodynamische Versuchsanstalt have tested many airfoil shapes in wind tunnels in an attempt to find some shapes that will produce high lift forces and at the same time have good lift-drag ratios. These groups found, however, that these are conflicting requirements. As the cambers increase, the lift normally increases, too, but the lift-drag ratio tends to decrease. Selection of airfoil shapes therefore will depend on the application. In a high-pressure vaneaxial fan, we will use a high-cambered airfoil, particularly near the blade root. On the other hand, if fan efficiency is more important than high static pressure, we will use a low-cambered airfoil shape.

You may wonder why the leading edge of an airfoil is blunt. Wouldn't the drag be smaller if the leading edge were pointed, like the trailing edge? This would indeed be the case, if the relative air velocity at the leading edge were exactly tangential. However, this tangential condition would exist for only one operating condition (i.e., for one flow rate and static pressure). Over most of the performance range, the relative air velocity would deviate from the tangential condition, and this would result in turbulence and in an increased drag. Another reason for the blunt leading edge is structural strength.

LIFT COEFFICIENT, DRAG COEFFICIENT

From the test data for lift L and drag D obtained from wind tunnel tests, we can calculate the corresponding coefficients as follows:

$$\text{Lift coefficient } C_L = \frac{844L}{AV^2} \Bigg\} \quad \text{for standard air density} \qquad (2.1)$$

$$\text{Drag coefficient } C_D = \frac{844D}{AV^2} \qquad\qquad\qquad\qquad (2.2)$$

where L and D are in pounds, A is the area of the tested airfoil plate in square feet, V is the relative air velocity in feet per second, and C_L and C_D are dimensionless coefficients. From these formulas we note that $C_L/C_D = L/D$. In other words, the lift-drag ratio is also the ratio of the corresponding coefficients.

CHARACTERISTIC CURVES OF AIRFOILS

As mentioned earlier, many airfoil shapes have been tested in wind tunnels. The airfoil plates tested by NACA usually have an airfoil chord of 5 in and a length of 30 in, as shown in Fig. 2.3. This is called an *aspect ratio* of 6.

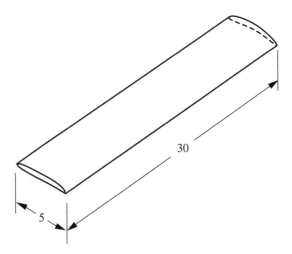

FIGURE 2.3 Airfoil plate for wind tunnel test, aspect ratio of 6.

Figure 2.4 shows the characteristic curves NACA obtained for its airfoil no. 6512 for an aspect ratio of 6. Note that the lift coefficient is much larger than the drag coefficient so that the lift-drag ratios are in the range of about 10 to 20.

For use in fan blades, the characteristic curves have to be converted from an aspect ratio of 6 to an infinite aspect ratio. This conversion further increases the lift-drag ratio so that the maximum lift-drag ratio will about triple. Table 2.2 shows the calculation for this conversion. Figure 2.5 shows the resulting characteristic curves for an infinite aspect ratio.

Please note the considerable difference between Figs. 2.4 and 2.5. Figure 2.5 (infinite aspect ratio) shows lower drag, resulting in higher lift-drag ratios. The reason for this considerable difference is turbulence at the two ends of the airfoil plate, which

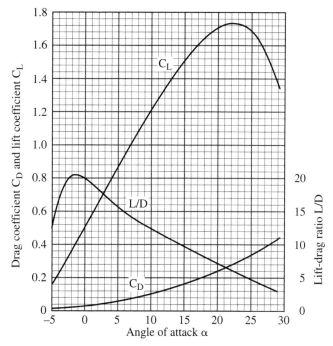

FIGURE 2.4 Characteristic curves for NACA airfoil no. 6512 (size of plate: 5 × 30 in; aspect ratio: 6).

occurs only for finite aspect ratios but is eliminated for an infinite aspect ratio and therefore also for axial-flow fans, where the fan blades are bordered by the fan hub and the housing wall so that no turbulence at the ends can develop. In designing axial-flow fans, therefore, one should use the airfoil's characteristic curves for an infinite aspect ratio.

Looking again at the characteristic curves of the NACA airfoil no. 6512 for an infinite aspect ratio, as shown in Fig. 2.5, we note the following points:

TABLE 2.2 Calculation for Converting the Characteristic Curves from an Aspect Ratio of 6 to an Infinite Aspect Ratio: NACA Airfoil No. 6512

α_6	L/D	C_L	C_{D6}	$\Delta\alpha$	α_∞	ΔC_D	$C_{D\infty}$	$C_L/C_{D\infty}$
−5	12.3	0.16	0.013	0.49	−5.5	0.001	0.012	13.3
−1	22.4	0.47	0.021	1.43	−2.4	0.012	0.009	52.2
3	17.8	0.75	0.042	2.28	0.7	0.030	0.012	62.5
7	13.8	1.01	0.073	3.07	3.9	0.054	0.019	53.2
15	9.5	1.52	0.160	4.62	10.4	0.123	0.037	41.1
23	5.9	1.74	0.295	5.29	17.7	0.161	0.134	13.0
29	3.2	1.40	0.43	4.26	24.7	0.104	0.326	4.3

Note: $\Delta\alpha = 18.24 C_L \times \frac{1}{6} = 3.04 C_L$; $\alpha_\infty = \alpha_6 - \Delta\alpha$; $\Delta C_D = C_L^2/\pi \times \frac{1}{6} = 0.05305 C_L^2$; $C_{D\infty} = C_{D6} - \Delta C_D$.

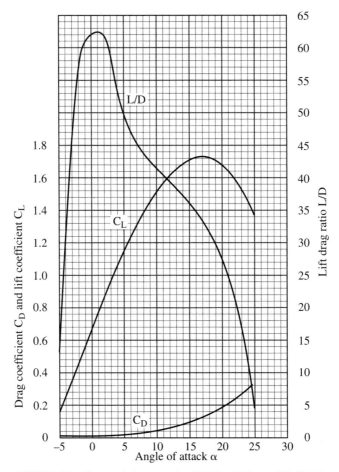

FIGURE 2.5   Characteristic curves for NACA airfoil no. 6512 (infinite aspect ratio).

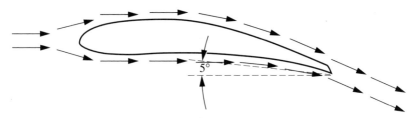

FIGURE 2.6   At a 5° angle of attack, the airflow is smooth and follows the contours of the airfoil. The direction of the airflow is deflected by 13.5°.

1. The lift coefficient is zero at an angle of attack of about −8°. (If this were a symmetric airfoil, the zero lift coefficient would occur at zero angle of attack.)

2. As the angle of attack is increased, the lift coefficient rises, until it reaches a maximum of about 1.7 at an angle of attack of about 15°. This is the top of the operating range for this airfoil.

3. The lift-drag ratio has its maximum of 62.4 at an angle of attack of 1°. The best operating range, then, will be at angles from 1° to about 10°, where the L/D ratio is still good (between 62 and 41) and the airflow is smooth, as shown in Fig. 2.6.

4. For angles of attack from 10° to 15°, the airflow can still follow the contour of the airfoil, but the fan efficiency will be somewhat impaired because of the lower lift-drag ratios.

5. For angles of attack larger than 15°, the airfoil will stall, resulting in a decrease in the lift coefficient. At these large angles of attack, the airflow can no longer follow the upper contour of the airfoil. It will separate from that contour, as shown in Fig. 2.7.

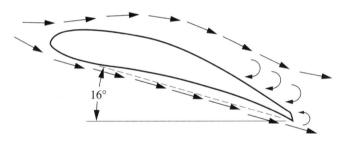

FIGURE 2.7 At a 16° angle of attack, the airfoil stalls, and separation of airflow takes place at the trailing edge and at the suction side of the airfoil, with small eddies filling the suction zones. The deflection of the airflow past the trailing edge is close to zero.

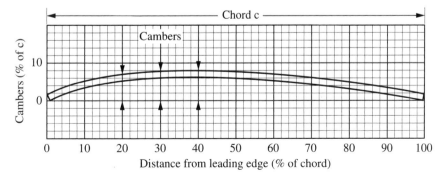

FIGURE 2.8 Dimensions of Göttingen single-thickness profile no. 417a.

Please note that angle of attack is not identical with blade angle. The blade angle of an axial-flow fan is much larger than the angle of attack, as will be discussed in Chap. 4.

SINGLE-THICKNESS SHEET METAL PROFILES

It sometimes is desirable to use single-thickness sheet metal blades rather than airfoil blades. The reason might be a dust-laden airflow or simply lower cost. The

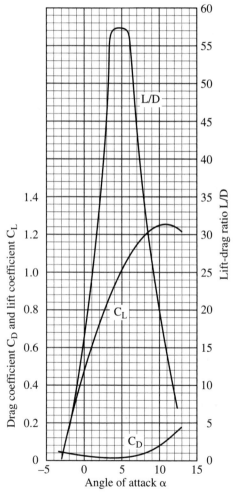

FIGURE 2.9 Performance of Göttingen single-thickness profile no 417a (infinite aspect ratio).

shape and characteristic curves for such a single-thickness profile are given in a Göttingen publication. Göttingen calls it profile no. 417a. Figure 2.8 shows the shape and Figure 2.9 shows the characteristic curves, converted to an infinite aspect ratio. Comparing these with the shape and characteristic curves for the NACA airfoil no. 6512, as shown in Figs. 2.2 and 2.5, we find the following similarities and differences:

1. The single-thickness profile has a maximum camber of 8 percent, which is about halfway between the upper (13.3 percent at 36 percent) and lower (1.3 percent at 36 percent) cambers of NACA airfoil no. 6512.

2. The maximum camber of the single-thickness profile is located at 38 percent of the chord from the leading edge, about the same as for NACA airfoil no. 6512.

3. The maximum lift coefficient for profile no. 417a is lower, 1.25 instead of 1.74.

4. The maximum lift-drag ratio is somewhat lower, 57 instead of 62.

5. The angle of attack at which the maximum lift coefficient occurs is considerably lower, 10° instead of 17°. This results in a narrower operating range and particularly in a narrower range for good lift-drag ratios. For example, the range for lift-drag ratios of 35 or more is 21° wide for NACA airfoil no. 6512 but only 6° wide for the single-thickness profile.

Despite these disadvantages, single-thickness profiles are often used in fan blades, especially in propeller fans and in tubeaxial fans.

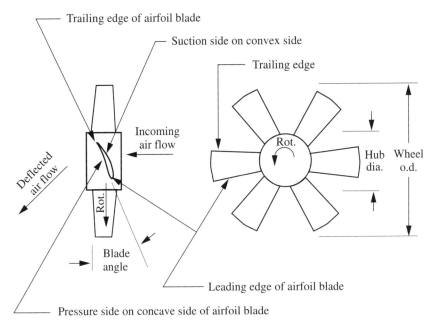

FIGURE 2.10   Airfoil as used in an axial-flow fan blade.

FUNCTION OF AIRFOIL BLADES IN AXIAL AND CENTRIFUGAL FANS

Let us now examine how the airfoil is used as the cross section of a fan blade. Figure 2.10 shows how an airfoil is used in an axial-flow fan blade. Here the concave side of the airfoil is the pressure side (just like in an airplane wing), so this is a normal condition.

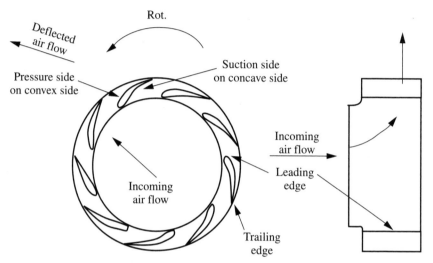

FIGURE 2.11   Airfoil as used in a backwardly curved centrifugal fan blade. Note that here the pressure side is on the *convex* side of the airfoil blade.

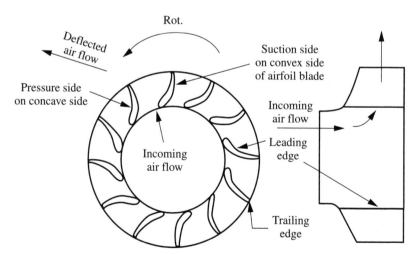

FIGURE 2.12   Airfoil as used in a centrifugal fan with radial-tip blades, a design that is rarely used but results in good efficiencies.

Figure 2.11 shows how the airfoil is used in a centrifugal fan with backwardly curved blades. Here the convex side of the airfoil is the pressure side, which, of course, is abnormal. This type of blade does not really have the function of an airfoil, since there is no lift force in the conventional sense. It is simply a backwardly curved blade with a blunt leading edge that helps broaden the range of good efficiencies.

Figure 2.12 shows how an airfoil could be used in a centrifugal fan with radial-tip blades. Here the concave side is the pressure side, as it should be. However, this configuration is rarely used, partly because of higher cost and partly because radial-tip blades are often used for handling dust-laden air and this is done better by thick, single-thickness blades.

CHAPTER 3
TYPES OF FANS, TERMINOLOGY, AND MECHANICAL CONSTRUCTION

SIX FAN CATEGORIES

This book will discuss the following six categories of fans:

1. Axial-flow fans
2. Centrifugal fans
3. Axial-centrifugal fans
4. Roof ventilators
5. Cross-flow blowers
6. Vortex or regenerative blowers

AXIAL-FLOW FANS

There are four types of axial-flow fans. Listed in the order of increasing static pressure, they are

1. Propeller fans (PFs)
2. Tubeaxial fans (TAFs)
3. Vaneaxial fans (VAFs)
4. Two-stage axial-flow fans

Propeller Fans

The *propeller fan,* sometimes called the *panel fan,* is the most commonly used of all fans. It can be found in industrial, commercial, institutional, and residential applications. It can exhaust hot or contaminated air or corrosive gases from factories, welding shops, foundries, furnace rooms, laboratories, laundries, stores, or residential attics or windows.

Sometimes several propeller fans are installed in the walls of a building, operating in parallel and exhausting the air. Figure 3.1 shows two propeller fans with

FIGURE 3.1 Two 21-in propeller fans with direct drive, mounted in a wall and exhausting air from a factory building.

direct drive mounted in the wall of a factory, near the ceiling where the hot air is located.

Figures 3.2 and 3.3 show the general configuration for a propeller fan with a belt drive from an electric motor. The units consist of the following eleven components: a spun venturi housing, a bearing base (plus braces), two bearings, a shaft, a motor base, an electric motor, two pulleys, a belt, and a fan wheel. In Fig. 3.2 the fan wheel

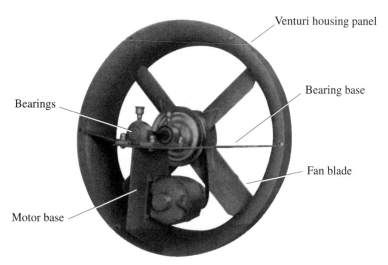

FIGURE 3.2 A 24-in propeller fan with belt drive. Note the location of the motor opposite the rotating blades.

has four blades, and the motor is mounted on a separate, vertical motor base. In Fig. 3.3 the fan wheel has six blades, and a horizontal base supports both the motor and the two bearings. In both figures, however, the motor is located opposite the rotating fan blades. This results in good motor cooling but some obstruction to the airflow.

Figure 3.4 shows the simpler configuration of a propeller fan with direct drive from an electric motor. This unit consists of only four components: a spun venturi housing, a motor base (plus braces), an electric motor, and a four-bladed fan wheel.

The belt-drive arrangement has the following three advantages:

1. It results in flexibility of performance, since any speed (rpms) can be obtained for the fan wheel by selection of the proper pulley ratio. However, when the speed is increased to boost the flow (cfm), the brake horsepower will increase even more, as the third power of the rpm ratio, as will be explained later.

2. In large sizes, belt drive is preferable, since it will keep the speed of the fan wheel low or moderate while keeping the motor speed high, for lower cost. (High-speed motors are less expensive than low-speed motors of the same horsepower.)

3. The motor will get good cooling from the air stream passing over it.

The direct-drive arrangement has the following five advantages:

1. It has a lower number of components, resulting in lower cost.

2. It requires no maintenance and regular checkups for adjustment of the belt.

3. It has a better fan efficiency, since a belt drive would consume an extra 10 to 15 percent of the brake horsepower.

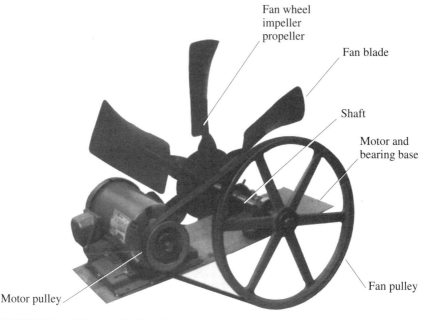

FIGURE 3.3 A 36-in propeller fan with belt drive. Note the large pulley ratio for a low speed of the fan wheel. *(Courtesy of Chicago Blower Corporation, Glendale Heights, Ill.)*

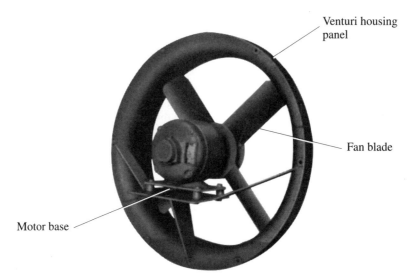

Venturi housing
panel

Fan blade

Motor base

FIGURE 3.4   A 24-in propeller fan with direct drive. Note the height adjustment of motor base for an even tip clearance.

4. It results in more flow (cfm) because the central location of the motor does not obstruct the airflow.

5. The performance flexibility of the belt-drive arrangement also can be obtained, but at an extra cost, by means of adjustable-pitch blades and by a variation in the number of blades. A 3° increase in the blade angle will result in a 10 to 15 percent increase in flow (15 percent in the range of small blade angles, 10 percent for larger blade angles). The static pressure can be boosted by an increase in the number of blades, up to a point.

Conclusion: Direct drive is less expensive and more efficient. It is preferable in small sizes. Belt drive is preferable in large sizes and results in better performance flexibility than direct drive, unless adjustable-pitch blades are used.

Figure 3.5 shows a 46-in propeller fan wheel of aluminum with a 13-in-diameter hub and with eight narrow airfoil blades welded to the hub. The hub-tip ratio is 0.28, a good ratio for a propeller fan. This is an efficient but expensive propeller-fan wheel. Most propeller-fan wheels have sheet metal blades riveted to a so-called spider, as shown in Fig. 3.6. This is a lightweight, lower-cost construction that is somewhat less efficient but adequate in small and medium sizes. Many propeller-fan wheels are plastic molds. In very small sizes, where cost is more important than efficiency, one-piece stampings are sometimes used.

Shutters.   Most propeller fans are used for exhausting from a space. They are mounted on the inside of a building, with the motor located on the inlet side, inside the building, and the air stream blowing outward. A shutter is mounted on the outside. There are two types of shutters: automatic shutters and motorized shutters.

Figure 3.7 shows an automatic shutter having three shutter blades linked together and mounted on hinged rods. The shutter will be opened by the air stream on start-up of the fan. It will be closed by the weight of the shutter blades when the fan is turned off. The motorized shutter, used mainly in larger sizes, is opened and

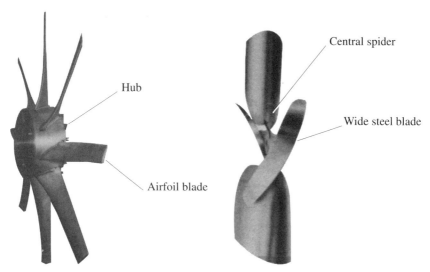

FIGURE 3.5 Cast-aluminum propeller-fan wheel with a 46-in o.d., 13-in hub diameter, and eight narrow airfoil blades welded to hub.

FIGURE 3.6 A 24-in propeller-fan wheel with four wide steel blades riveted to central spider.

closed by a separate small motor mounted on the shutter frame. When the shutter is closed, it will prevent heat losses due to backdraft and keep out wind, rain, and snow.

Screen Guards. Figure 3.8 shows a fan guard, sometimes fitted over the motor side of a propeller fan for safety, whenever the fan is installed less than 7 ft from the floor. It uses a steel mesh, designed for minimum interference with the air stream.

FIGURE 3.7 Automatic shutter having a square frame and three shutter blades linked together for simultaneous opening and closing.

FIGURE 3.8 Fan guard, used for safety and mounted on the inlet side of a propeller fan.

Man Coolers.   Another application for propeller fans (besides exhausting from a space) is for cooling people (as the term *man coolers* implies) or products or for supplying cool air to certain processes. These cooling fans are located in hot places, such as steel mills, foundries, and forge plants. They are also used for cooling down furnaces for maintenance work, for cooling electrical equipment (such as transformers, circuit breakers, and control panels), or for drying chemical coatings.

Figure 3.9 shows a man cooler mounted on a heavy pedestal for stability. It has a lug on top so that it can be moved easily to various locations. It has a 30-in propeller-fan wheel with six narrow blades and direct drive from a 3-hp, 1740-rpm motor.

Figure 3.10 shows a similar type of man cooler. It has a 30-in propeller-fan wheel with eight narrow blades and direct drive from a 3-hp, 1150-rpm motor. It has a bracket for mounting it on a wall, up high enough so that it cannot be damaged by trucks and the wires cannot be cut by wheels. As a special feature, this unit has a conical discharge nozzle that boosts the outlet velocity for deeper penetration. The nozzle contains some straightening vanes, almost like a vaneaxial fan, to prevent excessive air spin at the narrow end of the cone.

FIGURE 3.9 Man cooler mounted on a pedestal for cooling people, products, or processes. *(Courtesy of Coppus Engineering Division, Tuthill Corporation, Millbury, Mass.)*

FIGURE 3.10 Man cooler with a bracket for mounting it on a wall, with conical discharge nozzle for deeper penetration. *(Courtesy of Bayley Fan Group, Division of Lau Industries, Lebanon, Ind.)*

Tubeaxial Fans

Figure 3.11 shows a tubeaxial fan with direct drive from an electric motor. It has a cylindrical housing and a fan wheel with a 33 percent hub-tip ratio and with ten blades that may or may not have airfoil cross sections. The best application for tubeaxial fans is for exhausting from an inlet duct. A short outlet duct can be tolerated, but the friction loss there will be larger than normal because of the air spin. If no inlet duct is used, a venturi inlet is needed to prevent a 10 to 15 percent loss in flow and an increased noise level. Figure 3.11 shows the motor on the inlet side, but it could be located on the outlet side as well.

In case of belt drive, the motor is located outside the cylindrical housing, and a belt guard is needed. Direct drive has fewer parts and therefore lower cost, the same as for propeller fans, and the performance flexibility again can be obtained by means of adjustable-pitch blades.

Vaneaxial Fans

Figure 3.12 shows a vaneaxial fan with belt drive from an electric motor. It has a cylindrical housing (like a tubeaxial fan) and a fan wheel with a 46 percent hub-tip ratio and with nine airfoil blades. It also has eleven guide vanes, neutralizing the air spin, so that the unit can be used for blowing (outlet duct) as well as for exhausting (inlet duct). Again, direct drive is simpler and less expensive than belt drive. Also, performance flexibility for direct drive can be obtained by means of adjustable-pitch blades. Again, a venturi inlet is needed if no inlet duct is used.

Figure 3.13 shows an axial-flow fan wheel with a 42 percent hub-tip ratio and with eight single-thickness steel blades. It could be used in a tubeaxial fan or in a vaneaxial fan.

Cylindrical housing

Access door

Fan wheel
impeller
propeller

FIGURE 3.11   Tubeaxial fan with direct drive from an electric motor on the inlet side (in the background) and with a fan wheel having a 33 percent hub-tip ratio and with ten blades. *(Courtesy of General Resource Corporation, Hopkins, Minn.)*

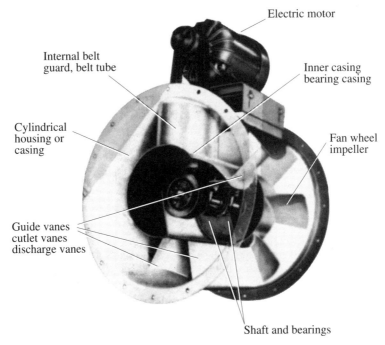

Electric motor

Internal belt
guard, belt tube

Inner casing
bearing casing

Cylindrical
housing or
casing

Fan wheel
impeller

Guide vanes
cutlet vanes
discharge vanes

Shaft and bearings

FIGURE 3.12 Vaneaxial fan with belt drive from an electric motor and with a fan wheel having a 46 percent hub-tip ratio and nine airfoil blades. *(Courtesy of General Resource Corporation, Hopkins, Minn.)*

Figure 3.14 shows a vaneaxial fan wheel with a 64 percent hub-tip ratio and with five wide airfoil blades. This hub-tip ratio would be too large for a tubeaxial fan but is quite common in vaneaxial fans.

Two-Stage Axial-Flow Fans

Two-stage axial-flow fans have the configuration of two fans in series so that the pressures will add up. This is an easy solution when higher static pressures are needed, but excessive tip speeds and noise levels cannot be tolerated. The two fan wheels may rotate in the same direction, with guide vanes between them. Or they may be counterrotating, without any guide vanes, as will be explained in more detail in Chap. 4.

CENTRIFUGAL FANS

There are six types of centrifugal fan wheels in common use. Listed in the order of decreasing efficiency, they are

1. Centrifugal fans with airfoil (AF) blades
2. Centrifugal fans with backward-curved (BC) blades

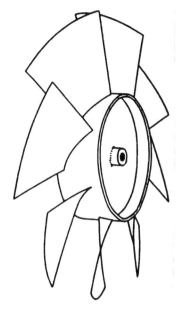

FIGURE 3.13   Axial-flow fan wheel with a 42 percent hub-tip ratio and with eight single-thickness steel blades for use as a tubeaxial fan wheel or as a vaneaxial fan wheel.

3. Centrifugal fans with backward-inclined (BI) blades

4. Centrifugal fans wirh radial-tip (RT) blades

5. Centrifugal fans with forward-curved (FC) blades

6. Centrifugal fans with radial blades (RBs)

These six types are used in a variety of applications, as will be discussed in more detail in Chap. 7.

Centrifugal Fans with AF Blades

The centrifugal fan with AF blades has the best mechanical efficiency and the lowest noise level (for comparable tip speeds) of all centrifugal fans. Figures 3.15 and 3.16 show two constructions for centrifugal fan wheels with AF blades. Figure 3.15 shows hollow airfoil blades, normally used in medium and large sizes. Figure 3.16 shows cast-aluminum blades, which are often used in small sizes and for testing and devel-

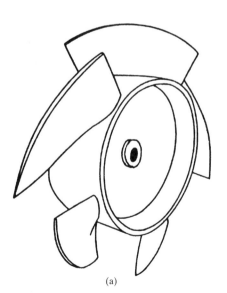
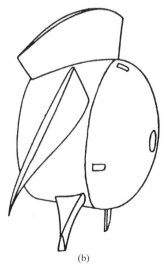

(a)   (b)

FIGURE 3.14   Vaneaxial fan wheel with a 64 percent hub-tip ratio and with five wide cast-aluminum airfoil blades: (*a*) without inlet hood; (*b*) with inlet hood.

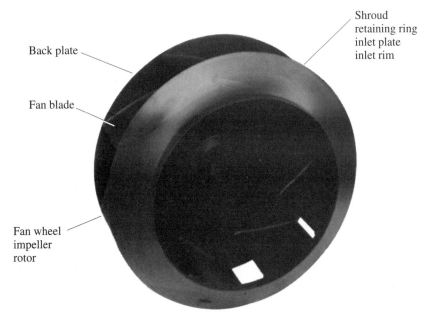

Back plate

Fan blade

Fan wheel
impeller
rotor

Shroud
retaining ring
inlet plate
inlet rim

FIGURE 3.15 Centrifugal fan wheel, SISW, with nine hollow airfoil steel blades welded to back plate and shroud. *(Courtesy of General Resource Corporation, Hopkins, Minn.)*

FIGURE 3.16 Experimental centrifugal fan wheel, SISW, with eleven cast-aluminum airfoil blades welded to the back plate but not yet welded to the shroud (held above the blades).

opment work, with the shroud held above the airfoil blades prior to welding it to the blades.

Centrifugal Fans with BC Blades

BC blades are single-thickness steel blades but otherwise are similar to AF blades with respect to construction and performance. They have slightly lower efficiencies but can handle contaminated air streams because the single-thickness steel blades can be made of heavier material than can be used for hollow airfoil blades.

Centrifugal Fans with BI Blades

Figure 3.17 shows a sketch of an SISW (single inlet, single width) centrifugal fan wheel with BI blades. These are more economical in production, but they are somewhat lower in structural strength and efficiency. Figure 3.18 shows the same fan wheel in a scroll housing. Figure 3.19 shows a BI centrifugal fan with scroll housing, noting the terminology for the various components.

Incidentally, scroll housings are not always used in connection with centrifugal fan wheels. Centrifugal fan wheels also can be used without a scroll housing, in such applications as unhoused plug fans, multistage units, and roof ventilators. An exception is FC centrifugal fan wheels. They require a scroll housing for proper functioning, as will be explained in Chap. 7.

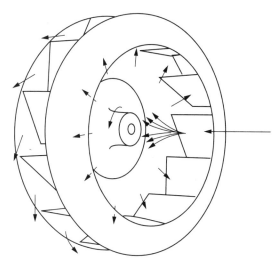

FIGURE 3.17 Centrifugal fan wheel, SISW, with BI blades welded to back plate and shroud.

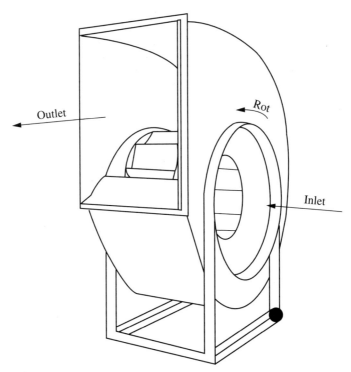

FIGURE 3.18 Angular view of centrifugal fan with BI wheel inside showing scroll housing, inlet side, and outlet side with cutoff all the way across housing width.

Centrifugal Fans with RT Blades

RT blades are curved, with good flow conditions at the leading edge. Only the blade tips are radial, as the term *radial-tip blades* indicates. Figure 3.20 shows a radial-tip centrifugal fan wheel. These RT wheels are used mainly in large sizes, with wheel diameters from 30 to 60 in, for industrial applications, often with severe conditions of high temperature and light concentrations of solids.

Centrifugal Fans with FC Blades

FC blades, as the name indicates, are curved forward, i.e., in the direction of the rotation. This results in very large blade angles and in flow rates that are much larger than those of any other centrifugal fan of the same size and speed. Figure 3.21 shows a typical SISW FC fan wheel, with many short blades and a flat shroud with a large inlet diameter for large flows. These fans are used in small furnaces, air conditioners, and electronic equipment, whenever compactness is more important than efficiency.

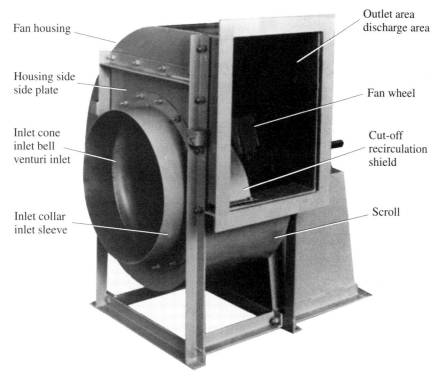

FIGURE 3.19 Angular view of centrifugal fan with BI wheel inside showing scroll housing, inlet side (inlet collar and inlet cone), and outlet side with recirculation shield opposite the inlet cone only. *(Courtesy of General Resource Corporation, Hopkins, Minn.)*

Centrifugal Fans with Radial Blades

Radial blades (RBs) are rugged and self-cleaning, but they have comparatively low efficiencies because of the nontangential flow conditions at the leading edge. Figure 3.22 shows an SISW RB fan wheel with a back plate but without a shroud. Sometimes even the back plate is omitted (open fan wheel), and reinforcement ribs are added for rigidity. These fans can handle not only corrosive fumes but even abrasive materials from grinding operations.

AXIAL-CENTRIFUGAL FANS

These fans are also called *tubular centrifugal fans, in-line centrifugal fans,* or *mixed-flow fans* (especially if the fan wheel has a conical back plate). The following two types of fan wheels are used in these fans:

1. A fan wheel with a flat back plate, as shown in Fig. 3.17, i.e., the same type as is used in a scroll housing. When used in an axial-centrifugal fan, however, the air

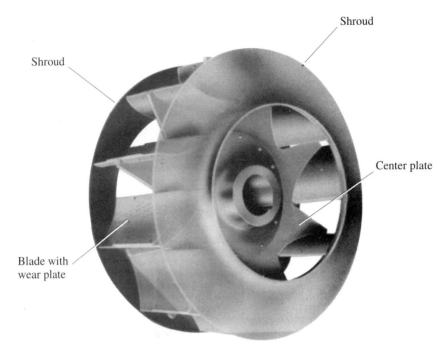

Shroud

Shroud

Center plate

Blade with
wear plate

FIGURE 3.20   DIDW (double inlet, double width) centrifugal fan wheel with ten RT blades. Note notched-out center plate and replaceable wear plates on pressure side of blades.

FIGURE 3.21   SISW centrifugal fan wheel with 52 FC blades fastened through corresponding slots in back plate and shroud.

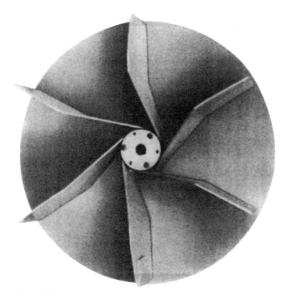

FIGURE 3.22 SISW centrifugal fan wheel with six radial blades welded to a back plate.

stream has to make two 90° turns, which, of course, results in some extra losses, especially if the diffuser ratio (housing i.d./wheel o.d.) is small.

2. A fan wheel with a conical back plate, as shown in Fig. 3.23. This fan wheel is more expensive to build, but the air stream here has to make only two 45° turns, a more efficient arrangement. In either case, the fan wheel usually has BI blades or occasionally AF or BC blades.

The following three types of housings are in common use in axial-centrifugal fans:

1. A cylindrical housing, as shown in Figs. 3.23 and 3.24.

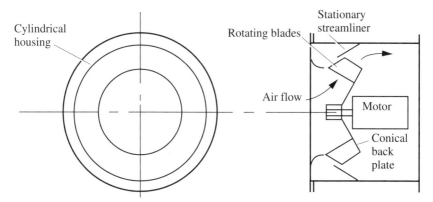

FIGURE 3.23 Mixed-flow fan showing direct motor drive, venturi inlet, fan wheel with conical back plate, and with a 45° diverging air stream discharging into a cylindrical housing.

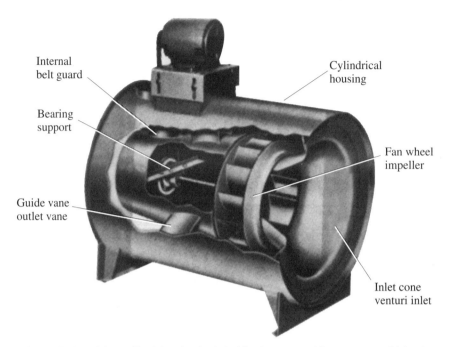

Internal belt guard

Cylindrical housing

Bearing support

Fan wheel impeller

Guide vane outlet vane

Inlet cone venturi inlet

FIGURE 3.24 Axial-centrifugal fan showing belt drive from an outside motor, venturi inlet, fan wheel with flat back plate, and outlet guide vanes, all assembled in a cylindrical housing. *(Courtesy of General Resource Corporation, Hopkins, Minn.)*

2. A square housing, as shown in Fig. 3.25.

3. A barrel-shaped housing, as shown in Figs. 3.26 and 3.27.

Various wheel and housing combinations are possible. Figure 3.27 shows a special type of barrel-shaped housing which is covered by my U.S. patent no. 3,312,386. It has a separate chamber for the motor so that direct drive can be used, even if hot or corrosive gases are handled. This fan has the trade name Axcentri Bifurcator, implying that the air stream is divided into two forks, flowing above and below the motor chamber but never coming into contact with the motor. The various types of axial-centrifugal fans will be discussed further in Chap. 9.

ROOF VENTILATORS

Figure 3.28 shows an exhaust roof ventilator with direct drive, a BI centrifugal fan wheel, and radial discharge using various spinnings. Various other models of roof ventilators are in common use. Some may have belt drive instead of direct drive, some may have axial fan wheels instead of centrifugal fan wheels, and some may be for upblast instead of radial discharge. While most models are for exhausting air from a building, some are for supplying air into a building. The various combinations of these features lead to ten different models, which will be illustrated and described in Chap. 10.

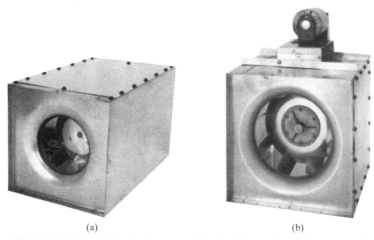

(a)   (b)

FIGURE 3.25   Mixed-flow fan in a square housing. Two models are shown, one for direct drive and one for belt drive. Both models have a venturi inlet, a fan wheel with a conical back plate, and an access door. The square housing results in lower cost and allows connection to either square or round ducts. *(Courtesy of FloAire, Inc., Bensalem, PA.)*

FIGURE 3.26   Mixed-flow fan with barrel-shaped spun housing for smaller diameters of inlet and outlet ducts. Direct drive. The fan wheel has a conical back plate. Outlet guide vanes (not shown) prevent excessive air spin at the small outlet diameter. *(Courtesy of FloAire, Inc., Bensalem, Pa.)*

FIGURE 3.27   Mixed-flow fan with barrel-shaped housing for smaller diameters of inlets and outlets. The fan wheel with conical back plate is directly driven by a motor in a separate chamber. Outlet vanes (not shown) prevent excessive outlet spin. *(Courtesy of Bayley Fan, Division of Lau Industries, Lebanon, Ind.)*

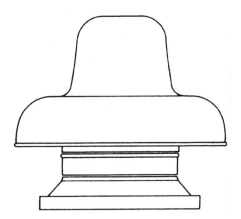

FIGURE 3.28   Schematic sketch of a centrifugal roof exhauster, direct drive, radial discharge, 15-in wheel diameter, 1 hp, 1725 rpm. *(Courtesy of Flo-Aire, Inc., Cornwells Heights, Pa.)*

CROSS-FLOW BLOWERS

A cross-flow blower is a unique type of centrifugal fan in which the airflow passes twice through a fan wheel with FC blading, first inward and then outward, as shown in Fig. 3.29. The main advantage of cross-flow blowers is that they can be made axially wider, in fact to any width desired. This makes them particularly suitable for certain applications such as air curtains, long and narrow heating or cooling coils, and dry blowers in a car wash. The flow pattern and principle of operation will be explained in Chap. 12.

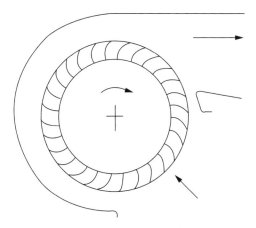

FIGURE 3.29 Cross-flow blower showing airflow passing twice through the rotating fan wheel.

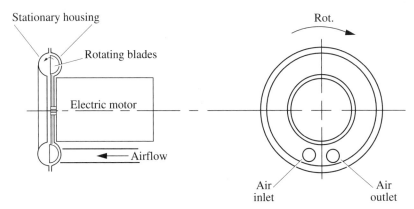

FIGURE 3.30 Vortex blower showing blades rotating in right half of the torus-shaped housing.

VORTEX OR REGENERATIVE BLOWERS

The vortex or regenerative blower is another unique type of centrifugal fan. Here the airflow circles around in an annular, torus-shaped space, similar to the shape of a doughnut. On one side of the torus are rotating fan blades, throwing the air outward, as shown in Fig. 3.30. The airflow then is guided back inward by the other side of the torus so that it must reenter the inner portion of the rotating blades. This results in a complicated flow pattern that will be discussed in detail in Chap. 9.

CONCLUSION

From the preceding we note that there are many different types of fans. Nevertheless, only two basic operating principles are used in all these fans: deflection of airflow and centrifugal force.

In *axial-flow fans,* the operating principle is simply deflection of airflow. Here the pressure is produced exclusively by the lift of the airfoil or of the single-thickness sheet metal profile used for the cross sections of the blades. Since an airfoil has a better lift-drag ratio over a wider range of angles of attack than a single-thickness profile (see Chap. 2), airfoil blades will result in better efficiencies than single-thickness profiles.

In *centrifugal fans* (including mixed-flow, cross-flow, and vortex fans), the operating principle is a combination of airflow deflection plus centrifugal force. This results in the following two differences between the performances of axial-flow fans and centrifugal fans:

1. Centrifugal fans normally produce more static pressure than axial-flow fans of the same wheel diameter and the same running speed. (Axial-flow fans, on the other hand, have the advantages of greater compactness and of easier installation.)

2. Since in centrifugal fans the airfoil lift contributes only a small portion of the pressure produced (while most of it is produced by centrifugal action), the improvement in performance due to airfoil blades (over sheet metal blades) is not as pronounced in centrifugal fans as it is in axial-flow fans.

CHAPTER 4
AXIAL-FLOW FANS

NOMENCLATURE

In past years, "fans and blowers" was a common expression. Axial-flow fans were called "fans," and centrifugal fans were called "blowers." Calling centrifugal fans blowers was misleading because centrifugal fans can be used for exhausting as well as for blowing. Axial-flow fans also can be used for either blowing or exhausting. Today the accepted terms are *axial-flow fans* and *centrifugal fans*.

The term *axial-flow fan* indicates that the air (or gas) flows through the fan in an approximately axial direction, as opposed to centrifugal fans (sometimes called *radial-flow fans*), where the air flows through the fan wheel approximately in a radially outward direction.

MATHEMATICAL FAN DESIGN VERSUS EXPERIMENTAL CUT-AND-TRY METHOD

The casual observer in general will not realize that the dimensions of an axial-flow fan (fan wheel outside diameter, hub diameter, number and width of the blades and vanes, blade and vane angles, curvature of blades and vanes, fan speed, and motor horsepower) can be calculated from the requirements (air volume in CFM and static pressure).

Let us dwell on the casual observer for a moment, and let us try to put ourselves in his or her place and look at the problem of fan design as he or she would do. The point of view of the casual observer will be about as follows: Fasten a number of blades somehow symmetrically around a hub. Put this fan wheel, together with a motor, into a housing, and run a test on it. If it does not give you "enough air," make certain changes on the unit. Make the hub diameter larger or smaller, and increase the number of blades, the blade angles, the blade widths, and the famous "scooping curvature." Keep on changing and retesting until you just about get what you want.

This viewpoint of the casual observer naturally is rather primitive. In some respects, it is even incorrect, since it overlooks certain limitations, such as the fact that the addition of blades or an increase in blade curvature does not always result in an increased air delivery. However, you cannot blame this casual observer if you stop and realize that many years ago this also was the conception of the fan designer. And you must admit one thing: As primitive as this purely experimental cut-and-try method may be, with a sufficient amount of persistence, time, and money, it often will be possible to obtain the desired air volume and static pressure by the application of this method. There are only three objections to this method:

1. The resulting units often will be larger, run at higher speeds, and consume more brake horsepower than necessary.
2. The method is too expensive because in general it will require the building and testing of three to five samples until the desired air volume and static pressure is obtained.

3. This method will practically always lead to units with uneven and turbulent airflow and with stalling effects in certain portions of the blade. As a result of all this, these units usually will be inefficient and noisy.

Once the necessity of fan design on a theoretical basis has been recognized, the first question is: Is it at all possible to determine the dimensions for a fan unit so that it will perform in accordance with a certain set of specifications, by pure calculation, this way completely eliminating the use of any experimental cut-and-try method? The answer to this question is: In most cases, this is possible, and even more than that, it is possible in more than one way. In other words, several designs are possible that will meet a certain set of requirements with respect to air delivery and pressure.

This naturally leads to the next question: If this problem of meeting the requirements can be solved by several designs, is it possible to find one optimal design? The answer is yes, providing that a definition for the word *optimal* can be agreed on.

The question "What is the optimal design?" is rather complex, and the answer to it will vary with the prospective application of the fan unit. In the majority of cases, it will include, among other things, the call for high efficiency and low sound level, both over the widest possible range of operation. Other requirements may be, for instance, a nonoverloading brake horsepower characteristic; or a flat pressure curve, which means a large free delivery; or a steep pressure curve, which means little variation in air delivery throughout the operating range; or a large pressure safety margin; or compactness; or some other supplementary requirement that may be desirable in a certain application. The combination of these requirements often results in interference problems and in conflicting specifications whose relative importance has to be considered before a decision is made.

AXIAL FLOW AND HELICAL FLOW

Let's be more specific about the statement that air flows through an axial-flow fan in an "approximately axial direction." On the inlet side, as the flow approaches the fan blades, the direction of the flow is axial, i.e., parallel to the axis of rotation, provided there are no inlet vanes or other restrictions ahead of the fan wheel. The fan blade then deflects the airflow, as shown in Fig. 2.10 (see Chap. 2).

The operating principle of axial-flow fans is simply deflection of airflow, as explained in Chap. 2 (page 2.1), on airfoils, and Chap. 3 (page 3.20), on types of fans. Past the blades, therefore, the pattern of the deflected airflow is of helical shape, like a spiral staircase. This is true for all three types of axial-flow fans: propeller fans, tubeaxial fans, and vaneaxial fans. Accordingly, the design procedures and design calculations are similar for all three types. As to both construction and performance, however, there are some differences.

We may anticipate one statement right here: The sequence propeller, tubeaxial, and vaneaxial also indicates the general trend of increasing weight, price, hub diameter, static pressure, aerodynamical load, and efficiency.

Coming back to the helical pattern of the airflow past the blades of an axial-flow fan, the air velocity there can be resolved into two components: an *axial velocity* and a *tangential or circumferential velocity.*

The axial velocity is the useful component. It moves the air to the location where we want it to go. In a propeller fan, the axial velocity moves the air across a wall or a partition. In a tubeaxial or vaneaxial fan, the axial velocity moves the air through a duct on the inlet side or on the outlet side or both.

The tangential or circumferential velocity component is an energy loss in the case of a propeller fan or a tubeaxial fan. In a vaneaxial fan, however, the tangential component is not a total loss; some of it is converted into static pressure, as will be explained later. This is the main reason why vaneaxial fans have higher efficiencies than propeller or tubeaxial fans.

BLADE TWIST, VELOCITY DISTRIBUTION

For good efficiency, the airflow of an axial-flow fan should be evenly distributed over the working face of the fan wheel. To be more specific, the axial air velocity should be the same from hub (or spider) to tip. The velocity of the rotating blade, on the other hand, is far from evenly distributed: It is low near the center and increases toward the tip. This gradient should be compensated by a twist in the blade, resulting in larger blade angles near the center and smaller blade angles toward the tip. This can be seen clearly in Fig. 4.1, which shows two views of a 30-in tube axial fan wheel with a 13-in hub diameter and eight single-thickness steel blades welded to the hub. Low-cost fan wheels (especially propeller-fan wheels) sometimes do not have this variation of the blade angle from hub to tip. They sometimes have the same blade angle from hub to tip (or worse, a slightly larger blade angle at the tip). This will result in a loss of fan efficiency because most of the airflow then will be produced by the outer portion of the blades, even at low static pressures. At higher static pressures, the blade twist is even more important, because without it, the inner portion of the blade will stall and permit reversed airflow, which, of course, will seriously affect the fan efficiency.

The *propeller fan,* as shown in Figs. 4.2 through 4.4, is the lightest, least expensive, and most commonly used fan. As mentioned, normally it is installed in a wall or in a partition to exhaust air from a building. This exhausted air, of course, has to be

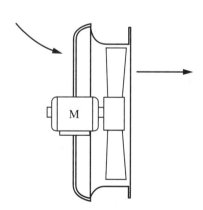

FIGURE 4.1 Two views of a fabricated tubeaxial fan wheel, 30-in o.d., 45 percent hub-tip ratio, and eight die-formed steel blades, with larger blade angles at the hub and smaller blade angles at the tip.

FIGURE 4.2 Propeller fan with motor on inlet side, 25 percent hub-tip ratio, and direct drive.

FIGURE 4.3 Two views of an 18-in propeller fan with direct drive from a ¼-hp, 1150-rpm motor, 40 percent hub-tip ratio, and five cast-aluminum airfoil blades that are backswept for a lower noise level.

FIGURE 4.4 A 48-in propeller fan with direct drive from a 7½-hp, 1150-rpm motor, 28 percent hub-tip ratio, and seven cast-aluminum airfoil blades.

replaced by fresh air, coming in through other openings. If these openings are large enough, the suction pressure needed is small. The propeller fan, therefore, is designed to operate in the range near free delivery, to move large air volumes against low static pressures. As can be seen from Figs. 4.2 through 4.4, the unit consists of a narrow mounting ring, a motor support, a motor, and a fan wheel. The mounting ring normally has a spun inlet bell that often is extended to a square mounting panel. The mounting panel carries some tubes or braces, which in turn support the motor base and the motor.

The motor is usually on the inlet side, as shown in Figs. 4.2 through 4.4, but in special applications it can be on the outlet side, which means slightly less noise. If the motor is on the inlet side, it is inside the building, protected from rain and snow by a shutter on the outside. When the fan is in operation, the shutter will be held open by the air stream. When the fan is not running, the shutter will close and prevent any backdraft from entering the building.

Large-size propeller-fan wheels usually run at low speeds (rpms) and therefore are belt driven. If the motor horsepower is large, good efficiency is desired, and to accomplish this, the fan wheel has a 20 to 40 percent hub-tip ratio and airfoil blades with a twist, resulting in blade angles between 30° and 50° at the hub and between 10° and 25° at the tip. As mentioned earlier, the larger blade angles at the hub will compensate for the lower blade velocities there, and this will result in a fairly even air velocity distribution over the face of the fan. This, as mentioned, is required for good fan efficiency.

Small propeller fans can be built with either direct drive or belt drive. Here the motor horsepower is small, and fan efficiency, therefore, is of minor importance. Lower cost is more important. These small propeller fans therefore do not have a hub and airfoil blades. Instead, they use a so-called spider with radial extensions to which sheet metal blades are riveted. The blade angles here are often constant from spider to blade tip. This results in most of the air being moved by the outer half of the blade. The inner portion of the blade produces mostly turbulence and, of course, consumes just as much power (or perhaps more) as it would if it were functioning properly. Sometimes there is some reverse flow near the spider when the static pressure increases. In other words, the tips of the blades move faster and produce say ½ in of static pressure, but the spider portion of the blades moves slower and may produce no static pressure. Therefore, some of the air past the blade tips will flow inward and then backward. Some small, low-cost fan wheels are stamped in one piece to keep the cost down.

The *tubeaxial fan,* as shown in Figs. 4.1 and 4.5, is a glorified propeller fan. It has a cylindrical housing, about one diameter long, containing a motor support, a motor, and a fan wheel. The motor can be located either upstream or downstream of the fan wheel. An upstream motor has the advantage that the airflow has a chance to smoothen out before it hits the fan blades, but this is only important if the venturi inlet is too small and therefore not effective. If the venturi inlet is adequate, the airflow will be smooth to begin with and does not need any extra space for smoothing. The upstream motor, on the other hand, has the disadvantage that some turbulence will be produced by the motor support ahead of the fan wheel. This may affect the efficiency and will result in a somewhat increased noise level.

In general, we should keep in mind that a blade, operating in turbulent airflow, will not function properly. Turbulence past the fan wheel, therefore, is not too harmful. It just increases the resistance of the system and therefore the static pressure against which the fan will operate. Turbulence ahead of the fan wheel, however, is harmful. It not only increases the static pressure required, but it also results in the blades operating in turbulent airflow and therefore with lower efficiency and a higher noise level.

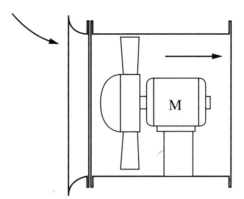

FIGURE 4.5 Tubeaxial fan with motor on outlet side and with separate venturi inlet, 43 percent hub-tip ratio, and direct drive.

The fan wheel of a tubeaxial fan can be similar to that of a propeller fan. It often has a medium-sized hub diameter, about 30 to 50 percent of the blade outside diameter, in the case of direct drive preferably not too much different from the motor diameter, for streamline flow conditions. The unit is designed to operate in the range of moderate static pressures, higher than for a propeller fan (due to the larger hub diameter) but not as high as for a vaneaxial fan (due to the smaller hub diameter and the lack of guide vanes, which in a vaneaxial fan convert some of the tangential air velocity into static pressure).

A tubeaxial fan can be connected to an inlet duct or an outlet duct or both. If there is no inlet duct, a spun venturi inlet is required, as shown in Fig. 1.13, to prevent vena contracta, as shown in Fig. 1.11. Small tubeaxial fans usually are designed with direct drive; large units are designed with belt drive.

The *vaneaxial fan,* as shown in Figs. 4.6 and 4.7, is a more elaborate unit. It has the outside appearance of a cylindrical housing at least one diameter long. As in a tubeaxial fan, this housing contains the motor support, the motor, and the fan wheel,

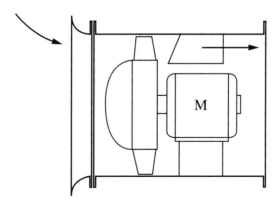

FIGURE 4.6 Vaneaxial fan with outlet vanes around the motor and with separate venturi inlet, 66 percent hub-tip ratio, and direct drive.

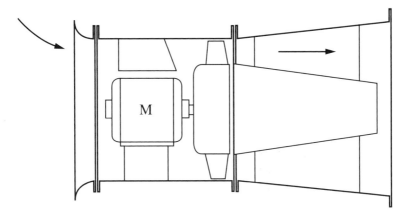

FIGURE 4.7 Vaneaxial fan with inlet vanes around the motor and with separate venturi inlet, 66 percent hub-tip ratio, direct drive, and outlet diffuser and tail piece for static regain.

but the vaneaxial fan housing contains in addition a set of guide vanes and sometimes an inner ring, a converging tail piece, and an expanding diffuser for static regain (see Chap. 1, page 1.12, on basics).

The guide vanes usually are arranged around the motor. This makes the unit more compact, since the same axial length is used for both the motor and the guide vanes. Motor and guide vanes can be located either past the fan wheel (see Fig. 4.6) or ahead of the fan wheel (see Fig. 4.7).

Hub Diameter *d* of Vaneaxial Fans

The hub diameter of a vaneaxial fan is larger (than that of a tubeaxial fan), usually between 50 and 80 percent of the wheel diameter, sometimes even slightly larger. The vaneaxial fan is designed to operate in the range of fairly high static pressures, and this requires a larger hub diameter. The customer usually specifies the required air volume, static pressure, fan diameter *D,* and speed (rpms). In designing a vaneaxial fan to meet these requirements, the first step will be to determine the hub diameter *d.* This can be done from the formula

$$d_{min} = (19,000/\text{rpm}) \sqrt{SP} \qquad (4.1)$$

where *d* is in inches and *SP* is in inches of water column. Hub diameter also can be determined from the graph shown in Fig. 4.8.

Suppose the customer requires that the vaneaxial fan should run at 1750 rpm and produce 12,000 cfm against 3 in of static pressure at the point of operation. Figure 4.8 indicates that the corresponding minimum hub diameter will be 18.8 in. (Please note that this is the hub diameter, regardless of the requirements for air volume and for wheel diameter *D.*) If for some reason a somewhat larger hub diameter *d* is desired, this will be acceptable. (It would merely result in a slight reduction in the annular area and therefore in the air volume, which could be compensated by a slight increase in the blade angles. It would not be critical.) A smaller hub diameter, on the other hand, could be critical. It might result in an inadequate performance of the inner blade portion, i.e., turbulence and possible reversed air flow near the hub. This inadequate performance is called *stalling.*

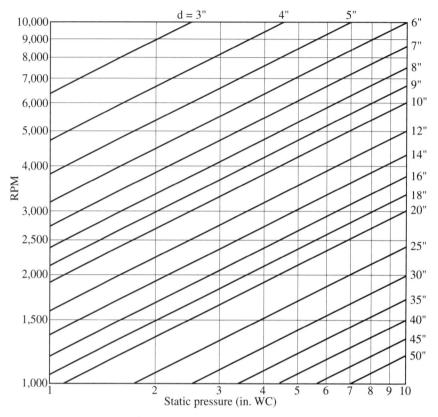

FIGURE 4.8   Minimum hub diameter d of a vaneaxial fan as a function of speed (rpms) and static pressure.

The static pressure SP produced by a vaneaxial fan can be calculated for each radius from the following formula:

$$SP = 3.43 \times 10^{-9} \times \text{rpm} \times z_B \times C_L \times l \times W \qquad (4.2)$$

where SP = static pressure, in inches of water column
    z_B = number of blades
    C_L = lift coefficient of airfoil at the angle of attack, used at this radius
    l = blade width at this radius, in inches
    W = air velocity relative to the rotating blade, in feet per minute (fpm) (a formula for W will be given later)

This formula indicates again that a larger hub diameter will result in a larger static pressure because, for a larger hub diameter,

1. The relative velocity W will be larger (due to the increased blade velocity).

2. There will be more room available for wider blades without overlapping.

For good efficiency, the static pressure produced should be the same for any radius from hub to tip. Since the relative air velocity W is smallest at the hub, this must be compensated by a larger $l \times C_L$ at the hub, but both l and C_L can be increased only up to a certain limit (l up to the point where the blades would overlap and C_L up to the maximum lift coefficient the airfoil can produce). This is why the hub diameter d can be slightly larger but not smaller than $d = (19,000/\text{rpm})\sqrt{SP}$. If d were too small, $l \times C_L$ could not be made large enough to compensate for the smaller W at the hub, and stalling would occur near the hub.

Wheel Diameter *D* of Vaneaxial Fans

After the hub diameter d has been determined, the next step is to check whether the wheel diameter D requested by the customer is acceptable. Obviously, it is not acceptable if it is smaller than the hub diameter d we just determined, but this is an extreme case that will happen rarely. However, even if D is larger than d, it may not be large enough.

In order to check whether the wheel diameter D requested by the customer is acceptable, we use the formula

$$D_{\min} = \sqrt{d^2 + 61 \ (\text{cfm/rpm})} \qquad (4.3)$$

or the graph shown in Fig. 4.9. Using the preceding values of $d = 18.8$ in and cfm/rpm $= 12,000/1750 = 6.86$, we find $D_{\min} = 27.2$ in. This is the minimum wheel diameter. It would result in a hub-tip ratio of $18.8/27.2 = 0.69$, a good hub-tip ratio for a vaneaxial fan. If the customer requested a wheel diameter of 28, 29, or 30 in, we would use this size. If the customer suggested, for example, a wheel diameter of 24 in or anything else smaller than 27 in, we would point out that this would be risky because the pressure safety margin would be too small and that we would prefer a larger wheel diameter D.

If the customer cannot accept a larger wheel diameter, a two-stage axial-flow fan may solve the problem. Then each stage has to produce only about one-half the static pressure, and the hub diameter d as well as the wheel diameter D can be reduced. This two-stage unit will be longer and more expensive. This may or may not be acceptable to the customer. If it is not acceptable, a centrifugal fan may have to be considered instead of a vaneaxial fan.

Summarizing, we found that the hub diameter d is a function of static pressure and speed and that the wheel diameter D is a function of d and of cfm/rpm.

Vaneaxial Fans of Various Designs

Best efficiencies for vaneaxial fans are obtained with airfoil shapes as cross sections of the blades because airfoils have large lift-drag ratios (see Chap. 2, page 2.9, on airfoils). Airfoils, then, result in higher static pressures (produced by the airfoil lift) and lower power consumption (produced by the airfoil drag) and therefore higher fan efficiencies.

Airfoil blades usually are made as aluminum castings and sometimes as steel castings, incorporating both features, the twist in the blade angles and the airfoil shape in the cross sections. Figures 4.10 through 4.17 show such fan wheels with cast-aluminum airfoil blades. Let us examine the different designs shown in these pictures.

Figures 4.10, 4.11, and 4.12 show three experimental fan wheels for the same vaneaxial fan. All three wheels have the same hub diameter and the same blade section (blade width, airfoil shape, and blade angle) at the hub, but the width at the blade

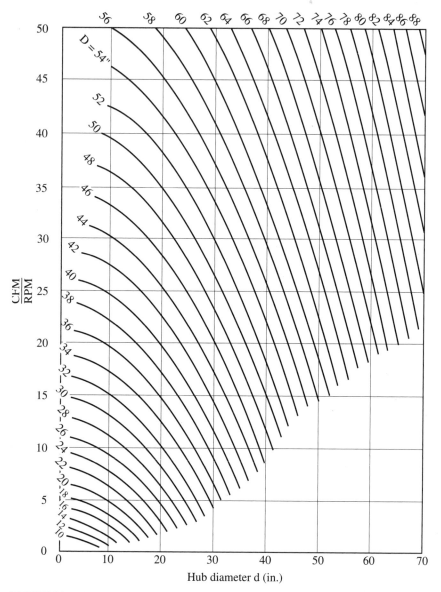

FIGURE 4.9   Minimum fan-wheel diameter D of a vaneaxial fan as a function of cfm, rpm, and hub diameter d.

tip is varied. Figure 4.10 shows a wide blade tip, Figure 4.11 shows a medium tip, and Figure 4.12 shows a narrow tip. I tested the three wheels and found the following:

1. Wide blade tips result in high pressure, high efficiency, and quiet operation, but they cause in considerable motor overload at the point of no delivery.

2. Medium tips reduce the maximum static pressure and the no-delivery overload.

FIGURE 4.10 A 31-in vaneaxial fan wheel, of cast aluminum, with a 63 percent hub-tip ratio, direct drive, and seven airfoil blades with wide tips for high pressure and quiet operation at high efficiency.

FIGURE 4.11 A 31-in vaneaxial fan wheel, of cast aluminum, with a 63 percent hub-tip ratio, direct drive, and seven airfoil blades with medium-wide tips for medium pressure and less overload at no delivery.

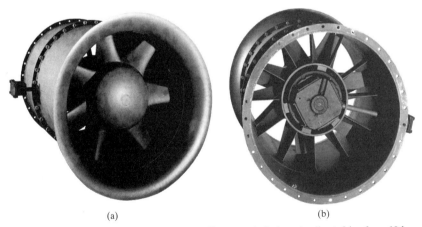

(a)   (b)

FIGURE 4.12 A 31-in vaneaxial fan with a 63 percent hub-tip ratio, direct drive from 10-hp, 1750-rpm motor, and seven airfoil blades with narrow tips for nonoverloading brake horsepower characteristics. Twelve outlet vanes go well with seven blades. (a) Inlet side. (b) Outlet side.

3. Narrow tips result in a nonoverloading brake horsepower characteristic. This was the fan wheel I adopted for production, even though the efficiency was slightly lower and the noise level was slightly higher, but still acceptable. A compromise between conflicting performance features had to be made.

FIGURE 4.13 Inlet view of a 5-in vaneaxial fan, of cast aluminum, with a 78 percent hub-tip ratio, seven airfoil blades, eight outlet vanes, belt drive from 1/12-hp shaded-pole motor, and a fan speed of 3800 rpm.

FIGURE 4.14 A 5-in vaneaxial fan for projector lamp cooling showing belt drive in assembled projector. Fan pulley is part of the fan-wheel casting.

Figures 4.13 and 4.14 are views from the inlet side of a 5-in vaneaxial fan that was designed for cooling a projector lamp. A large 78 percent hub-tip ratio was needed in order to produce sufficient static pressure to force the cooling air through some narrow passages. Not as many guide vanes are needed because of the small size. Eight vanes go well with seven blades.

FIGURE 4.15   A 34-in tank engine cooling fan wheel, of aluminum casting, 35 hp, 2400 rpm. View from the inlet side showing teeth inside the hub for gear drive from engine.

FIGURE 4.16   Angular side view of the same 34-in cast-aluminum tank engine cooling fan wheel showing the airfoil cross sections of the blades and the variation in the blade angles from hub to tip.

FIGURE 4.17   The same 34-in cast-aluminum tank engine cooling fan wheel, view from the outlet side showing a 63 percent hub-tip ratio and twelve blades with blade width slightly decreasing from hub to tip.

An interesting design is shown in Figures 4.15, 4.16, and 4.17. These are three views of a cast-aluminum fan wheel with airfoil blades. This 34-in fan wheel was designed to cool the engine of an Army tank. It used gear drive from the engine, as shown in Fig. 4.15, a view from the inlet side. Figure 4.16 is an angular side view showing the airfoil sections of the blades and the variation of the blade angles from hub to tip. Figure 4.17 is a view from the outlet side showing the 63 percent hub-tip ratio and the twelve blades.

Figures 4.18 through 4.20 are three views of the 34-in fan wheel redesigned in steel, by request of the customer, to replace the cast-aluminum fan wheel. Here, some single-thickness steel blades with slots were welded to a heavy steel disk, even though the aerodynamic conditions of the steel disk were not ideal. Furthermore, certain other difficulties had to be addressed, such as the lower lift coefficients of the single-thickness blades and the risk of blade vibration due to the lesser rigidity of the single-thickness blades. These difficulties were overcome by an increase in the hub-tip ratio and by using more blades of narrower

FIGURE 4.18 A 34-in tank engine cooling fan wheel redesigned in steel, 35 hp, 2400 rpm. View from the inlet side showing the welds on the slots and on the tabs of the blades.

width. Various shapes of welding tabs were tried out. The steel blades were subjected to vibrations in order to determine which shape would have the best resistance to fatigue failure. Figure 4.21 shows the test setup, in which six blades (differing only in the way they were attached to the heavy steel disk) were vibrated several million times by connecting rods from solenoids. The solenoids were kept cool by air streams from a separate pressure blower.

Figure 4.22 shows another vaneaxial fan wheel with cast-aluminum airfoil blades. Figure 4.23 shows the corresponding vaneaxial fan housing. This unit was designed for grain drying. It had an 18-in wheel outside diameter and direct drive from a 10-hp, 3450-rpm motor.

Single-thickness blades, as shown in Figs. 4.1, 4.18, 4.19, and 4.24 also have the desired variation in the blade angles from hub to tip. For accuracy, uniformity, and economy of production, single-thickness steel blades are press-formed in a die that takes care of the springback of the material.

In small sizes, the entire fan wheel can be molded in plastic, as shown in Fig. 4.25. This is a good and efficient fan wheel, even though it has only single-thickness blades. The figure shows a good blade twist from hub to tip.

FIGURE 4.19 Angular side view of the same 34-in tank engine cooling fan wheel of welded steel showing single-thickness steel blades welded to a heavy disk.

FIGURE 4.20 The same 34-in tank engine cooling fan wheel. View from the outlet side showing a 68 percent hub-tip ratio and 16 steel blades of constant width from hub to tip and the heavy disk with mounting holes and balancing holes.

FIGURE 4.21   Setup for testing single-thickness steel blades for fatigue failure due to vibration of the blades.

Inlet Bell

Figures 4.12 through 4.14 and Fig. 4.24 showed vaneaxial fans with inlet bells attached. In vaneaxial fans, the inlet bell (also called the *venturi inlet*) is even more important than in propeller fans or tubeaxial fans because, owing to the larger hub diameter the annular area between the hub outside diameter and the housing inside

FIGURE 4.22   Front view of an 18-in vaneaxial fan wheel, of cast aluminum, with a 48 percent hub-tip ratio, six airfoil blades with medium-wide tips for grain drying, and direct drive from 10-hp, 3450-rpm motor.

FIGURE 4.23 Inlet view of housing for an 18-in vaneaxial fan, direct drive. Motor and fan wheel were removed so that the inner ring, motor base, and outlet guide vanes can be seen.

diameter is smaller and the acceleration of the entering air stream is therefore greater. Without an inlet bell, the vena contracta would be worse and would affect a larger portion of the blades (to operate in turbulent air and to be starved for air), particularly if the fan wheel is located near the housing inlet, as in Fig. 4.6. The use of an inlet bell, therefore, will boost the flow rate by 10 to 15 percent. It also will increase the fan efficiency and reduce the noise level considerably.

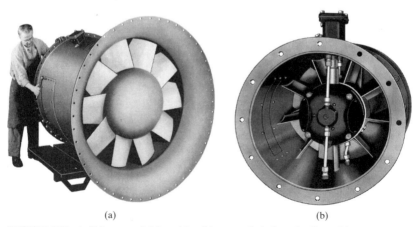

(a)   (b)

FIGURE 4.24 A 48-in vaneaxial fan with a 54 percent hub-tip ratio, direct drive from 25-hp, 1150-rpm motor, ten single-thickness steel blades welded to the fabricated hub, and nine outlet vanes: (a) inlet side; (b) outlet side.

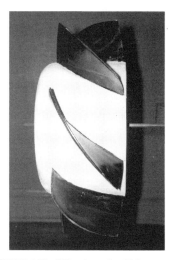

FIGURE 4.25 Side view of a 10-in vaneaxial fan wheel, plastic mold, with a 64 percent hub-tip ratio, seven single-thickness blades, and blade angles varying from 50° at the hub to 34° at the tip, for a 16° twist. *(Courtesy of Coppus Engineering Division, Tuthill Corporation, Millbury, Mass.)*

For the shape of the inlet bell, an elliptic contour is ideal, but a circular contour is a good approximation and reduces the depth of the spinning. As mentioned in Chap. 1 (page 1.14), on basics, the radius of curvature of the inlet bell should be at least 14 percent of the housing inside diameter. This is graphically shown in Fig. 4.26. For example, for a 42-in throat inside diameter (or housing inside diameter), the radius of curvature should be at least 5⅞ in.

Figures 4.27 and 4.28 show an oversize inlet bell that was built for experimental purposes to confirm by test that not much would be gained if the radius of curvature were made much larger than 14 percent of the housing inside diameter. Figure 4.27 shows the fan blades stationary. Figure 4.28 shows a curved string being drawn by the air stream while the fan is running; note the shadow of the string, indicating that it did not touch the surface of the inlet bell but followed the curvature of the converging air stream.

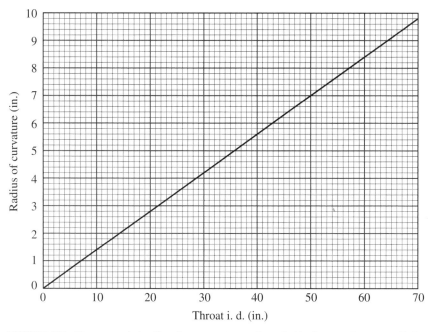

FIGURE 4.26 Recommended radius of curvature versus throat inside diameter for a venturi inlet.

FIGURE 4.27 An oversize inlet bell shown on a vaneaxial fan with five wide-tip airfoil blades in stationary position.

Direct Drive, Belt Drive, Duct Connection

As in tubeaxial fans, direct drive generally is preferable in small vaneaxial fans so that obstructions to airflow (from the belt housing), belt losses, maintenance, and the extra expense for bearings, brackets, and belt housings is avoided. Belt drive, on the other hand, is preferable in large sizes so that the running speed can be kept low without the use of expensive low-speed motors. Belt drive also permits better cover-

FIGURE 4.28 The same inlet bell with the fan wheel running and a string being drawn in by the air stream entering the unit.

FIGURE 4.29 Inlet view of a 54-in vaneaxial fan (930 rpm, belt drive from 25-hp, 1750-rpm motor) with a 45 percent hub-tip ratio, six cast-aluminum airfoil blades, and the equivalent of eleven outlet vanes. *(Courtesy of Ammerman Division, General Resource Corporation, Hopkins, Minn.)*

age of the range so that requirements for air volume and static pressure can be met more closely. Figure 4.29 shows a belt-driven vaneaxial fan.

The cylindrical shape of vaneaxial fans makes them suitable for straight-line installation. Like tubeaxial fans, they can be connected to an inlet duct, an outlet duct, or both.

Guide Vanes

As mentioned previously, the airflow past an axial-flow fan wheel has a helical pattern. This means that the air moves along cylindrical surfaces with practically no radial component, only an axial and a rotational or circumferential component.

It is the function of the guide vanes to eliminate or at least reduce the air spin past the fan blades. There are two ways in which guide vanes can be provided to perform their function of reducing the rotational energy loss: They can be located on the outlet side or on the inlet side of the fan blades. In the case of outlet vanes, the static pressure is produced partly by the blades and partly by the vanes. In the case of inlet vanes, the vanes do not produce any static pressure; they merely prepare the airflow for the blades.

In order to study the two types of guide vanes, let us make a schematic sketch of the flow pattern along a cylindrical surface for each type of vane. In order to do this, we have to "unroll" or develop this cylindrical surface into a flat plane. This is done in Fig. 4.30 for outlet vanes and in Fig. 4.31 for inlet vanes. These two figures show not only the shapes of the blades and vanes but also the different velocities of the air flowing past the blades and vanes.

Outlet Guide Vanes

The function of outlet vanes is easier to understand. Figure 4.30 shows how the airflow will pass first through the rotating blade section and then through the stationary guide vane section. The airflow approaches the blades with an air velocity V_0 of

$$V_0 = V_a = \frac{\text{cfm}}{\text{annular area}} \tag{4.4}$$

The airflow then gets deflected by the blades and leaves the blades with velocity V_1. This velocity V_1 has an axial component V_0 that, of course, has to be retained for continuity (see Chap. 1, page 1.10, on basics). V_1 also has a rotational component V_r, resulting in

$$V_1 = \sqrt{V_0^2 + V_r^2} \tag{4.5}$$

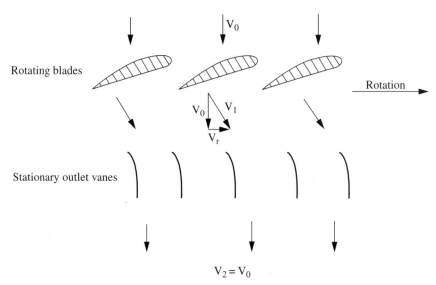

FIGURE 4.30 Function of outlet vanes. They guide the helical airflow, produced by the rotating blades, back to an axial direction, thereby decelerating the air velocity from V_1 to V_2.

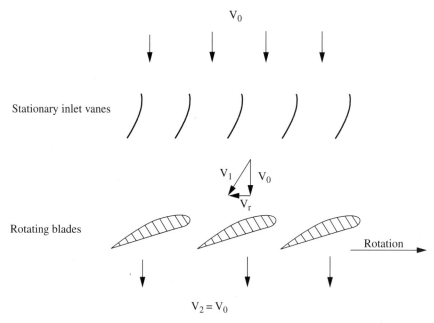

FIGURE 4.31 Function of inlet vanes. They guide the approaching axial airflow into a helical motion, with the rotational component opposite to the fan rotation. The rotating blades then deflect the airflow back into a more or less axial direction, thereby decelerating the air velocity from V_1 to V_2.

so that V_1 is about 20 to 30 percent larger than V_0 at the hub. At the tip, the increase will be smaller, about 10 to 15 percent. In other words, the deflection of the airflow will be largest at the hub, according to the following formula:

$$V_r = \frac{233 \times 10^5}{\text{rpm}} \times \frac{SP}{r} \qquad (4.6)$$

where V_r = rotational component of the helical airflow, in fpm
  SP = static pressure, in inches of water column
   r = radius of the rotating blade section, in inches

The airflow then passes through the stationary vane section. These outlet vanes guide the air stream back into the axial direction. The air leaves the guide vane section with a velocity V_2 that again must be the same as V_0 for continuity. In the vane section, therefore, the airflow is decelerated from V_1 to V_0. Some of this difference $V_1 - V_0$ is converted to static pressure. This phenomenon of static recovery was explained in Chap. 1. Figure 4.24 shows a vaneaxial fan housing with the outlet guide vanes visible.

Inlet Guide Vanes

The function of inlet vanes is different. It is illustrated in Fig. 4.31. Here, the air is first drawn through the stationary vane section. These inlet vanes are curved in such a way that they guide the airflow into a helical motion whose rotational component is opposite to the fan rotation. This rotational component should be just sufficient to neutralize the subsequent deflection in the direction of the fan rotation that is imposed on the airflow by the rotating blades. As a result, the air should leave the blades in an approximately axial direction. In this arrangement, the airflow approaches the guide vanes with air velocity

$$V_0 = V_a = \text{cfm/annular area} \qquad (4.4)$$

The airflow then is guided into a spin by the stationary inlet vanes and leaves them with velocity

$$V_1 = \sqrt{V_0^2 + V_r^2} \qquad (4.5)$$

Again, V_1 is larger than V_0. Finally, as the air passes through the rotating blade section, it is deflected back into an approximately axial direction and thereby is slowed down to V_0 again, with some of the difference $V_1 - V_0$ converted to static pressure.

Shape of the Guide Vanes

The shape of the guide vanes can be determined from the requirement of tangential conditions at the leading edge of the outlet vanes and the trailing edge of the inlet vanes. The width and spacing of the vanes follow considerations similar to those generally used in the design of turning vanes in duct elbows. The considerations are, however, not quite the same because the airflow in elbows usually does not change its velocity (except for the direction), whereas outlet vanes operate in decelerated airflow and inlet vanes in accelerated airflow.

Since the guide vanes are stationary, the relative air velocities and therefore the losses here are much smaller than in the blade section. From this it might be surmised that the shape of the guide vanes is not quite as critical as that of the blades. This is actually true for the outlet vanes. The shape of inlet vanes, however, is still critical, not because of the comparatively small losses occurring in them, but because of the indirect effect on the subsequent blades. If the helical air motion entering the blade section is not in accordance with the calculation, the blades will not operate under the precalculated conditions and naturally will not perform as expected. This is one of the disadvantages of inlet vanes and an advantage of outlet vanes (others will be discussed later).

The cross section of the guide vanes—like that of the blades—can be either an airfoil shape or a single line (as shown in Figs. 4.30 and 4.31) of proper curvature. Airfoil vanes are made as aluminum castings (or sometimes of hollow-steel construction), whereas single-sheet steel vanes are fabricated. The single-sheet steel construction is used more commonly and will result in satisfactory performance, so the extra expense of the airfoil construction is seldom justified. Only for certain performance features (such as a wider efficiency curve) will airfoil vanes be useful, especially in the case of the more critical inlet vanes.

The Pros and Cons of Guide Vanes

Now that the function of the guide vanes has been clarified, the first question to be asked naturally is: In which cases is it worthwhile to provide such guide vanes? Or in other words, in which cases should the more elaborate and more expensive vaneaxial fan be used instead of the tubeaxial fan? The answer is: The vaneaxial fan should be used whenever higher pressure and higher efficiency are requested or desired. The vaneaxial fan is indicated for high-pressure requirements because it converts some of the rotational component V_r back into static pressure, thereby producing more static pressure than the tubeaxial fan. The vaneaxial fan is indicated whenever high efficiency is desired because—owing to the higher static pressure—the vaneaxial fan produces a larger air horsepower and therefore a higher efficiency.

The next question to be asked is: Which is the better location for the stationary guide vanes, ahead or past the rotating blades? The answer is: Outlet vanes will be preferable in most cases. Let's analyze this answer. Inlet vanes (ahead of the blades) are occasionally (but not too often) the preferred configuration. For example, if the motor should be on the inlet side (in order to be more accessible) and the guide vanes should be around the motor (to save space), this would be a good reason to use inlet vanes. Another example: If strong inlet turbulence (due to an inlet damper or an inlet elbow or due to an inadequate venturi inlet) can be expected, the inlet vanes will give the air stream an opportunity to somewhat smooth out before it hits the rotating blades, so this would be another reason for inlet vanes. On the other hand, inlet vanes have the following five disadvantages:

1. The shape of inlet vanes is more critical, as was discussed earlier.
2. By giving the air stream a spin opposite the fan rotation, inlet vanes obviously result in larger relative air velocities (relative to the blades), almost as if the fan wheel were to rotate at a higher speed. This results in a higher noise level, however.
3. Inlet vanes, while reducing a strong inlet turbulence, will themselves produce some slight inlet turbulence, again resulting in a slightly increased noise level. In other words, outlet vanes will be quieter if the blades operate in a fairly smooth airflow.
4. Inlet vanes are designed to compensate for the subsequent deflection of the air stream by the rotating blades, but they can do this only for one point on the per-

formance curves. At lower static pressures, there will still be some air spin opposite the fan rotation, even past the blades. At higher static pressures, there will be an air spin in the fan rotation despite the inlet spin from the inlet guide vanes. The reason for this is that the deflection by the rotating blades (the rotational component V_r) is not constant but is proportional to the static pressure, as can be seen from the formula for V_r, given earlier. A remaining air spin could be quite harmful, particularly if there is a converging cone past the fan, which would amplify the spin, as illustrated in Fig. 1.8 in Chap. 1. Outlet vanes, on the other hand, will do a better job of removing the air spin past the blades. They will do this for any point on the performance curves, by brute force, so to speak, even if tangential conditions do not prevail at the leading edge of the outlet vane.

5. The deceleration from V_1 to V_2, with simultaneous partial conversion of kinetic energy (velocity pressure) to potential energy (static pressure), of the airflow is done by the rotating blades in the case of inlet vanes but by the stationary guide vanes in the case of outlet vanes. Obviously, the air velocity relative to the stationary outlet vanes is considerably smaller and therefore has smaller losses due to turbulence than the air velocity relative to the rotating blades. This conversion, called *static regain*, therefore will be more efficient for outlet vanes. In other words, outlet vanes will provide more static regain.

On balance, we can say that outlet vanes will be preferable in most applications.

Velocity Diagrams, Relative Air Velocities, Blade Angles $\beta + \alpha$

Let us repeat the formula for the static pressure SP produced at each radius of a vaneaxial fan:

$$SP = 3.43 \times 10^{-9} \times \text{rpm} \times z_B \times C_L \times l \times W \tag{4.2}$$

This formula indicates that the static pressure produced is proportional to W, the air velocity relative to the blade. Figure 4.32 shows the velocity diagram for outlet vanes and explains the meaning of the various velocities.

From Fig. 4.32, we note that W consists of two components: V_a in axial direction and $V_B - \frac{1}{2}V_r$ in circumferential direction. Thus we get

$$W = \sqrt{V_a^2 + \left(V_B - \frac{1}{2}V_r\right)^2} \tag{4.7}$$

and that the corresponding air angle β can be calculated from

$$\tan \beta = \frac{V_a}{V_B - \frac{1}{2}V_r} \tag{4.8}$$

The corresponding blade angle must be made somewhat larger, namely, $\beta + \alpha$, where α is the angle of attack (see Figs. 2.1 and 2.4, Chap. 2, on airfoils). The angle α is determined from the characteristic curves (for an infinite aspect ratio) for the airfoil used. The angle α is largest at the hub (about 2° to 5°) and smallest at the tip (about 0° to 3°).

The angle δ indicates the air angle past the blades, but as can be seen from Fig. 4.32, δ is measured from the axial direction (not from the circumferential direction

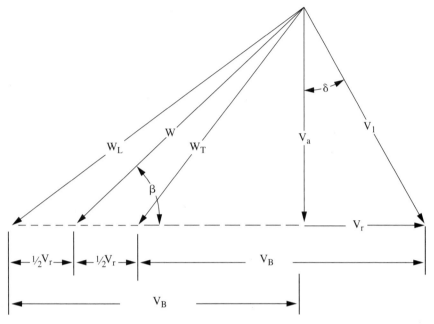

FIGURE 4.32 Velocity diagram for outlet vanes.

$$A_a = \frac{(D^2 - d^2)\pi}{576}$$ annular area in square feet if D and d are in inches

$V_a = \text{cfm}/A_a$ axial air velocity in feet per minute through the annulus

$$V_r = \frac{233 \times 10^5}{\text{rpm}} \times \frac{SP}{r}$$ rotational component in feet per minute of the helical air velocity past the blades

SP = static pressure in inches of water column

r = radius in inches of the rotating blade section under consideration

V_1 = helical air velocity in feet per minute past the blades, that is, $V_1 = \sqrt{V_a^2 + V_r^2}$

W_L = air velocity in feet per minute relative to the leading edge of the blade

W_T = air velocity in feet per minute relative to the trailing edge of the blade

W = air velocity in feet per minute relative to the blade, average of W_L and W_T

$V_B = (2r\pi/12) \times \text{rpm}$ blade velocity in feet per minute of the rotating blade section

$$W = \sqrt{V_a^2 + \left(V_B - \frac{1}{2}V_r\right)^2}$$

$$\tan \beta = \frac{V_a}{V_B - \frac{1}{2}V_r}$$ average direction of the relative air velocity W while passing the blade section

$$\tan \delta = \frac{V_r}{V_a}$$ direction of the air velocity V_1 past the blade, entering the guide vane section; also angle of the leading edge of the guide vane

like β). The angle δ is also the angle at the leading edge of the outlet vane. It can be calculated from

$$\tan \delta = \frac{V_r}{V_a} \tag{4.9}$$

Figure 4.33 shows the velocity diagram for inlet vanes and again explains the meaning of the various velocities. Here, the formula for the relative air velocity W is different:

$$W = \sqrt{V_a^2 + \left(V_B + \frac{1}{2} V_r\right)^2} \tag{4.10}$$

The corresponding air angle β now can be calculated from

$$\tan \beta = \frac{V_a}{V_B + \frac{1}{2} V_r} \tag{4.11}$$

Again, the angle of attack α is added to β to obtain the blade angle $\beta + \alpha$. The angle δ here is the air angle past the vane section, i.e., the angle at the trailing edge of the guide vane. Again, it can be calculated from

$$\tan \delta = \frac{V_r}{V_a} \tag{4.9}$$

Number of Blades z_B

Let's discuss a question that has been asked many times: Is there such a thing as an optimal number of blades for a vaneaxial fan, and if there is an optimal number, what is it? Let's analyze this interesting question.

From the formula for the static pressure produced, we already know that this pressure is proportional to the product $z_B \times l$, the number of blades times the blade width. This means that a certain design can be modified by, for instance, doubling the number of blades and reducing their width to one-half without any appreciable change in the fundamental design and in the resulting performance of the unit, at least as far as air volume and static pressure are concerned. But what about turbulence and noise? They are—and this is an important point—mostly produced by the edges (both leading and trailing edges) and not by the blade surface. Therefore, fewer and wider blades will result in a better fan efficiency and a lower noise level. On the other hand, if the number of blades becomes too small and the blade width, therefore, too large, the fan hub becomes too wide axially and thus heavy, bulky, expensive, and hard to balance. We are facing two conflicting requirements: fewer blades for better efficiency and less noise but more blades for less weight, etc.

Aerodynamically, the optimal number of blades would be one very wide blade, draped around the entire hub, because this would keep the number of blade edges and the turbulence losses to a minimum. One wide blade, therefore, would result in the best efficiency and in the lowest noise level but in an impractical and costly fan wheel. As a compromise between efficiency and cost, five to twelve blades are good practical solutions.

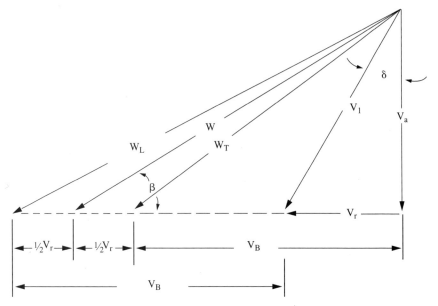

FIGURE 4.33   Velocity diagram for inlet vanes.

$$A_a = \frac{(D^2 - d^2)\,\pi}{576} \qquad \text{annular area in square feet}$$

$V_a = \text{cfm}/A_a \qquad$ axial air velocity in feet per minute through the annulus

$$V_r = \frac{233 \times 10^5}{\text{rpm}} \times \frac{SP}{r} \qquad \text{rotational component in feet per minute of the helical air velocity past the vanes}$$

$SP = $ static pressure in inches of water column

$r = $ radius in inches of the rotating blade section under consideration

$V_1 = \sqrt{V_a^2 + V_r^2} \qquad$ helical air velocity in feet per minute past the vanes

$W_L = $ air velocity in feet per minute relative to the leading edge of the blade

$W_T = $ air velocity in feet per minute relative to the trailing edge of the blade

$W = $ air velocity in feet per minute relative to the blade, average of W_L and W_T

$V_B = (2r\pi/12) \times \text{rpm} \qquad$ blade velocity in feet per minute of the rotating blade section

$$W = \sqrt{V_a^2 + \left(V_B - \frac{1}{2}V_r\right)^2}$$

$$\tan \beta = \frac{V_a}{V_B + \frac{1}{2}V_r} \qquad \text{average direction of the relative air velocity } W \text{ while passing the blade section}$$

$$\tan \delta = \frac{V_r}{V_a} \qquad \text{direction of the air velocity } V_1 \text{ past the vane, entering blade section; also angle of the trailing edge of the guide vane}$$

Width of Blades *l*

The width of the blades is measured along a curve or, to be exact, along the intersection of a cylindrical surface with the blades. However, measuring the blade width straight across will result in almost the same dimension, except for very wide blades. The pressure produced, as mentioned before, is proportional to the product $z_B \times l$.

After we have decided on the number of blades z_B, we must now determine the width of the blades.

At the hub, the blades must be nonoverlapping, for two reasons: Overlapping blades might choke the airflow, and—if a simple sand casting is used—they could not be drawn from the sand. Usually, overlapping blades will be avoided if the blade width l is made equal to or smaller than $1 \le 3.4d/z_B$, where d is the hub diameter and z_B is the number of blades. This, then, will be the blade width at the hub. In some designs, the blade width is constant all the way from the hub to the tip of the blades, but often it varies.

As far as the point of design is concerned, the designer has a certain amount of freedom in selecting the blade width for the outer portion of the blades, since variations in blade width at each radius can be compensated by corresponding variations in the lift coefficient C_L of the profile used in that section. This freedom can be used to control certain performance features, not just concerning the point of design. The first and most natural idea would be to make the blade narrower toward the tip because of the greater blade velocity V_B at the tip. In fact, many fan wheels are designed this way. Sometimes, however, conditions are such that wide-tip blades have certain advantages, such as a significantly lower noise level, a steeper pressure curve, and a higher maximum pressure. However, wide-tip blades result in a deeper stalling dip (especially for large blade angles) and in a larger brake horsepower at the no-delivery point. The fan, of course, should never operate at the no-delivery point, but if by an unforeseen accident it does, the brake horsepower overload might damage the motor. Narrow-tip blades, on the other hand, result in a flatter pressure curve, a larger free-delivery air volume, a lower no-delivery brake horsepower, and less sensitivity of the sound level to inlet turbulence.

Number of Guide Vanes z_V

We have already discussed the number of blades z_B and have reached the conclusion that five to twelve blades will be an acceptable compromise between good fan efficiency and a reasonably low cost. Now we wonder what the number of guide vanes z_V should be. There are only two simple rules for the determination of z_V:

1. z_V should be larger than z_B because the guide vanes should be closer to each other. The risk of choking the airflow due to overlapping is remote.

2. The numbers for z_B and z_V should have no common divisor; otherwise, two blades would pass two vanes simultaneously, thereby increasing the noise level. In other words, the occurrence of blades passing vanes should be staggered rather than concentrated. Suppose that $z_B = 6$. This would rule out the numbers 8, 9, 10, 12, 14, and 15 for z_V. Seven vanes would be spaced too far apart, except in a very small fan, say, a fan of 5-in diameter or less. Thus 11 or 13 vanes would be a good selection.

Suppose that $z_B = 7$. This would be a good selection for the number of blades. It would make the fan wheel fairly narrow in axial direction and easy to balance. And it would go with any number of vanes z_V except 14. In fact, 10, 11, 12, 13, and 15 vanes would all go well with seven blades.

Suppose that we have 13 guide vanes, but 2 or 3 of them have to be skipped because they would interfere with a belt housing or a motor support. The remaining 10 or 11 guide vanes then would be spaced as if there were 13 guide vanes.

TWO-STAGE AXIAL-FLOW FANS

Two-stage axial-flow fans are sometimes a good solution for applications where higher static pressure is required or (as mentioned on page 4.9) where the static pressure per stage should be reduced so that a smaller hub diameter can be used. There are two ways to design a two-stage axial-flow fan: (1) with two fan wheels rotating in the same direction and with guide vanes placed between the two stages (Fig. 4.34) or (2) with two counterrotating fan wheels and no guide vanes at all (Fig. 4.35). By either method, the two-stage unit will approximately double the static pressure.

Two-Stage Axial-Flow Fan with the Same Rotation for Both Stages

In the first configuration, a double-shaft extension motor can be placed between the two fan wheels, as shown in Fig. 4.34, or two separate motors can be used. The sta-

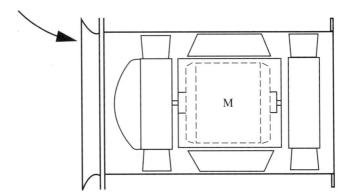

FIGURE 4.34   Two-stage axial-flow fan with a 66 percent hub-tip ratio and direct drive from a double-shaft extension motor. Guide vanes between the two fan wheels act as outlet vanes for the first stage and as inlet vanes for the second stage.

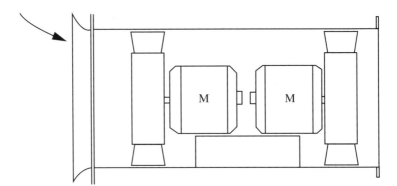

FIGURE 4.35   Two-stage axial-flow fan with a 66 percent hub-tip ratio, two counterrotating fan wheels, direct drive from two motors, and no guide vanes.

tionary guide vanes also will be between the two fan wheels. Their function is shown in Fig. 4.36. They pick up the helical airflow produced by the rotating blades of the first stage and—due to their curvature—reverse the rotational component V_r to the opposite rotation, whereby the air velocity first decelerates and then accelerates again. In other words, they act as outlet vanes for the first stage and as inlet vanes for the second stage, as shown in Figs. 4.34 and 4.36. The rotating blades of the second stage then deflect the airflow back into a more or less axial direction, thereby decelerating the air velocity from V_2 to $V_3 = V_0$. This configuration has the advantage that the same fan wheel usually is used for both stages, even though theoretically the second-stage fan wheel should have slightly smaller blade angles.

Two-Stage Axial-Flow Fan with Counterrotating Stages

In the second configuration, as illustrated in Fig. 4.35, the two fan wheels run in opposite directions and are driven by two separate motors. The air spin produced by

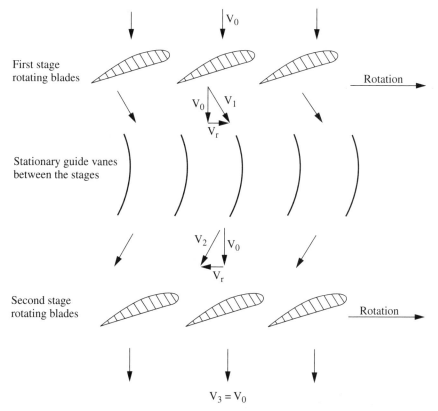

FIGURE 4.36 Function of two-stage axial-flow fan with both stages rotating in the same direction and with guide vanes between the two stages. The stationary vanes reverse the helical airflow from the first stage into the opposite rotation, whereby the velocity first decelerates and then accelerates again. The rotating blades of the second stage then deflect the airflow back into a more or less axial direction, thereby decelerating the air velocity from V_2 to V_3.

the first stage is more or less neutralized by the deflection produced by the second stage, as shown in Fig. 4.37. As a result, no guide vanes are needed, which reduces the manufacturing cost somewhat and compensates for the possible extra expense of two motors instead of one. Another advantage of this configuration is that in case that one of the two motors should fail, the unit can still deliver some air with only one stage running.

Performance of Axial-Flow Fans

The performance of fans is determined by laboratory tests. The test methods are described in a manual prepared jointly by the Air Movement and Control Association, Inc. (AMCA) and the American Society of Heating, Refrigeration and Air Conditioning Engineers, Inc. (ASHRAE). Various standard setups (such as test ducts and test chambers, both for blowing and for exhausting) are specified in the manual. They will be discussed in detail in Chap. 18. After a test has been run, the test data are processed (subjected to certain calculations), and the results are plotted on a graph, such as shown in Figs. 4.38 and 4.39.

Pressure Curve

Figure 4.38 shows the shape of a typical static pressure versus air volume curve. It certainly has a strange shape, first going up, then going down, and then going up again. What causes this obviously inconsistent behavior of an axial-flow fan? Let's

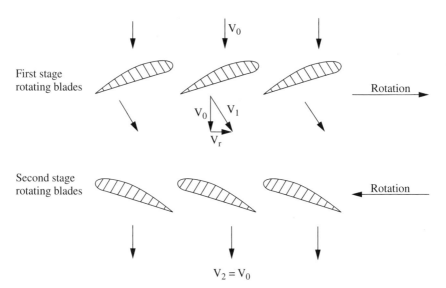

FIGURE 4.37 Function of two-stage axial-flow fan with two counterrotating fan wheels and no guide vanes. The rotating blades of the first-stage fan wheel deflect the airflow into a helical motion, thereby accelerating the velocity from V_0 to V_1. The rotating blades of the second-stage fan wheel then deflect the airflow back into a more or less axial direction, thereby decelerating the air velocity from V_1 to V_2.

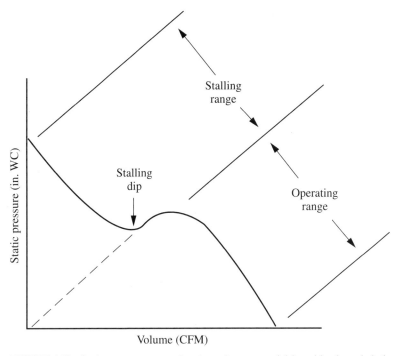

FIGURE 4.38 Static pressure versus air volume for a vaneaxial fan with a large hub-tip ratio and with large blade angles.

analyze it. Starting at the free-delivery ($SP = 0$) point, the static pressure rises to a peak value. This is the good operating range of the fan, from free delivery to the peak pressure. As the air volume—due to increasing restrictions—gradually becomes smaller in this operating range, the axial air velocity V_a gradually decreases, too. As a result, both the angle of attack (between the relative velocity W and the airfoil or single-thickness profile) and therefore the lift coefficient C_L will increase (see Fig. 2.5). This increase in the lift coefficient is the reason why the static pressure increases as the air volume decreases. When the maximum lift coefficient is reached, the angle of attack has become so large that the airflow is no longer able to follow the upper contour of the airfoil or profile, and it separates from that upper contour. The fan then stalls. From here on, the lift coefficient starts to decrease, as shown in Fig. 2.5, and the static pressure decreases with it. If nothing else would happen, the static pressure would go all the way down to zero, as indicated by the dashed line in Fig. 4.38. In other words, the static pressure versus air volume curve would look similar to the C_L versus α curve in Fig. 2.5. It would first rise and then fall. However, something else happens. After a more or less pronounced stalling dip in the static pressure curve, the axial-flow fan starts acting like an inefficient and noisy mixed-flow fan. (Low efficiency and high noise level usually go together in fans because they both are the result of air turbulence and eddies.) As the airflow approaches the fan inlet, the blades throw the air outward by centrifugal force and in this way produce the static pressures of the stalling range, which keep increasing until the point of no delivery (zero cfm) is reached. Figures 4.40 and 4.41 show a comparison of the airflow in the operating range and in the stalling range. In the

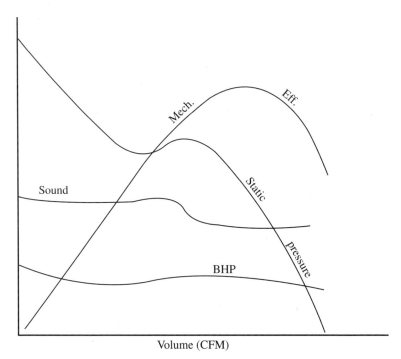

FIGURE 4.39 Static pressure, brake horsepower, mechanical efficiency, and sound level versus air volume for a vaneaxial fan with a large hub-tip ratio and with large blade angles.

operating range, the airflow passing through the unit is smooth and quiet, without any eddies. In the stalling range, the airflow is turbulent, inefficient, and noisy. It crowds toward the tips of the blades, and some of it even forms eddies past the blades and returns to the roots of the blades to be thrown outward again, especially as the point of no delivery is approached.

The depth of the stalling dip is minor if the hub-tip ratio and the blade angles are small. For larger hub-tip ratios and larger blade angles, the stalling dip becomes deeper.

Pressure Safety Margin

Precaution must be taken in the selection of axial-flow fans so that the unit, when installed in the field, will not operate in any part of the inefficient and noisy stalling range. For this reason, a pressure safety margin of 30 to 50 percent should be provided. This means that the maximum operating pressure of the selected unit, i.e., the peak pressure of the operating range, should be 30 to 50 percent higher than the pressure required for the application. This pressure safety margin will allow for possible errors that may have been made in the determination of system resistance and to allow for possible fluctuations of the system.

Using a 30 to 50 percent pressure safety margin is good practice. If the safety margin were too small, there would still be the risk that the installed unit might operate in the stalling range. On the other hand, if the safety margin were larger than really

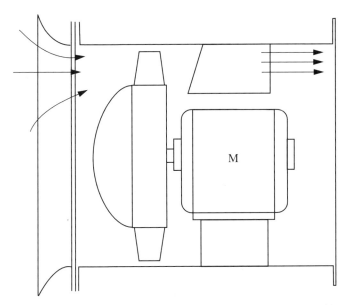

FIGURE 4.40 In the good operating range, the airflow passing through a vaneaxial fan is smooth and has no radial components.

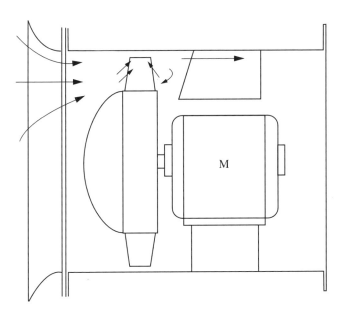

FIGURE 4.41 In the stalling range, the airflow passing through a vane-axial fan is turbulent, with eddies and with radially outward components.

necessary, this would increase the motor horsepower required. Thus 30 to 50 percent is a good compromise between risk and excessive brake horsepower.

Curves for Brake Horsepower, Efficiency, and Sound Level

Coming back to Fig. 4.39, it shows the general shape of the performance curves for an axial-flow fan. Four curves are plotted against volume output (cfm), showing the variation of the static pressure, brake horsepower, mechanical efficiency, and sound level for varying air volumes. We have already discussed the shape of the pressure curve. The following characteristics are to be noted in the shape of the three other curves.

The brake horsepower curve also first goes up, then goes down, and then goes up again, but the variations are less than in the pressure curve. The brake horsepower at the no-delivery point may be higher or lower than the maximum brake horsepower in the operating range, depending on the design. As mentioned previously, wide-tip blades have the advantage of a lower noise level but the disadvantage of motor overload at the no-delivery point.

The efficiency curve has its maximum point at 65 to 80 percent of the peak pressure. As can be seen in Fig. 4.39, the curve drops off on both sides of this maximum point.

The sound-level curve has its minimum somewhere in the operating range, usually near the point of maximum efficiency. It shows little variation within the operating range but shows a sudden increase as the stalling point is approached, where the airflow becomes turbulent. After this sudden increase, which in the range of higher tip speeds is often as much as 25 dB, the unit stays noisy throughout the stalling range and again shows little variation within this range. We might say that the sound-level curve has two levels: a low level in the operating range and a high level in the stalling range. The quality of the sound also changes noticeably from a predominantly musical note in the operating range to a low-pitch rumbling noise in the stalling range. The frequency of the musical note is equal to the number of blades times the revolutions per second. This is so accurate that one can even use this as a method to determine the revolutions per minute (speed) of an unaccessible axial-flow fan installed in a system. We simply check the musical note with a pitch pipe, look up its frequency f in Table 4.1, and calculate the speed from the frequency f and the number of blades z_B using the following formula:

$$\text{rpm} = 60f/z_B \tag{4.12}$$

Suppose the fan has seven blades and the musical note is A_3, having a frequency of 220 vibrations per second. The running speed of the fan then will be $60 \times 220/7 = 1886$ rpm.

Influence of Hub-Tip Ratio on Performance

We have already established that a larger hub diameter enables a vaneaxial fan to produce more static pressure. In the section on minimum hub diameter we introduced the formula

$$d_{\min} = (19{,}000/\text{rpm})\sqrt{SP} \tag{4.1}$$

Solving this equation for static pressure, we get

$$SP = (d_{\min} \times \text{rpm}/19{,}000)^2 \tag{4.13}$$

TABLE 4.1 Musical Notes and Their Frequencies (vibrations per second)

Musical note	Frequency f	Musical note	Frequency f	Musical note	Frequency f
A_2	110	A_3	220	A_4	440
$A_{\#2}$	117	$A_{\#3}$	233	$A_{\#4}$	466
B_2	123	B_3	247	B_4	494
C_3	131	C_4	262	C_5	523
$C_{\#3}$	139	$C_{\#4}$	277	$C_{\#5}$	554
D_3	147	D_4	294	D_5	587
$D_{\#3}$	156	$D_{\#4}$	311	$D_{\#5}$	622
E_3	165	E_4	330	E_5	659
F_3	175	F_4	349	$F_{\#5}$	698
$F_{\#3}$	185	$F_{\#4}$	370	$F_{\#5}$	740
G_3	196	G_4	392	G_5	784
$G_{\#3}$	208	$G_{\#4}$	415	$G_{\#5}$	831
				A_5	880

This indicates that the static pressure produced increases as the square of the hub diameter or, for a constant wheel diameter D, as the square of the hub-tip ratio. In order to illustrate this, let's compare the performances of two vaneaxial fans having the same wheel diameter and running at the same speed but having different hub-tip ratios. Such a comparison of performances is shown in Fig. 4.42. I tested two 29-in vaneaxial fans with outlet vanes and plotted their performance curves. Both fans had five blades and 16° blade angles at the tip, and both ran at 1750 rpm, but their hub-tip ratios were 52 and 68 percent, respectively. Analyzing the graph shown in Fig. 4.42, we find the following differences between the two static pressure curves:

1. The first fan, with the 52 percent hub-tip ratio, produced a maximum static pressure of 3.05 inWC. The second fan, with the 68 percent hub-tip ratio, produced a

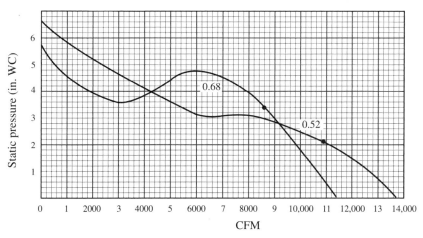

FIGURE 4.42 Comparison of the performances of two 29-in vaneaxial fans at 1750 rpm with five blades, eleven outlet vanes, 16° tip angles, and hub-tip ratios of 52 and 68 percent.

considerably higher maximum static pressure (as could be expected), namely, 4.75 inWC.

2. According to the formula, the second fan should produce $(0.68/0.52)^2 \times 3.05 = 5.22$ inWC of maximum static pressure. The reason why it actually produced about ½ inWC less than this is that the hub diameter is not the only parameter influencing the static pressure. The blade width is the other parameter. It so happened that the second fan had a slightly narrower blade tip, and this resulted in a slightly lower maximum static pressure.

3. On each curve, the point of maximum efficiency is indicated. It occurs at 2.1 inWC for the first fan and at 3.4 inWC for the second fan. According to the formula, the second fan should produce $(0.68/0.52)^2 \times 2.1 = 3.6$ inWC at the point of maximum efficiency. Again, the slightly lower static pressure actually produced is a result of the slightly narrower tip.

4. While the first fan produces less static pressure, it delivers more air volume and therefore has a flatter pressure curve. The reason for this is the larger annular area resulting from the smaller hub diameter.

5. The second fan has a pronounced stalling dip, whereas the first fan has just a slight reversed curvature in the pressure curve. Vaneaxial fans with larger hub-tip ratios have deeper stalling dips.

Influence of the Blade Angle on Performance

Figure 4.43 shows the performance of a typical 36-in vaneaxial fan running at 1750 rpm. This fan has a 23-in hub diameter, corresponding to a 64 percent hub-tip ratio, and ten blades with sheet metal profiles (not airfoils), with the blade widths varying from 7½ in at the hub to 11 in at the tip. The blades had a 12° twist from hub to tip, and the blade angles were varied over a wide range, resulting in tip angles from 13° (10 hp) to a maximum of 33° (50 hp). In Fig. 4.43, we note the following:

1. Seven static pressure curves are shown, for tip angles of 13° (10 hp), 16° (15 hp), 19° (20 hp), 22° (25 hp), 25° (30 hp), 29° (40 hp), and 33° (50 hp).

2. Let's analyze the performance shown for the middle curve, which uses a 22° tip angle and a 25-hp motor. It has a maximum static pressure of 6.2 inWC. The maximum efficiency occurred at a static pressure of 4.5 inWC, so the pressure safety margin was 38 percent. The volume at this point was 22,700 cfm. Calculating the minimum hub diameter, we get $d_{min} = (19,000/\text{rpm})\sqrt{SP} = (19,000/1750)\sqrt{4.5} = 23.0$ in. This is the hub diameter we actually used. Calculating the minimum wheel diameter, we get $D_{min} = \sqrt{d^2 + \text{cfm/rpm}} = \sqrt{13^2 + 61 \times 22,700/1750} = \sqrt{529 + 791} = \sqrt{1320} = 36.3$ in. We actually used a 36-in wheel diameter.

3. As the blade angles were increased from 13° to 33°, the air volume (cfm) doubled, while the maximum static pressure increased by about 50 percent and the stalling dip deepened.

Some manufacturers make axial-flow fan wheels with adjustable-pitch airfoil blades. This angular adjustment can be accomplished in various ways. My U.S. patent no. 4,610,600 describes one method that maintains good fan efficiency. Usually, the adjustment is done with the power shut off, but some designs permit automatic "in-flight" adjustment.

From the performance curves shown for this 36-in vaneaxial fan at 1750 rpm, one could calculate the performance curves for other running speeds and for other sizes as long as they are in geometric proportion with the 36-in size. These conversions for

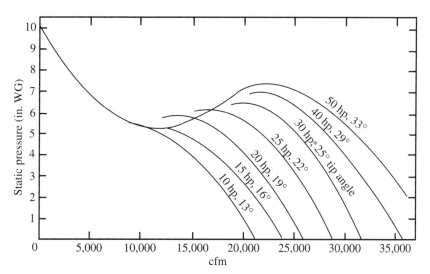

FIGURE 4.43 Performance of a 36-in vaneaxial fan at 1750 rpm with tip angles varied from 13° (10 hp) to 33° (50 hp). *(From Bleier, F. P., Fans, in Handbook of Energy Systems Engineering. New York: Wiley, 1985.)*

other speeds and sizes are done by using certain formulas known as *fan laws*. These will be discussed in Chap. 5.

Comparison of Performance for the Four Types of Axial-Flow Fans

Figure 4.44 shows a comparison of the static pressure curves for the four types of axial-flow fans, all having the same wheel diameter and the same running speed. While the general shape of the four curves is similar, the range of static pressures is quite different. The propeller fan delivers the largest air volume but produces the lowest static pressure and therefore has the flattest pressure curve. The tube-axial fan is in the middle. The vaneaxial fan delivers less volume in the low-pressure range but produces the highest static pressure and therefore has the steepest pressure curve.

Figure 4.44 also shows some parabolic curves that intersect the pressure curves. These parabolic curves are called *system characteristics* because they characterize the system (ducts, coils, elbows, dampers, etc.), the same as the pressure curves characterize the fan. System characteristics will be discussed in more detail in Chap. 5. For the time being, let's just say that they show the static pressure needed to overcome the resistance of the system and that this static pressure increases as the square of the air volume to be blown or drawn through the system. The point of intersection of the fan's pressure curve with the system characteristic will be the point at which the fan will operate in this system.

Table 4.2 shows the highlights of this discussion of the four types of axial-flow fans: propeller fans, tubeaxial fans, vaneaxial fans, and two-stage axial flow fans. This table again shows that the sequence propeller, tubeaxial, and vaneaxial indicates the general trend of increasing hub diameter, static pressure, and efficiency. The efficiencies shown on the last line of the table are the maximum mechanical efficiencies that can be obtained with a superdeluxe model. Practical fan efficiencies usually are slightly lower. Belt-driven units, again, have slightly lower efficiencies.

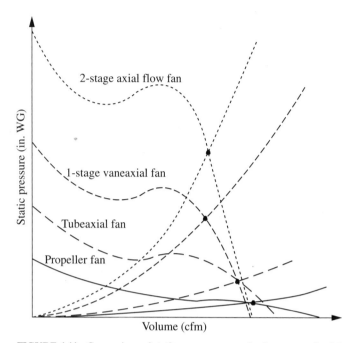

FIGURE 4.44 Comparison of static pressure curves for four types of axial-flow fans having the same wheel diameter and the same running speed. Parabolic curves indicate four system characteristics through the maximum efficiency point of each fan. (*From Bleier F. P., Fans, in Handbook of Energy Systems Engineering. New York: Wiley, 1985.*)

INFLUENCE OF THE TIP CLEARANCE ON THE PERFORMANCE OF VANEAXIAL FANS

Figures 4.45 through 4.47 show the results of a series of tests I conducted on a 29-in vaneaxial fan running at 1750 rpm. I started with a unit having a ¹⁄₁₆-in tip clearance and then machined the fan wheel outside diameter down by ¹⁄₁₆ in on the radius for a ¹⁄₈-in tip clearance and retested the unit. I repeated this process several times until I finally had a 1-in tip clearance. Not that I ever expected to use such a large tip clearance. I just wanted to determine exactly what the performances at these different tip clearances would be. These curve sheets indicate that for increasing tip clearances, the following changes in performance occur:

1. The air volume decreases, but not too much.

2. The maximum static pressure in the operating range decreases considerably, from 2.5 in at a ¹⁄₁₆-in tip clearance to 1 in at a 1-in tip clearance.

3. The brake horsepower in the operating range decreases considerably, but not as much as the static pressure.

4. As a result, the mechanical efficiencies decrease considerably, from a maximum of 82 percent at a ¹⁄₁₆-in tip clearance, to 72 percent at ⁵⁄₁₆ in, to 60 percent at ¹³⁄₁₆ in, to 58 percent at 1 in.

TABLE 4.2 Typical Features of Axial-Flow Fans with Direct Drive

Type of fan	Propeller fan	Tubeaxial fan	Single-stage vaneaxial fan	Two-stage axial-flow fan
Casing	Mounting ring or Mounting panel	Short cylindrical housing	Cylindrical housing	Long cylindrical housing
Motor support	Inlet side of panel preferred	Inside housing, outlet side preferred	Inside housing, outlet side preferred	Inside housing, between the two stages
Guide vanes	None	None	Past fan wheel preferred	Between the two stages or none
Hub-tip ratio	0–40%	30–50%	45–80%	50–80%
SP (inWC)	0–1	½–2½	1–9	4–18
Blade angle at hub	30–50°	30–50°	30–60°	30–60°
Blade angle at tip	10–25°	10–25°	10–35°	10–35°
Maximum mechanical efficiency	70%	75%	90%	70%

5. The noise level at free delivery increases somewhat, from 73.5 dB at a ¹⁄₁₆-in tip clearance to 76 dB at a ¾-in tip clearance.

6. The noise level in the stalling range goes up and down periodically, fluctuating between 96 and 108 dB. This seems to be the result of a resonance condition. It reaches maxima at tip clearances of ¼, ½, and ¾ in. This may be an interesting finding, but it is of no practical value, since the fan should never be used in the stalling range.

What is of practical value is what we found about the performance in the operating range, that the smallest possible tip clearance will result in optimal performance in all respects: pressure, efficiency, and noise level. In small vaneaxial fans, it would be worthwhile to machine the wheel outside diameter and the housing inside diameter so that a small tip clearance can be held. In larger fans, this would be too expensive, and as a compromise between cost and performance, a somewhat larger tip clearance will have to be accepted.

Tubeaxial and propeller fans produce lower pressures and efficiencies than vaneaxial fans. A larger tip clearance, therefore, is less harmful in these fans.

VANEAXIAL FANS WITH SLOTTED BLADES

Figure 2.6 showed how the airflow will pass smoothly over an airfoil blade, thereby producing a lift (see Fig. 2.1), most of which is the result of negative suction pressure on the top side of the airfoil. This lift results in the static pressure produced by the fan. As the angle of attack gets larger and larger, the lift and therefore the static pres-

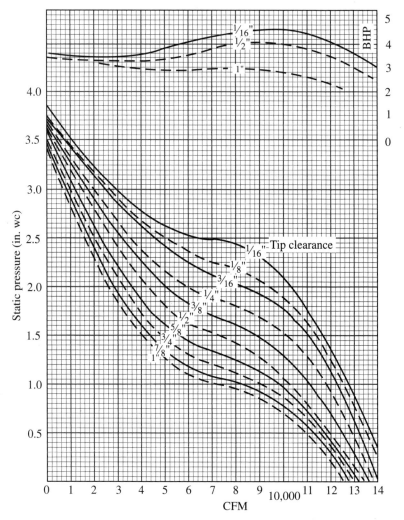

FIGURE 4.45 Influence of tip clearance on the pressure and brake horsepower curves of a 29-in vaneaxial fan running at 1750 rpm.

sure become larger too, but there comes a point where the angle of attack becomes so large that the airflow can no longer follow the airfoil contour and will separate from the upper (suction) side of the airfoil, as shown in Fig. 2.7. In a typical airfoil, this will happen when the angle of attack reaches about 10° to 15°.

Please note that this 10° to 15° is not the blade angle. The blade angle is much larger. The blade angle is $\beta + \alpha$, where β is the air angle, which can be calculated from the velocity diagrams (Fig. 4.32 for outlet vanes or Fig. 4.33 for inlet vanes):

$$\tan \beta = \frac{V_a}{V_B - \frac{1}{2}V_r} \qquad \text{for outlet vanes} \qquad (4.8)$$

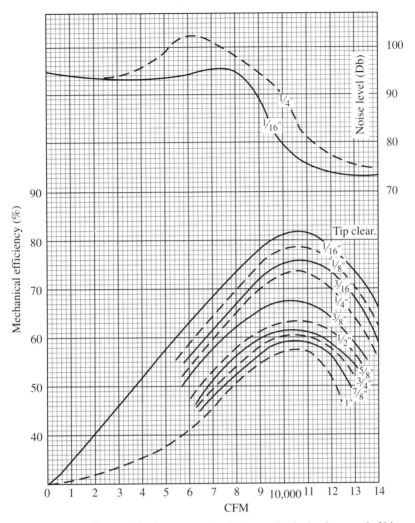

FIGURE 4.46 Influence of tip clearance on the efficiency and noise-level curves of a 29-in vaneaxial fan running at 1750 rpm.

$$\tan \beta = \frac{V_a}{V_B + \frac{1}{2}V_r} \qquad \text{for inlet vanes} \qquad (4.11)$$

As mentioned previously, outlet vanes are preferable in most applications.

In other words, whenever the angle of attack exceeds about 15°, separation of airflow will occur, and the stalling range of the vaneaxial fan will start.

The so-called slotted blade was patented by Dr. Herman E. Sheets in 1943 and is manufactured in small sizes by General Dynamics Corporation. The slotted blade will delay separation of the airflow and thereby extend the good operating range of

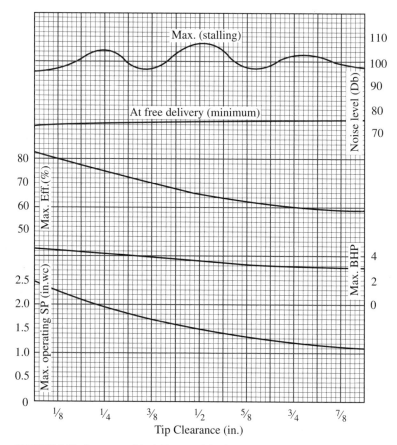

FIGURE 4.47 Summary of the influence of tip clearance on the performance of a 29-in vaneaxial fan running at 1750 rpm.

a vaneaxial fan. It will permit larger blade angles and larger angles of attack, resulting in higher lift coefficients and thereby in higher maximum static pressures before stalling starts. How does the slotted blade accomplish this?

Figure 4.48 shows how the slotted blade functions. It really is not one slotted blade but consists of two slightly overlapping airfoil blades with a slot between the trailing edge of the first blade and the leading edge of the second blade. Figure 4.48 shows this configuration and the air velocities relative to the rotating blades. As the airflow approaches the leading edge of the first blade, it divides into an upper airflow and a lower airflow, the same as in a normal (not slotted) blade.

The lower airflow passes under the bottom of the first blade and then again divides into two branches. Most of it will pass under the second blade and past the second blade, but a small portion will pass through the slot between the two blades with a high velocity because the slot is narrow. After passing through the narrow slot, this small portion of the lower airflow then will flow over the top of the second blade and join the rest of the lower airflow.

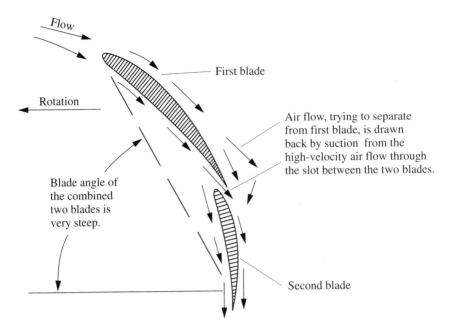

FIGURE 4.48 Function of the slotted blade. The flow arrows shown represent relative air velocities (relative to the moving blades). The absolute air velocities would be the vector sum of the relative air velocities plus the blade velocity. The absolute air velocities past the blades would have a strong spin to the left, which then would be neutralized by the outlet guide vanes, the same as for a solid blade.

The upper airflow will pass over the top of the first blade and then over the top of the second blade, as long as the angle of attack at the first blade is small. However, as the resistance of the system increases, the angle of attack at the first blade increases too, until a point is reached where the angle of attack at the first blade becomes too large and the airflow can no longer follow the upper contour of the first airfoil so that the airflow would separate from the top contour of the first airfoil, if nothing else happens.

However, something else does happen that will delay the separation of the airflow from the top of the first airfoil. Actually, two things happen:

1. One is the high-velocity air stream through the slot. According to Bernoulli, this high-velocity air stream will produce a low-pressure area—in fact, a negative suction pressure past the slot. This suction pressure will draw the airflow coming from the top of the first blade back to the top contour of the second blade.

2. The second thing is the negative pressure produced by the suction side of the second airfoil. This negative pressure assists in drawing the airflow back. Both phenomena together will prevent flow separation for awhile.

This means that the slotted blade can go to steeper blade angles, larger lift coefficients, and higher static pressures. Furthermore, because of the larger blade angles, there is room for more blades, which again results in higher static pressures. The maximum static pressure thus obtained easily can be double that of a normal blade.

The air volume also is increased due to the larger blade angles. This means that a certain requirement for air volume and static pressure can be fulfilled with a lower speed or with a smaller fan. This will make up for some of the increased cost of the slotted blade.

The high static pressure produced by the slotted blade results in a deep stalling dip and in a no-delivery static pressure that is smaller than the maximum static pressure in the operating range. With this unconventional appearance of the pressure curve, the user must be doubly careful that the fan will not operate in the stalling range.

APPLICATIONS WITH FLUCTUATING SYSTEMS

Figure 4.39 showed the usual shape of the performance curves for a vaneaxial fan. The brake horsepower curve shown has a no-delivery value just slightly above the maximum brake horsepower in the operating range. This is still acceptable. In some designs, however, the no-delivery brake horsepower rises considerably above the maximum brake horsepower in the operating range. This happens whenever the blade width at the tip is much wider than the blade width at the hub and the blade angles are small. As mentioned, wider tips have the advantage of lower noise levels.

Even though the fan should never operate in the stalling range, there are installations where the fluctuations of the system become so great that—despite the pressure safety margin—the operating point is temporarily switched into the stalling range, sometimes even to the no-delivery point. This may happen, for example, when wind pressure works against the outlet of a system, when intake dampers for the makeup air of an exhaust system are accidentally closed, or when the cooling coils of a refrigerating system freeze up.

In applications where this could happen, the unit should be designed in such a way that the no-delivery brake horsepower is kept below the motor horsepower. In other words, wide-tip blades may have to be avoided in such applications. The problem of controlling the no-delivery brake horsepower is hardest in cases of small aerodynamic load, such as propeller fans and fans with small blade angles. The problem is relatively easier in cases of heavy aerodynamic load such as vaneaxial fans with large blade angles or with slotted blades.

Since the operating range is the most important range, it occasionally may happen that the no-delivery brake horsepower would overload the motor. In such cases, the motor should be equipped with overload protection.

NOISE LEVEL

The noise level produced by well-designed axial-flow fans is lower than that of centrifugal fans of the same tip speed, but it is more sensitive to the effect of turbulent airflow (which will raise the noise level). Here is a list of factors (in the order of importance) that produce noise and therefore should be avoided when quiet operation is desired:

1. Operation in the stalling range
2. High tip speed
3. Lack of an inlet bell if installed without an inlet duct

4. Obstructions in the air stream ahead of and close to the blades (support arms, belt housing, conduit pipes)

5. Elbows in the duct work ahead of and close to the fan inlet

6. Inlet guide vanes as opposed to outlet guide vanes

7. Obstructions in the air stream past and close to the fan blades

8. Vibrations due to poor balance or due to a resonance condition

9. Single-thickness blades as opposed to airfoil blades.

10. Many narrow blades as opposed to fewer and wider blades (because as mentioned previously, it is the blade edges, particularly the trailing edges, rather than the blade surfaces that produce turbulence and therefore noise)

Figure 4.49 shows a comparison of the noise-level curves for two vaneaxial fans of the same size and speed, one having outlet vanes (solid line) and one having inlet vanes (dotted line). The curve for outlet vanes has the typical shape that was shown in Fig. 4.39. The pronounced difference between the operating and stalling levels indicates that the airflow in the operating range is smooth and stable. If it were turbulent, the operating level would be almost as high as the stalling level.

The curve for inlet vanes has the same basic shape, but it shows a higher operating noise level than the curve for outlet vanes. The difference between the two ranges, therefore, is smaller for inlet vanes. This is due to the increase in the relative air velocities caused by the inlet vanes, which naturally has the strongest effect in the operating range and practically no effect in the stalling range.

OUTLET DIFFUSER AND OUTLET TAIL PIECE

An outlet diffuser and an outlet tail piece, as shown in Fig. 4.7, are sometimes used in large vaneaxial fans. Their function is to provide for a gradual increase in the annular area toward the outlet and consequently for a gradual decrease in the air velocity, thereby avoiding abrupt changes and improving the flow conditions with

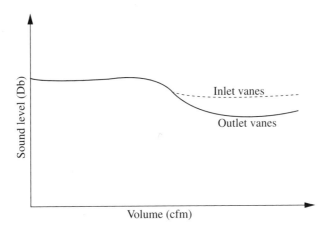

FIGURE 4.49 Sound-level curves for vaneaxial fans.

respect to static regain, i.e. (see page 1.12), the reconversion of useless to useful energy. For best results, the two cones, one diverging and one converging, are long and have small angles, about 6° with the center line. Such a diffuser and tail piece will increase the fan efficiencies by about 4 percent, due to a slight increase in air volume and static pressure. This may permit a slight blade angle reduction and consequently a slight reduction in the brake horsepower. In large units, such as those used for mine ventilation, this brake horsepower reduction may be considerable.

The diffuser and tail piece, on the other hand, have the disadvantages that they make the unit longer and more expensive, and in case of direct motor drive, they reduce the motor cooling effect of the air stream and the accessibility of the motor. They are therefore often omitted in small and medium-sized units. In very large units, however, where belt drive is used and the motor is located outside the housing, the outlet diffuser and tail piece are usually included.

SELECTION OF AXIAL-FLOW FANS

Table 4.2 indicated that the sequence propeller, tubeaxial, vaneaxial, and two-stage axial indicates the general trend of increasing static pressure. For even higher pressures, centrifugal fans are used, as will be discussed in Chap. 7. The ranges of the various types overlap. For example, the borderline between tubeaxial and vaneaxial fans is quite flexible. In addition, the range for centrifugal fans widely overlaps with the range for two-stage axial-flow fans and even with the range for single-stage vaneaxial fans. Keeping all this in mind, I will give four examples of how to select axial-flow fans. This section will explain certain procedures used in fan selection. The same procedures then can be applied in other fan selections.

The problem of fan selection usually presents itself in the following manner: A certain air volume and static pressure are required for a certain system. Often the customer will even make a fan selection, by suggesting the type and size of fan, the fan speed, and the motor horsepower. The first step will be to check whether all these data are acceptable and whether perhaps a better selection could be made.

Example 1: Suppose the requirements call for 20,600 cfm at a static pressure of 2 inWC to be produced by a 30-in vaneaxial fan with a belt drive at 2000 rpm from a 10-hp motor. Let's check these data. A 30-in vaneaxial fan will have an outlet area OA of 5.1 ft². As mentioned in Chap. 1, the outlet velocity OV will be 20,600/5.1 = 4039 fpm, and the corresponding velocity pressure VP will be $(4039/4005)^2 = 1.02$ inWC. The total pressure will be $TP = SP + VP = 2.00 + 1.02 = 3.02$ in WC, and the air horsepower *ahp* will be (cfm × TP)/6356 = (20,600 × 3.02)/6356 = 9.29. With a 10-hp motor, as required by the customer, this fan would have to have a 93 percent mechanical efficiency, which is more than can be expected. This means that a 10-hp motor would be overloaded. A 15-hp motor will be needed in order to get 20,600 cfm at a static pressure of 2 inWC from a 30-in vaneaxial fan.

Please note that the fan speed did not enter into this calculation. In other words, in order to get 20,600 cfm at a static pressure of 2 inWC from a 30-in vaneaxial fan, a 15-hp motor will be required, regardless of what the fan speed will be.

Next, let's calculate the minimum hub diameter d_{min} and the minimum wheel diameter D_{min} for these requirements, using our formulas (see pages 4.7 and 4.9). The minimum hub diameter d_{min} will be $d_{min} = (19,000/\text{rpm})\sqrt{SP} = (19,000/2000)\sqrt{2} = 13.44$ in. We decide to make $d = 14$ in. This will give us a hub-tip ratio of 14/30 = 0.47. Table 4.2 shows that vaneaxial fans have hub-tip ratios from 45 to 80 percent, so our 47 percent is within this range.

Next, we calculate the minimum wheel diameter D_{min}. It will be

$$D_{min} = \sqrt{d^2 + 61(\text{cfm/rpm})} = \sqrt{14^2 + 61(20,600/2000)}$$

$$= \sqrt{196 + 628} = 28.71 \text{ in}$$

Our 30-in wheel diameter is larger than the 28.71-in minimum wheel diameter we obtained by calculation. The 30-in wheel diameter suggested by the customer, therefore, is a good figure.

Now we are ready to study the rating tables published in vaneaxial fan catalogs. Table 4.3 shows such a rating table for a 30-in vaneaxial fan with belt drive, taken from the vaneaxial fan catalog of the Ammerman Division of the General Resource Corporation. Figure 4.29 shows a photograph of this belt-driven vaneaxial fan. From table 4.3 we note the following:

1. This is the customary format to present the performance of belt-driven vaneaxial fans, with the static pressures shown on top, the volumes and outlet velocities shown on the left side, and the speed (rpms) and brake horsepowers shown at the cross points. Later we will explain how these performance tables can be calculated from performance tests.

2. For this 30-in vaneaxial fan, we find figures rather close to our requirements. For 20,629 cfm and a static pressure of 2 inWC, this fan will run at 2061 rpm and consume 14.1 bhp. A 15-hp motor would be all right.

As a rule, we can obtain the same air volume and static pressure with a larger fan at a lower speed. This larger fan will have two advantages:

1. It will have a lower power consumption (bhp) and therefore a lower operating cost.
2. It will have a lower tip speed and therefore a lower noise level.

The smaller fan, on the other hand, also will have two advantages:

1. It will be more compact.
2. It will have a lower first cost.

To check these four points, let us look at the rating table for the next larger size in the Ammerman catalog, i.e., the 36-in size. It is shown in Table 4.4. We find the following comparative data for these two sizes of vaneaxial fans:

Size (in)	30	36
Speed (rpm)	2061	1395
Volume (cfm)	20,629	20,625
SP (inWC)	2.00	2.00
Brake horsepower (bhp)	14.1	11.1
Tip speed (fpm) =	16,187	13,148
$= (D/12)\pi \text{rpm}$		

Comparing these figures for the 30- and 36-in sizes, we find the following:

1. For the 36-in size, the tip speed will be 19 percent lower, for a reduced noise level, an important consideration.

TABLE 4.3 Performance Table for a 30-in Vaneaxial Fan with Belt Drive from the Catalog of the Ammerman Division of General Resource Corporation

Vaneaxial fan Size 30

Belt-driven Vaneaxial fan - Size 30

Inlet and outlet diameter = 30 3/4 in
Inlet and outlet area = 5.16 ft²
Tip speed = 7.969 × rpm
Max. bhp = 1.680$(rpm/1000)^3$

Volume, cfm	OV, fpm	1/4-in SP rpm	bhp	1/2-in SP rpm	bhp	3/4-in SP rpm	bhp	1-in SP rpm	bhp	1 1/4-in SP rpm	bhp	1 1/2-in SP rpm	bhp	1 3/4-in SP rpm	bhp
5,157	1000	593	0.35	749	0.69	905	1.13								
6,189	1200	656	0.46	788	0.81	917	1.24	1036	1.80						
7,220	1400	724	0.62	835	0.98	945	1.40	1056	1.91						
8,252	1600	797	0.81	894	1.18	991	1.61	1089	2.11	1166	2.50	1279	3.18	1379	4.18
9,283	1800	876	1.04	960	1.44	1046	1.89	1132	2.40	1188	2.70	1287	3.37	1395	4.37
10,314	2000	955	1.32	1031	1.76	1109	2.26	1187	2.77	1220	2.99	1309	3.65	1421	4.68
11,346	2200	1034	1.65	1105	2.14	1176	2.66	1246	3.21	1264	3.37	1343	4.01	1457	5.10
12,377	2400	1115	2.04	1181	2.58	1245	3.09	1310	3.70	1317	3.82	1387	4.44	1502	5.64
13,409	2600	1199	2.51	1258	3.05	1318	3.65	1378	4.26	1374	4.33	1438	4.96	1555	6.29
14,440	2800	1282	3.04	1338	3.61	1393	4.26	1450	4.89	1436	4.89	1495	5.57	1613	7.00
15,472	3000	1366	3.64	1419	4.29	1470	4.94	1523	5.61	1503	5.55	1558	6.28	1675	7.84
16,503	3200	1451	4.34	1500	5.01	1548	5.66	1597	6.38	1574	6.31	1624	7.07	1740	8.75
17,535	3400	1536	5.14	1581	5.79	1626	6.48	1672	7.21	1646	7.14	1693	7.98	1808	9.66
18,566	3600	1618	5.99	1664	6.68	1707	7.42	1750	8.19	1719	8.05	1764	8.89	1879	10.7
19,598	3800	1702	6.90	1746	7.70	1788	8.47	1828	9.31	1793	9.03	1836	9.87	1950	11.9
20,629	4000	1789	7.98	1830	8.82	1869	9.59	1907	10.5	1869	10.2	1910	11.0	2023	13.2
21,660	4200	1877	9.24	1913	10.0	1950	10.9	1987	11.8	1947	11.3	1985	12.3	2097	14.5
22,692	4400	1962	10.6	1998	11.3	2033	12.3	2069	13.2	2024	12.7	2061	13.6	2174	16.0
23,723	4600	2048	12.0	2083	12.9	2116	13.8	2150	14.7	2103	14.1	2139	15.1	2252	17.6
24,755	4800	2136	13.7	2168	14.5	2200	15.4	2232	16.4	2183	15.7	2218	16.7		
25,786	5000	2224	15.3	2254	16.2	2285	17.2	2316	18.1	2265	17.4	2297	18.4		

Volume, cfm	OV, fpm	2-in SP		2½-in SP		3-in SP		3½-in SP		4-in SP		4½-in SP		5-in SP	
		rpm	bhp	rpm	bhp	rpm	bhp	rpm	bhp	rpm	bhp	rpm	bhp	rpm	bhp
8,252	1600	1469	4.89												
9,283	1800	1478	5.08	1650	6.85										
10,314	2000	1497	5.40	1656	7.14										
11,346	2200	1527	5.85	1672	7.56	1807	9.17								
12,377	2400	1568	6.41	1698	8.05	1814	9.45	1953	11.5						
13,409	2600	1615	7.07	1734	8.61	1830	9.87	1960	11.9	2089	14.1	2214	16.6	2335	19.5
14,440	2800	1667	7.77	1778	9.38	1855	10.4	1977	12.5	2097	14.6	2217	17.0	2338	20.1
15,472	3000	1726	8.61	1830	10.3	1891	11.2	2004	13.2	2116	15.3	2226	17.6	2350	20.9
16,503	3200	1787	9.52	1883	11.2	1934	12.1	2039	14.0	2143	16.1	2246	18.3	2375	21.8
17,535	3400	1852	10.5	1942	12.2	1981	13.1	2079	15.0	2176	17.0	2275	19.3	2407	22.8
18,566	3600	1920	11.6	2005	13.4	2034	14.1	2126	16.1	2218	18.1	2312	20.4	2440	24.0
19,598	3800	1991	12.8	2072	14.7	2091	15.3	2179	17.4	2265	19.5	2353	21.6		
20,629	4000	2061	14.1	2140	16.0	2154	16.7	2235	18.8	2316	20.9	2397	23.0		
21,660	4200	2134	15.5	2210	17.4	2218	18.0	2293	20.2	2371	22.3				
22,692	4400	2210	17.0	2281	19.0	2284	19.5	2355	21.7						
23,723	4600	2286	18.7			2352	21.2								

Note: bhps shown do not include bell-drive losses.

TABLE 4.4 Performance Table for a 36-in Vaneaxial Fan with Belt Drive from the Catalog of the Ammerman Division of General Resource Corporation

Vaneaxial Fan Size 36

Belt-driven vaneaxial fan - Size 36

Inlet and outlet diameter = 36¾ in
Inlet and outlet area = 7.37 ft²
Tip speed = 9.523 × rpm
Max. bhp = 4.094(rpm/1000)³

Volume, cfm	OV, fpm	¼-in SP		½-in SP		¾-in SP		1-in SP		1¼-in SP		1½-in SP		1¾-in SP	
		rpm	bhp	rpm	bhp	rpm	bhp	rpm	bhp	rpm	bhp	rpm	bhp	rpm	bhp
7,366	1000	496	0.50	627	0.98	757	1.62								
8,839	1200	549	0.66	659	1.16	767	1.77	867	2.57						
10,312	1400	606	0.88	699	1.40	791	2.00	884	2.73	976	3.57	1070	4.54		
11,786	1600	667	1.15	748	1.69	829	2.30	911	3.02	994	3.86	1077	4.82	1154	5.97
13,259	1800	733	1.48	803	2.06	875	2.70	947	3.43	1021	4.27	1095	5.22	1167	6.24
14,732	2000	799	1.88	863	2.52	928	3.23	993	3.96	1058	4.81	1124	5.73	1189	6.68
16,205	2200	865	2.35	925	3.06	984	3.80	1043	4.59	1102	5.45	1161	6.34	1219	7.28
17,678	2400	933	2.92	988	3.68	1042	4.42	1096	5.29	1150	6.18	1203	7.08	1257	8.05
19,152	2600	1003	3.58	1053	4.36	1103	5.21	1153	6.09	1202	6.99	1251	7.95	1301	8.98
20,625	2800	1073	4.34	1120	5.16	1166	6.08	1213	6.99	1258	7.93	1304	8.97	1350	10.0
22,098	3000	1143	5.20	1187	6.12	1230	7.05	1274	8.01	1317	9.01	1359	10.1	1402	11.2
23,571	3200	1214	6.20	1255	7.16	1295	8.09	1336	9.11	1377	10.2	1417	11.4	1456	12.5
25,044	3400	1285	7.34	1323	8.27	1361	9.26	1399	10.3	1438	11.5	1476	12.7	1513	13.8
26,518	3600	1354	8.56	1392	9.54	1428	10.6	1464	11.7	1500	12.9	1536	14.1	1572	15.3
27,991	3800	1424	9.85	1461	11.0	1496	12.1	1530	13.3	1564	14.5	1598	15.7	1632	17.0
29,464	4000	1497	11.4	1531	12.6	1564	13.7	1596	15.0	1629	16.2	1661	17.5	1693	18.8
30,937	4200	1571	13.2	1601	14.3	1632	15.5	1663	16.8	1694	18.1	1725	19.4	1755	20.7
32,410	4400	1642	15.1	1672	16.2	1701	17.5	1731	18.8	1760	20.2	1790	21.5	1819	22.8
33,884	4600	1714	17.2	1743	18.4	1771	19.7	1799	21.0	1827	22.4	1856	23.8	1884	25.2
35,357	4800	1787	19.5	1814	20.7	1841	22.0	1868	23.4	1895	24.8	1922	26.3		
36,830	5000	1861	21.9	1886	23.2	1912	24.5	1938	25.9						

Volume, cfm	OV, fpm	2-in SP rpm	2-in SP bhp	2½-in SP rpm	2½-in SP bhp	3-in SP rpm	3-in SP bhp	3½-in SP rpm	3½-in SP bhp	4-in SP rpm	4-in SP bhp	4½-in SP rpm	4½-in SP bhp	5-in SP rpm	5-in SP bhp
11,786	1600	1229	6.99												
13,259	1800	1237	7.26	1381	9.79										
14,732	2000	1253	7.71	1386	10.2	1512	13.1								
16,205	2200	1278	8.35	1399	10.8	1518	13.5								
17,678	2400	1312	9.15	1421	11.5	1531	14.1	1634	16.4						
19,152	2600	1351	10.1	1451	12.3	1552	14.9	1640	17.0	1748	20.1	1853	23.7		
20,625	2800	1395	11.1	1488	13.4	1582	16.0	1654	17.8	1755	20.9	1855	24.3	1954	27.8
22,098	3000	1444	12.3	1531	14.7	1618	17.3	1677	18.8	1771	21.8	1863	25.1	1956	28.7
23,571	3200	1495	13.6	1576	16.0	1658	18.7	1706	20.0	1793	23.0	1879	26.2	1966	29.8
25,044	3400	1550	15.0	1625	17.4	1702	20.2	1740	21.4	1821	24.3	1904	27.5	1987	31.1
26,518	3600	1607	16.6	1678	19.1	1750	21.9	1779	23.0	1856	25.9	1935	29.1	2014	32.6
27,991	3800	1666	18.3	1734	21.0	1802	23.8	1823	24.8	1895	27.8	1969	30.9	2042	34.3
29,464	4000	1725	20.1	1791	22.9	1856	25.7	1870	26.8	1938	29.8	2006	32.9		
30,937	4200	1786	22.1	1849	24.9	1911	27.8	1919	28.8	1984	31.9				
32,410	4400	1849	24.3	1909	27.2	1968	30.3	1971	31.0						
33,884	4600	1913	26.7												

Note: bhps shown do not include belt-drive losses.

4.51

2. For the 36-in size, the brake horsepower will be 21 percent lower, so the operating cost will be considerably lower. This will offset the higher first cost of the 36-in fan. With an 11.1 bhp, we might even get by with a 10-hp motor. This, again, will compensate for the higher first cost of the 36-in fan.

In view of these conditions, the 36-in vaneaxial fan probably will be preferable.

Example 2: An inexpensive fan should be selected that will deliver 5100 cfm or preferably more against a static pressure of ¼ inWC and that will be quiet enough for installation as an exhaust fan in a record store. Since the static pressure is only ¼ inWC, it will be a propeller fan, as pictured in Fig. 4.50. We look at Table 4.5, showing an Ammerman catalog sheet for propeller fans, both for direct drive and for belt drive. We select a 36-in propeller fan with belt drive at 575 rpm from a ¾-hp motor. This fan will deliver 5142 cfm at a static pressure of ¼ inWC. It has a low tip speed of 5419 fpm, so it will have a low noise level.

If this fan were for installation in a factory, where quiet operation is not required, we would select from the same Ammerman catalog sheet a 24-in propeller fan with direct drive from a ¾-hp, 1140-rpm motor. This fan will deliver 5811 cfm at a static pressure of ¼ inWC. It will have a tip speed of 7163 fpm (32 percent larger) and therefore will have a higher noise level, but it will deliver 13 percent more volume and it will be less expensive due to the smaller size and the direct-drive arrangement, which eliminates two pulleys, two bearings, a bearing base, and a belt. The ¾-hp, 1140-rpm motor will be slightly more expensive than the ¾-hp, 1750-rpm motor used for the belt-drive arrangement in the 36-in propeller fan.

Example 3: An axial-flow fan should be selected that will deliver 25,000 cfm at a static pressure of 3½ inWC, a considerable static pressure that probably will require a vaneaxial fan but possibly could be produced by a less expensive tubeaxial fan. We first look at an old catalog by International Engineering, Inc., Dayton, Ohio, covering their line of tubeaxial fans. International calls them "duct boosters." Figure 4.51 shows the fan wheel. It has a 52 percent hub-tip ratio and eight wide blades. Table 4.6 shows the performance for International's 15 direct-drive models. The ninth model

Direct drive

Belt drive

FIGURE 4.50 Propeller fans with sheet metal blades for low cost. *(Courtesy of Ammerman Division of General Resource Corporation, Hopkins, Minn.)*

has a 36-in diameter and is driven by a 30-hp, 1750-rpm motor. By interpolation between static pressures of 3 and 4 inWC, we find that this tubeaxial fan will deliver the required 25,000 cfm at a static pressure of 3½ inWC.

Looking again at Table 4.4, showing the performance for Ammerman's 36-in vaneaxial fan, we find that the required 25,000 cfm at a static pressure of 3½ inWC can be obtained with their 36-in vaneaxial fan at 1779 rpm and that this fan will consume only 23.0 bhp, due to the fact that a vaneaxial fan has a higher efficiency than a tubeaxial fan and therefore consumes less brake horsepower for the same duty. This means that the 36-in vaneaxial fan will have two advantages over the 36-in tubeaxial fan:

1. The 36-in vaneaxial fan will have a lower operating cost because of lower power consumption.
2. The 36-in vaneaxial fan will have a lower first cost too (an unusual combination), since it will need only a 25-hp motor instead of a 30-hp motor.

The 36-in vaneaxial fan therefore will be a better selection.

Example 4: A vaneaxial fan for straight-line installation should be selected that will deliver 4000 cfm against a static pressure of 4 inWC. Table 4.7 shows the performance in an old Aerovent catalog for belt-driven vaneaxial fans. For the 15-in size, we find that we can obtain 3970 cfm at a static pressure of 4 inWC with belt drive at 3486 rpm, with a power consumption of 4.77 bhp. We could use direct drive from a 5-hp, 3450-rpm motor. This would eliminate the belt losses and reduce the power consumption to about 4.5 bhp. The outlet area will be 1.27 ft^2, so the outlet velocity will be $OV = 3970/1.27 = 3126$ cfm and the corresponding velocity pressure will be $VP = (3126/4005)^2 = 0.61$ inWC, resulting in a total pressure $TP = SP + VP = 4.00 + 0.61 = 4.61$ inWC, with an air horsepower ahp = cfm $\times$ $TP/6356 = 3970 \times 4.61/6356 = 2.88$ and a mechanical efficiency of $2.88/4.5 = 0.64 = 64$ percent.

If good accessibility to the motor is important, we would use belt drive instead of direct drive, or since the static pressure of 4 inWC is fairly high relative to the small 4000 cfm, we might consider using a centrifugal fan if it would fit into the geometric configuration of the system.

OVERLAPPING PERFORMANCE RANGES

The four examples presented for the selection of an axial-flow fan tend to indicate that the requirements for air volume and static pressure often determine what type of fan should be used for a specific application. If the static pressure is low, say, less than ¾ inWC, it probably should be a propeller fan (if it can be mounted in a wall or partition) or a tubeaxial fan (if it should exhaust from a duct). If the static pressure is between ½ and 3 inWC and the fan should exhaust from a duct, it probably should be a tubeaxial fan; if the fan should blow into a duct, a vaneaxial fan would be preferable because the tubeaxial fan would produce a spin in the outlet duct, and this would increase the friction path and therefore the static pressure beyond the estimated pressure loss in the outlet duct. If the static pressure is more than 1½ inWC and good efficiency is desired, it probably should be a vaneaxial fan rather than a tubeaxial fan, regardless of whether the fan is blowing or exhausting. If the static pressure is more than 6 inWC, it could be a two-stage axial-flow fan, but then the efficiency would be somewhat lower; for better efficiency, it should be a centrifugal fan.

TABLE 4.5 Performance Table for Propeller Fans, Direct and Belt Drive

Fan size WFD	rpm	WFD Direct Drive 0 in cfm	0 in bhp	1/8 in cfm	1/8 in bhp	1/4 in cfm	1/4 in bhp	3/8 in cfm	3/8 in bhp	1/2 in cfm	1/2 in bhp	hp
8D	1550	320	0.02	240	0.02	158	0.02	40	0.03			1/2
8E	1050	225	0.01	100	0.01							1/20
10D	1550	650	0.03	500	0.04	340	0.04	190	0.05	70	0.06	1/12
10E	1050	420	0.01	230	0.01							1/20
12A	1750	1479	0.14	1357	0.14	1221	0.15	886	0.16	734	0.18	1/6
12B	1140	963	0.04	726	0.04	418	0.05	200	0.06			1/12
12E	1050	887	0.03	600	0.03	321	0.04	40	0.05			1/20
14A	1750	2180	0.22	2049	0.23	1885	0.24	1706	0.25	1209	0.26	1/4
14B	1140	1420	0.061	1151	0.069	682	0.074	410	0.09			1/12
14C	860	1071	0.026	611	0.031	210	0.042					1/12
16A	1750	2771	0.27	2582	0.27	2385	0.3	2138	0.32	1784	0.32	1/3
16B	1140	1805	0.07	1502	0.08	1013	0.09					1/12
16C	860	1362	0.03	850	0.04	500	0.06					1/12
18A	1750	3844	0.50	3731	0.52	3575	0.55	3276	0.57	2877	0.58	1/2
18B	1140	2570	0.14	2263	0.15	1672	0.16	1136	0.19			1/6
18C	860	1939	0.06	1397	0.07	765	0.09					1/12
20A	1750	5406	0.82	5246	0.83	5073	0.87	4873	0.90	4628	0.93	1
20B	1140	3522	0.23	3255	0.24	2870	0.27	2187	0.29	1529	0.31	1/3
20C	860	2657	0.10	2287	0.11	1375	0.13					1/6
22A	1750	7195	1.33	7020	1.33	6843	1.39	6616	1.43	6414	1.46	1 1/2
22B	1140	4687	0.37	4400	0.39	4040	0.42	3272	0.45	2561	0.47	1/2
22C	860	3536	0.16	3120	0.18	2308	0.20	1400	0.21	895	0.24	1/4
24A	1750	10382	2.24	10120	2.32	9889	2.38	9553	2.41	9187	2.50	3
24B	1140	6763	0.62	6352	0.66	5811	0.71	5057	0.77	4021	0.81	3/4
24C	860	5102	0.27	4498	0.30	3400	0.34					1/3
30A	1750	17807	4.73	17390	4.82	16921	4.99	16470	5.12	16010	5.28	5
30B	1140	11600	1.31	10925	1.40	10193	1.49	9127	1.59	7504	1.6	1 1/2
30C	860	8751	0.56	7843	0.63	6243	0.69	4455	0.74			3/4
36B	1140	21373	3.82	20502	3.56	19618	4.12	18620	4.27	17400	4.46	5
36C	860	16123	1.64	14996	1.75	13646	1.87	11750	1.99	9000	2.06	2

(bhp does not include any belt losses)

	rpm	0 in cfm	0 in bhp	1/8 in cfm	1/8 in bhp	1/4 in cfm	1/4 in bhp	3/8 in cfm	3/8 in bhp	1/2 in cfm	1/2 in bhp
24 J	1550	8965	1.50	8624	1.53	8283	1.56	7939	1.60	7582	1.67
K	1355	7837	1.00	7447	1.03	7058	1.05	6651	1.11	6143	1.18
L	1230	7114	0.75	6685	0.77	6250	0.81	5763	0.86	4984	0.92
M	1075	6218	0.50	5726	0.32	5219	0.56	4416	0.61	3178	0.64
N	860	4695	0.29	4093	0.32	2783	0.34	1989	0.38	—	—
30 K	1655	16840	4.00	16325	4.13	15910	4.26	15446	4.39	14922	4.54
L	1140	11600	1.31	10925	1.40	10193	1.49	9127	1.59	7504	1.60
M	1030	10480	0.96	9734	1.04	8868	1.14	7205	1.19	5693	1.24
N	860	8751	0.56	7843	0.63	6243	0.69	4455	0.74	3364	0.84
36 L	1050	19685	2.98	18746	3.11	17740	3.25	16686	3.40	15157	3.57
M	815	17154	1.97	16077	2.09	14877	2.21	13274	2.36	11009	2.41
N	720	13498	0.96	12073	1.05	10128	1.17	6724	1.23	5173	1.34
P	660	12373	0.74	10799	0.83	8221	0.90	5287	0.97	3597	1.12
Q	575	10780	0.49	8857	0.57	5142	0.62	3217	0.74	1578	0.88
42 M	1255	31790	7.48	30835	7.48	29917	7.58	29044	2.29	28159	8.15
N	1095	27737	4.97	26643	4.97	25625	5.21	24615	5.40	23563	5.74
P	925	23431	3.00	22153	3.02	20964	3.21	19690	3.49	18196	3.66
Q	860	21784	2.41	20423	2.45	19138	2.66	17661	2.87	16005	3.01
R	800	20264	1.94	18814	1.99	17427	2.21	15723	2.37	13676	2.44
48 N	1054	38455	7.47	37273	7.86	36069	8.20	34792	8.43	33404	8.69
P	860	31372	4.06	29930	4.37	28384	4.58	26598	4.81	24584	5.06
Q	775	28276	2.97	26662	3.25	24914	3.41	22748	3.65	20204	3.81
R	677	24701	1.98	22820	2.20	20589	2.38	17809	2.53	13574	2.68
S	535	19520	0.98	17052	1.13	13585	1.26	8708	1.33	5866	1.43
54 Q	745	38265	4.93	36440	5.42	34307	5.72	32350	6.02	29845	6.20
R	690	35440	3.92	33420	4.35	31176	4.62	28808	4.85	25581	5.04
S	630	32358	2.98	30092	3.35	27699	3.60	24716	3.77	19946	4.05
T	550	28249	1.98	25579	2.28	22644	2.47	17771	2.69	12505	2.75
60 Q	745	38265	4.93	36440	5.42	34307	5.72	32350	6.02	29845	6.20
R	695	50886	7.45	47915	7.84	44944	8.23	41973	8.62	38877	8.99
S	605	44296	4.92	40883	5.25	37471	5.59	34020	5.93	27383	5.85
T	510	37340	2.94	33292	3.23	29254	3.52	21810	3.58	17004	3.67
U	450	32947	2.02	28359	2.27	22780	2.44	15404	2.51	11093	2.75

Source: From the Catalog of the American Division of the General Resources Corp.

FIGURE 4.51   High-pressure tubeaxial fan wheel with a 52 percent hub-tip ratio and eight wide blades. *(Courtesy of International Engineering, Inc., Dayton, Ohio.)*

As mentioned previously, the performance ranges for the various types overlap, so the decision on the type of fan can be based on other considerations, such as those listed below.

The vaneaxial fan has the following advantages over the centrifugal fan:

1. Greater compactness
2. Lower first cost
3. Straight-line installation, resulting in a lower installation cost
4. Lower sound level at the same tip speed

The centrifugal fan, on the other hand, has the following advantages over the vaneaxial fan:

1. Natural adaptability to installations requiring a 90° turn of the air stream
2. Better accessibility of the motor compared with direct-drive vaneaxial fans
3. Better protection of the motor against hot or contaminated gases than for a vaneaxial fan with direct drive
4. Greater assurance for operation in the efficient and quiet performance range, particularly for systems with fluctuating flow resistance

SAMPLE DESIGN CALCULATION FOR A 27-IN VANEAXIAL FAN

Suppose a customer wants several 27-in vaneaxial fans with direct drive from a 7½-hp, 1750-rpm motor to deliver 11,500 cfm against a static pressure of 2.5 inWC. A noise level not more than 86 dB is desired; therefore, we will use

TABLE 4.6 Performance of High-Pressure Tubeaxial Fans, Direct Drive

Catalogue no.	Propeller diameter, in	No. of blades	Speed (rpm)	Motor hp	"Code" capacity in cfm free air	Capacity "code" against static, in H₂O									Domestic approx. weight LDX	Export approx. weight LDX	Domestic approx. weight DBY	Export approx. weight DBY
						0.50	0.70	1.00	1.50	2.00	2.50	3.00	4.00	5.00				
LDX-DBY 2	16	8	1740	½	3,640	2,800	2,350	995							228	272	246	295
LDX-DBY 3	20	8	1740	1½	7,150	5,970	5,600	4,600	2,450						233	299	260	330
LDX-DBY 4	24	8	1740	5	12,350	10,800	10,350	9,700	8,300						368	432	400	464
LDX-DBY 4A	24	8	1160	1	8,100	6,050	5,200	2,050							263	326	293	356
LDX-DBY 6	28	8	1740	10	19,550	17,700	17,100	16,000	14,990	13,000	9,250	4,930	2,100		518	584	575	643
LDX-DBY 6A	28	8	1160	3	12,800	10,500	9,700	7,500	2,300						362	422	422	490
LDX-DBY 8	32	8	1740	15	28,900	26,650	25,950	25,100	23,400	22,400	20,000	16,100	7,600		752	865	815	923
LDX-DBY 8A	32	8	1160	5	19,100	16,200	15,370	13,400	8,000						578	682	618	725
LDX-DBY 11	36	8	1740	30	41,500	38,500	37,800	37,000	34,800	33,000	32,000	29,150	21,000	10,500	1024	1165	1048	1236
LDX-DBY 11A	36	8	1160	7½	27,200	23,400	22,000	21,030	17,500	9,950					846	982	896	1035
LDX-DBY 14	42	8	1750	60	66,000	61,850	61,000	59,000	58,000	56,000	53,500	52,000	46,800	39,250	1690	1940	1740	2015
LDX-DBY 14A	42	8	1160	20	43,300	39,000	37,400	32,500	29,000	10,850					1250	1500	1300	1575
LDX-DBY 15	42	8	870	10	27,200	23,400	22,000	21,030	17,500	9,950					1076	1260	1125	1310
LDX-DBY 18	48	8	1160	40	64,500	59,000	57,500	54,800	51,500	49,000	44,000	35,000	16,000	7,500	1995	2220	2070	2295
DX-DBY 18A	48	8	870	15	47,700	42,000	40,000	37,000	30,000	15,000					1563	1780	1625	1860

Source: Courtesy of International Engineering, Inc., Dayton, Ohio.

TABLE 4.7 Performance of Belt-Driven Vaneaxial Fans, from an Old Aerovent Catalog

Size 15

Maximum rpm = 4590

$$\text{Maximum bhp} = .124\left(\frac{\text{rpm}}{1000}\right)^{3}$$

CFM	OUTLET VEL.	1/4" SP		1/2" SP		3/4" SP		1" SP		1-1/4" SP		1-1/2" SP		1-3/4" SP		2" SP		2-1/4" SP		2-1/2" SP		2-3/4" SP		3" SP		3-1/2" SP		4" SP	
		RPM	BHP	RPM	BHP	RPM	BHP	RPM	BHP	RPM	BHP	RPM	BHP	RPM	BHP	RPM	BHP	RPM	BHP	RPM	BHP	RPM	BHP	RPM	BHP	RPM	BHP	RPM	BHP
1730	1400	1126	.18	1322	.28	1516	.40																						
1990	1600	1226	.23	1411	.34	1580	.46	1752	.61																				
2240	1800	1334	.30	1500	.41	1654	.54	1804	.69	1953	.84																		
2490	2000	1470	.39	1596	.50	1746	.65	1878	.79	2024	.97	2150	1.13																
2740	2200	1606	.51	1702	.61	1835	.76	1960	.91	2085	1.08	2207	1.26	2333	1.45	2450	1.63												
2990	2400	1742	.66	1813	.74	1938	.90	2057	1.06	2174	1.24	2276	1.40	2398	1.62	2499	1.80	2622	2.04										
3220	2600	1873	.82	1922	.88	2024	1.03	2136	1.19	2244	1.37	2354	1.58	2454	1.76	2563	1.98	2670	2.20	2773	2.43	2884	2.67						
3470	2800	2005	1.00	2058	1.08	2143	1.22	2244	1.39	2337	1.57	2438	1.77	2537	1.98	2645	2.20	2738	2.43	2838	2.67	2891	2.79	3032	3.16				
3720	3000	2142	1.22	2192	1.31	2290	1.49	2338	1.58	2432	1.77	2534	1.99	2624	2.20	2720	2.43	2812	2.66	2910	2.93	3002	3.18	3085	3.39	3266	3.93		
3970	3200	2279	1.47	2332	1.57	2370	1.65	2453	1.83	2540	2.02	2624	2.23	2714	2.45	2811	2.69	2903	2.96	2988	3.19	3066	3.43	3161	3.71	3323	4.23	3486	4.77
4220	3400	2415	1.75	2462	1.85	2503	1.94	2562	2.08	2643	2.29	2710	2.45	2800	2.70	2908	3.02	2976	3.21	3073	3.51	3137	3.71	3217	3.96	3390	4.59	3550	5.13
4470	3600	2550	2.06	2600	2.18	2643	2.29	2670	2.35	2762	2.61	2830	2.79	2912	3.04	2999	3.30	3080	3.58	3140	3.77	3229	4.07	3295	4.33	3452	4.89	3607	5.51
4720	3800	2685	2.40	2728	2.51	2773	2.63	2810	2.75	2866	2.92	2926	3.10	3020	3.40	3086	3.62	3167	3.89	3240	4.16	3303	4.39	3382	4.68	3523	5.26	3677	5.87

Maximum rpm = 3820 Size 18 Maximum bhp = $.309\left(\dfrac{rpm}{1000}\right)^3$

CFM	OUTLET VEL.	1/4" SP		1/2" SP		3/4" SP		1" SP		1-1/4" SP		1-1/2" SP		1-3/4" SP		2" SP		2-1/4" SP		2-1/2" SP		2-3/4" SP		3" SP		3-1/2" SP		4" SP	
		RPM	BHP	RPM	BHP	RPM	BHP	RPM	BHP	RPM	BHP	RPM	BHP	RPM	BHP	RPM	BHP	RPM	BHP	RPM	BHP	RPM	BHP	RPM	BHP	RPM	BHP	RPM	BHP
2500	1400	939	.25	1103	.40	1264	.57																						
2860	1600	1023	.33	1177	.49	1317	.66	1461	.87																				
3220	1800	1113	.43	1251	.60	1379	.78	1504	.99	1628	1.21																		
3580	2000	1225	.57	1331	.72	1455	.93	1566	1.14	1688	1.40	1793	1.63																
3940	2200	1339	.74	1419	.88	1530	1.09	1634	1.31	1738	1.55	1840	1.82	1946	2.08	2042	2.35												
4300	2400	1453	.94	1511	1.06	1615	1.29	1715	1.53	1812	1.79	1898	2.02	2000	2.34	2084	2.59	2186	2.93										
4640	2600	1561	1.17	1603	1.27	1688	1.48	1781	1.72	1871	1.98	1963	2.28	2046	2.53	2137	2.84	2226	3.16	2312	3.49								
5000	2800	1672	1.44	1716	1.56	1787	1.76	1871	2.00	1948	2.25	2032	2.54	2115	2.85	2205	3.17	2283	3.49	2367	3.85	2411	4.02	2528	4.54				
5360	3000	1786	1.76	1828	1.88	1910	2.14	1950	2.28	2028	2.55	2112	2.87	2188	3.17	2268	3.50	2345	3.83	2427	4.22	2503	4.57	2572	4.88	2723	5.65		
5720	3200	1900	2.12	1944	2.26	1976	2.38	2045	2.63	2118	2.91	2188	3.20	2263	3.52	2344	3.87	2421	4.27	2491	4.59	2556	4.93	2635	5.34	2771	6.10	2907	6.86
6080	3400	2014	2.52	2053	2.66	2087	2.79	2136	3.00	2204	3.29	2259	3.53	2334	3.89	2425	4.34	2481	4.62	2562	5.05	2615	5.34	2682	5.70	2827	6.60	2959	7.39
6440	3600	2126	2.96	2168	3.14	2204	3.29	2226	3.39	2302	3.75	2359	4.02	2428	4.38	2501	4.75	2538	5.15	2618	5.42	2692	5.86	2747	6.24	2878	7.04	3008	7.93
6800	3800	2238	3.45	2274	3.62	2312	3.80	2343	3.95	2390	4.21	2439	4.47	2518	4.89	2573	5.21	2641	5.60	2701	6.00	2754	6.32	2820	6.74	2937	7.57	3066	8.46

Ratings based on tests in accordance with Bulletin 110, Plate VII with entrance orifice and belt drive construction.
Brake horsepower includes belt drive and bearing losses.

When tapered outlet is used, cfm will be increased 5% and bhp reduced 2% at constant rpm and SP. For higher SP or cfm consult factory.

1. Airfoil blades rather than single-thickness blades
2. Blades that are somewhat wider at the tip than at the hub
3. Outlet guide vanes rather than inlet guide vanes

A 7½-hp, 1750-rpm motor will have an outside diameter of 10½ to 11⅛ in, depending on the type of motor. We can calculate the dimensions of this 27-in vaneaxial fan using the various formulas discussed previously.

As a first step, we will calculate the hub diameter d using Eq. (4.1):

$$d_{min} = (19,000/rpm)\sqrt{SP}$$

$$= (19,000/1750)\sqrt{2.5}$$

$$= 17.17 \text{ in}$$

We decide to make the hub diameter $d = 17\frac{1}{4}$ in.

Next, we will check whether the desired 27-in wheel diameter is acceptable, using Eq. (4.3):

$$D_{min} = \sqrt{d^2 + 61(cfm/rpm)} = \sqrt{17.25^2 + 61\,(11500/1750)}$$

$$= \sqrt{297.56 + 400.86} = \sqrt{698.42}$$

$$= 26.43 \text{ in}$$

The 27-in wheel diameter will be all right.

We will use a ⅛-in tip clearance, small enough for good performance yet large enough for easy assembly. This will make the housing inside diameter 27¼ in. We will use nine airfoil blades ($z_B = 9$) and 14 outlet guide vanes ($z_V = 14$) of single-thickness sheet metal.

Next, let's check whether the 7½-hp motor will be adequate. The outlet area of the fan housing will be

$$OA = \left(\frac{27.25}{24}\right)^2 \pi = 4.050 \text{ ft}^2$$

so the outlet velocity at the specified point of operation will be

$$OV = \frac{11,500}{4.050} = 2840 \text{ fpm}$$

the velocity pressure will be

$$VP = \left(\frac{2840}{4005}\right)^2 = 0.503 \text{ inWC}$$

the total pressure will be

$$TP = 2.50 + 0.503 = 3.003 \text{ inWC}$$

and the air horsepower will be

$$ahp = \frac{cfm \times TP}{6356} = \frac{11,500 \times 3.003}{6356} = 5.43$$

so the mechanical efficiency will be

$$ME = \frac{\text{ahp}}{\text{bhp}} = \frac{5.43}{7.5} = 0.724 = 72.4 \text{ percent}$$

at the specified point. This is acceptable, so the 7½-hp motor will be adequate. In fact, we can expect a higher efficiency than this.

The annular area A_a between the between the hub outside diameter and the housing inside diameter will be

$$A_a = \left(\frac{27.25}{24}\right)^2 \pi - \left(\frac{17.25}{24}\right)^2 \pi = 4.050 - 1.623 = 2.427 \text{ ft}^2$$

and the axial velocity V_a through the annular area will be

$$V_a = \frac{11,500}{2.427} = 4738 \text{ fpm} \quad \text{and} \quad V_a^2 = 2245 \times 10^4 \text{ (fpm)}^2$$

We will need this figure to calculate the relative air velocity.

Now we can go ahead with the calculation of the various dimensions for three radii: hub (8.625 in radius), middle (11.0625 in radius), and blade tip (13.500 in radius). Table 4.8 shows the dimensions we obtained and the various formulas we used to calculate them.

For the first nine lines of Table 4.8, the formulas we used are shown. For line 10, we have some freedom in selecting the blade width l, as long as l is below a certain value for each radius. The formula for this value is shown on line 10. We selected a tip width 46 percent larger than the blade width at the hub. As mentioned, a wider blade tip will result in a lower noise level. Unfortunately, it also will result in an increased no-delivery brake horsepower, but this will be acceptable if the overload is not too much. The tests later showed a no-delivery overload of 20 percent, which is acceptable.

Line 11 shows the airfoil section selected for each radius. Figure 4.52 shows the shape of the three airfoil sections selected for this 27-in vaneaxial fan. The section at the bottom is for the 85/8-in radius, adjacent to the hub, the next one is for the 11¹⁄₁₆-in radius, in the middle of the blade, and the section on top is for the 13½-in radius, at the blade tip. The hub section is the NACA airfoil no. 6512. The two other sections are thinner airfoils. They are modifications of the NACA no. 6512, with the cambers reduced to 72 and 53 percent, respectively. There are two reasons why the hub section of the blade is made thicker than the other sections:

1. The hub section has to have more mechanical strength.

2. The hub section needs a higher lift coefficient C_L because of the lower blade velocity there. Airfoils with larger upper cambers normally will have higher lift coefficients.

Line 12 shows the upper maximum camber for each of the three airfoil shapes: 13.2 percent (NACA no. 6512), 9.5 percent (72 percent of 13.2 percent), and 7.0 percent (53 percent of 13.2 percent).

Lines 13 and 14 show the angle of attack α and the corresponding lift coefficient C_L for each section. These data can be obtained from the graph shown in Fig. 4.53.

Line 15 shows the static pressure that will be produced at each radius, calculated from the formula shown. We selected l and C_L in such a way that this static pressure is about 2.5 inWC, as was requested by the customer.

TABLE 4.8 Calculations of the Dimensions for a 27-in Vaneaxial Fan at 1750 rpm

Description	Symbol	Unit	Formula	Hub	Middle	Tip
Radius	r	in		8.625	11.0625	13.500
Blade velocity	V_B	fpm	$(2\pi/12) \times \text{rpm} = 916.3r$	7,903	10,137	12,370
Rotational component of helical air velocity past blades	V_r	fpm	$(233 \times 10^5/\text{rpm}) \times$ $SP/r = 33{,}286/r$	3,859	3,009	2,466
tan (average relative air angle)	$V_B - \tfrac12 V_r$	fpm		5,973.5	8,632.5	11,137
	$\tan \beta$	1	$V_d/(V_B - \tfrac12 V_r) =$ $4738/(V_B - \tfrac12 V_r)$	0.7932	0.5489	0.4254
Average relative air angle	β	degrees and minutes		38°25'	28°45'	23°3'
	$(V_B - \tfrac12 V_r)^2$	(fpm)^2		3568×10^4	7452×10^4	12403×10^4
	W^2	(fpm)^2	$V_a^2 + (V_B - \tfrac12 V_r)^2$	5813×10^4	9697×10^4	14648×10^4
Average relative air velocity	W	fpm	$\sqrt{W^2}$	7,624	9,847	12,103
Blade width	l	in	$\leq 6.8\, r/z_B = 0.756r$	6.5	8.0	9.5
Airfoil selected				NACA no. 6512	NACA no. 6512 modified	NACA no. 6512 modified
Maximum upper camber		%		13.2	9.5	7.0
Angle of attack	α	degrees		3.0	2.5	2.0
Lift coefficient	C_L	1		0.94	0.59	0.41
Static pressure	SP	in WC	$3.43 \times 10^{-7} \times \text{rpm} \times$ $z_B \times C_L \times l \times W =$ $0.54 \times 10^{-4} \times C_{L \times l \times W}$	2.52	2.51	2.55
Blade angle	$\beta + \alpha$	degrees		41.5	31.2	25.0
tan (air angle past trailing edge)	$\tan \delta$	1	$V_r/V_a = V_r/4738$	0.8145	0.6351	0.5205
Air angle past trailing edge	δ	degrees and minutes		39°10'	32°25'	27°30'

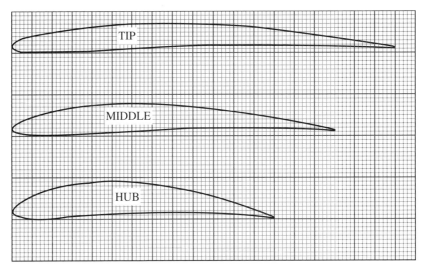

FIGURE 4.52   Three airfoil sections (reduced size) for the 27-in vaneaxial fan wheel.

Line 16 shows the blade angle $\beta + \alpha$. Line 17 shows how the absolute air angle δ past the blade is calculated. Line 18 shows this absolute air angle. This also will be the angle at the leading edge of the outlet guide vane.

Figure 4.54 shows how the principal blade dimensions vary from hub to tip. The blade width increases uniformly (in a straight line), while the blade angle and the

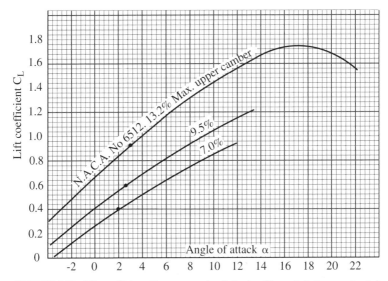

FIGURE 4.53   Lift coefficient C_L versus angle of attack α (for an infinite aspect ratio) for the three airfoils used in the 27-in vaneaxial fan blade.

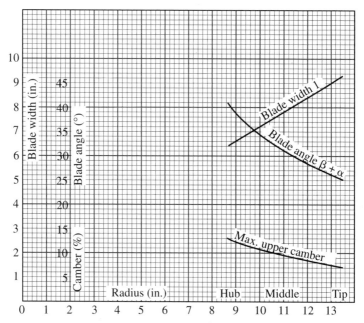

FIGURE 4.54 Blade width, blade angle, and maximum upper camber versus radius for the 27-in vaneaxial fan blade.

maximum upper camber decrease, but not uniformly. They decrease at a faster rate near the hub and at a slower rate near the tip.

Figure 4.55 shows the vector diagrams for the various absolute and relative velocities, which were explained in detail in Fig. 4.41. Figure 4.55 shows three velocity diagrams, for the three blade sections in Table 4.8, at the hub, in the middle of the blade, and at the tip. We note the following:

1. The axial air velocity V_a through the annulus between the hub and the housing is constant from hub to tip.

2. The blade velocity V_B and the air velocity W relative to the blade increase from hub to tip.

3. The angle β of the relative air velocity W decreases from hub to tip.

4. The rotational component V_r of the air velocity decreases from hub to tip.

Figure 4.56 shows a sketch of the 27-in cast-aluminum vaneaxial fan wheel at a scale of 1/10. This sketch was made using the dimensions we obtained from our calculations. A sketch of the corresponding vaneaxial fan housing would look similar to Figs. 4.6 and 4.40.

Figure 4.57 shows the performance obtained when this 27-in vaneaxial fan was tested. We note the following:

1. The *SP* curve passes through the point of requirement, 11,500 cfm at a static pressure of 2½ inWC.

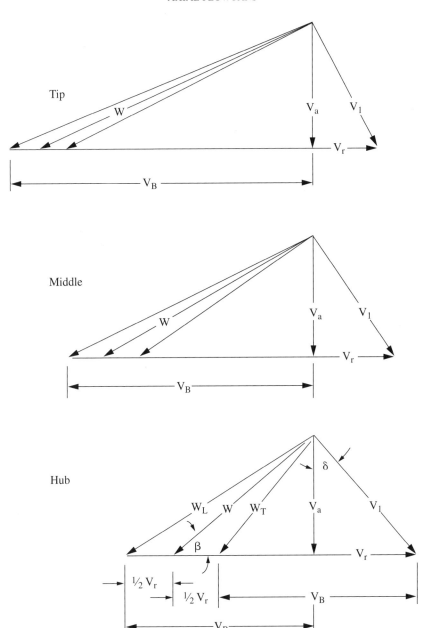

FIGURE 4.55 Velocity diagrams for 27-in vaneaxial fan blade.

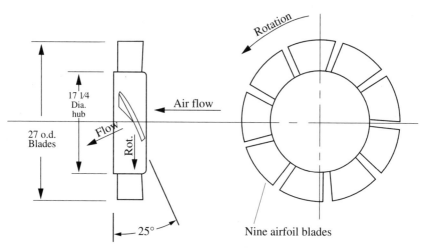

FIGURE 4.56 27-in vaneaxial fan wheel with 17¼-in-diameter hub, nine airfoil blades, and 25° blade angle at the tip.

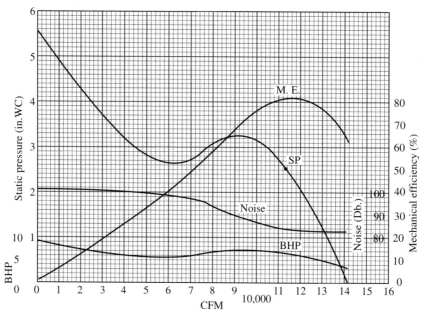

FIGURE 4.57 Performance obtained by test of the 27-in vaneaxial fan at 1750 rpm.

Air flow

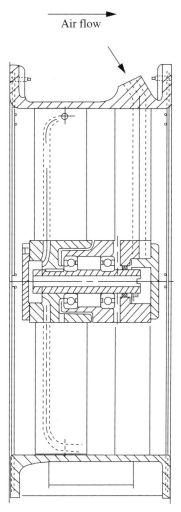

FIGURE 4.58 A 15-in tubeaxial fan with direct drive from a compressed-air turbine mounted above the inlet side of the fan housing. Note the space provided between the fan housing and the turbine, allowing the airflow to enter without excessive obstruction. *(Courtesy of Coppus Engineering Division, Tuthill Corporation, Millbury, Mass.)*

FIGURE 4.59 Schematic sketch of 20-in tubeaxial fan assembly driven by compressed-air jets, showing compressed-air inlet, hollow shaft, ball bearings, and two-bladed fan wheel. *(Courtesy of Coppus Engineering Division, Tuthill Corporation, Millbury, Mass.)*

2. The maximum static pressure is 3.3 inWC. This provides a 32 percent pressure safety margin over the specified 2.5 inWC.

3. The maximum mechanical efficiency is 83 percent. It occurs at the point of requirement.

4. The operating range is from 14,200 cfm at free delivery to 9400 cfm at a static pressure of 3.3 inWC.

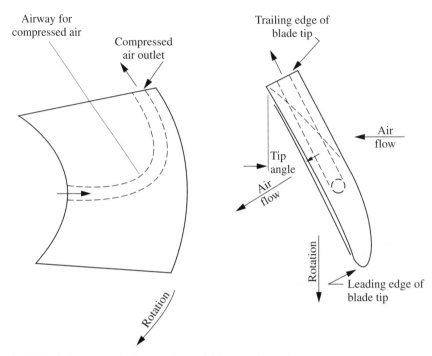

FIGURE 4.60 Schematic diagram of an axial-flow fan blade showing the airway for the compressed air, first flowing radially outward and then turning 90° to a circumferential direction for propulsion.

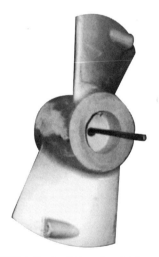

FIGURE 4.61 An 11-in tubeaxial fan wheel driven by compressed-air jets with a 34 percent hub-tip ratio and two airfoil blades. View from inlet side showing the compressed-air nozzles at the blade tips. *(Courtesy of Coppus Engineering Division, Tuthill Corporation, Millbury, Mass.)*

FIGURE 4.62 An 11-in tubeaxial fan assembly. View from the inlet side, again showing the two-bladed fan wheel with the compressed-air nozzles at the blade tips. *(Courtesy of Coppus Engineering Division, Tuthill Corporation, Millbury, Mass.)*

FIGURE 4.63 A 20-in vaneaxial fan wheel driven by compressed-air jets with a 31 percent hub-tip ratio and four airfoil blades. View from inlet side showing the compressed-air nozzles on two of the four blades. *(Courtesy of Texas Pneumatic Tools, Inc., Reagan, Texas.)*

5. The maximum power consumption in the operating range is 7.7 bhp. It occurs at 9600 cfm, close to the point of maximum static pressure.

6. The no-delivery brake horsepower overloads the motor by a moderate 20 percent.

7. The sound-level curve has two plateaus, one around 84 dB in the operating range and one around 100 dB in the stalling range, for a considerable 16-dB difference.

AXIAL-FLOW FANS DRIVEN BY COMPRESSED AIR

Most fans are driven by electric motors, either by direct drive or by belt drive or occasionally by gear drive. However, there are three other methods for driving a fan:

1. It can be driven by an *engine* (gasoline, propane, or diesel). As an example, a multistage turbo blower, driven by a diesel engine, will be described in Chap. 7.

2. A fan can be driven by a *turbine* (compressed air, steam, or water). Figure 4.58 shows a 15-in tubeaxial fan directly driven by a turbine (compressed air or steam) located on the inlet side of the fan housing. Turbine-driven fans are used in locations of hazardous atmospheres or where electric power is not available. A typical application would be a marine ventilator used to exhaust the gas from the inside of an oil tanker.

3. A fan can be driven by *reaction* to compressed-air jets discharging circumferentially through small openings at the blade tips and shooting toward the trailing edge. Figure 4.59 shows how this can be accomplished. The path of the com-

pressed air is as follows: First, it is ducted from the outside into a chamber inside a stationary hub that is supported by some struts or possibly by some guide vanes. From this chamber, the compressed air flows into the inside of a hollow, rotating shaft and then passes through some cored airways in the rotating hub and in the blades, first flowing radially outward and then turning 90° to a circumferential direction. Finally, a high-velocity jet of compressed air is ejected through an opening at or near the trailing edge of the blade tip, thereby pushing the blade forward and producing the fan rotation, just like a jet plane is pushed forward by the high-velocity jet ejected from the rear of the engine.

FIGURE 4.64 Same 20-in vaneaxial fan wheel driven by compressed-air jets. View from outlet side, again showing the compressed-air nozzles on two of the four blades. *(Courtesy of Texas Pneumatic Tools, Inc., Reagan, Texas.)*

Figures 4.59 through 4.66 show examples of axial-flow fans driven by compressed-air jets. Some have two blades, and some have four blades. Some are tubeaxial fans, and some are vaneaxial fans. They all have discharge nozzles at the blade tips but not necessarily on all blades. The discharge nozzles protrude above the surface of the blades and therefore will somewhat impair fan efficiency. To minimize this effect, the four-bladed fan wheels provide discharge noz-

FIGURE 4.65 A 20-in vaneaxial fan assembly. View from the inlet side showing the four-bladed fan wheel and the outlet guide vanes, partly covered by the fan blades. *(Courtesy of Texas Pneumatic Tools, Inc., Reagan, Texas.)*

FIGURE 4.66 Same 20-in vaneaxial fan assembly. View from the outlet side showing the outlet guide vanes and the four-bladed fan wheel behind them. *(Courtesy of Texas Pneumatic Tools, Inc., Reagan, Texas.)*

zles on only two blades. The ratio of the volume delivered divided by the volume of the compressed air consumed is called the *delivery ratio*. It is a measure of the fan efficiency.

Compressed-air–driven fans can be used for exhaust as well as for supply. They are particularly well suited for the ventilation of confined spaces, for the removal of hazardous fumes or of contaminated air from welding and sandblasting operations, and for locations where compressed air is available but electric power is not.

CHAPTER 5
FAN LAWS

CONVERSION OF FAN PERFORMANCE

There are certain general fan laws that are used to convert the performance of a fan from one set of variables (such as size, speed, and gas density) to another. Suppose a fan of a certain size and speed has been tested and its performance has been plotted for for the standard air density of 0.075 lb/ft³. We then can compute the performance of another fan of geometric similarity by converting the performance data in accordance with these fan laws without running a test on this second fan.

We call them *general fan laws* because they apply to any type of fan: axial-flow, centrifugal, and mixed-flow fans, roof ventilators, cross-flow blowers, and vortex blowers.

We have already seen that the performance of a fan can be presented in the catalog in two ways:

1. In the form of a *performance graph,* such as Fig. 4.39, showing static pressure, brake horsepower, efficiency, and noise level versus air volume.
2. In the form of a *rating table,* such as Table 4.3, showing air volume, fan speed, and brake horsepower at certain static pressures.

The performance graph is the original presentation. It is the result of a test on the fan. The rating table then was derived from the performance graph by using the fan laws. The calculation of rating tables is one important application of the fan laws. However, the fan laws also are used for other purposes, such as the conversion of a customer requirement from high temperature to the standard air density of the catalog or the prediction of the performance of a new fan design.

VARIATION IN FAN SPEED

In order to convert the performance of a fan at one speed to another speed, we take a number of points on the performance graph and convert the corresponding data for air volume, static pressure, brake horsepower, efficiency, and noise level from the speed of the graph to the desired speed using the following rules:

The air volume (cfm) varies directly with the speed:

$$\frac{\text{cfm}_2}{\text{cfm}_1} = \frac{\text{rpm}_2}{\text{rpm}_1} \qquad (5.1)$$

The pressures vary as the square of the speed:

$$\frac{SP_2}{SP_1} = \left(\frac{\text{rpm}_2}{\text{rpm}_1}\right)^2 \qquad (5.2)$$

The brake horsepower varies as the cube of the speed:

$$\frac{bhp_2}{bhp_1} = \left(\frac{rpm_2}{rpm_1}\right)^3 \tag{5.3}$$

The efficiency remains constant but, of course, shifts to the new air volume values.

The noise level is increased (or decreased) by 50 times the logarithm (base 10) of the speed ratio:

$$N_2 - N_1 = 50 \log_{10}\frac{rpm_2}{rpm_1} \tag{5.4}$$

At this point, three things should be noted:

1. All pressures vary as the square of the speed: the static pressures, the velocity pressures, and the total pressures.
2. All powers vary as the cube of the speed: the brake horsepower and the air horsepower.
3. The exponents add up: 1 (for air volume) plus 2, (for static pressure) equals 3 (for brake horsepower). This is no coincidence. The air horsepower is proportional to the product air volume times total pressure (see Eq. 1.7). Therefore, if air volume varies as the first power of the speed and total pressure varies as the second power of the speed, their product (air horsepower) must vary as the third power of the speed, and if air horsepower varies that way, brake horsepower must vary that way, too.

Here is an example of how these rules are applied. Let's take the performance graph shown in Fig. 4.57 for our 27-in vaneaxial fan at 1750 rpm and replot in Fig. 5.1 the four performance curves (static pressure, brake horsepower, efficiency and noise level) in dashed lines. What will be the performance of this 27-in vaneaxial fan if we run it, say, at a 50 percent higher speed, i.e., at 2625 rpm? Let's take the performance data for six points and enter them in the top half of Table 5.1. Our conversion factors will be as follows:

$$\text{Air volume (cfm) conversion factor} = \frac{rpm_2}{rpm_1} = \frac{2625}{1750} = 1.5$$

$$\text{Static pressure conversion factor} = \left(\frac{rpm_2}{rpm_1}\right)^2 = \left(\frac{2625}{1750}\right)^2 = 2.25$$

$$\text{Brake horsepower conversion factor} = \left(\frac{rpm_2}{rpm_1}\right)^3 = \left(\frac{2625}{1750}\right)^3 = 3.375$$

$$\text{Efficiency conversion factor} = 1$$

For conversion of the noise level, we do not use a factor, but we add an amount that can be calculated as

$$50 \log_{10}\frac{2650}{1750} = 50 \log_{10} 1.5$$

$$= 50 \times 0.17609 = 8.8 \text{ dB}$$

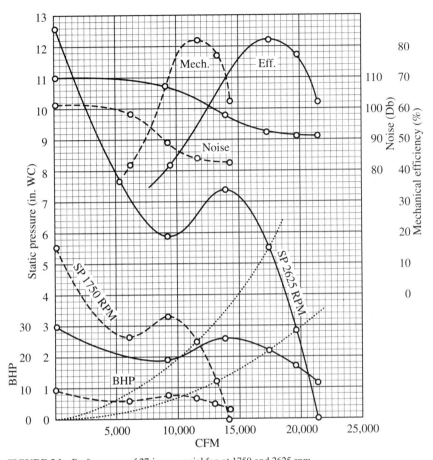

FIGURE 5.1   Performance of 27-in vaneaxial fan at 1750 and 2625 rpm.

We apply these conversion factors to obtain the performance data for this 27-in vaneaxial fan at 2625 rpm and enter these data in the bottom half of Table 5.1. We now plot these new data in solid lines in Fig. 5.1. We notice the following:

1. The increased speed results in a steeper static pressure curve because the conversion factor for static pressure is larger than the conversion factor for air volume. In other words, an increased speed boosts the static pressure more than the air volume.

2. The power-consumption brake horsepower becomes considerably higher because the conversion factor equals the third power of the speed ratio. At 1750 rpm a 7½-hp motor is adequate, but at 2625 rpm a 25-hp motor is needed.

3. The efficiency remains the same.

4. The noise level increases by 8.8 dB.

Figure 5.1 also shows two parabolic curves (dotted lines) connecting corresponding points on the pressure curves. These parabolic curves indicate how the perfor-

TABLE 5.1   Conversion of Fan Performance to a Different Speed

		Free delivery	In-between Point	Maximum efficiency	Maximum SP	Stalling dip	No delivery
27-in	cfm	14,200	13,020	11,500	9,200	6,100	0
vaneaxial	SP	0	1.25	2.50	3.30	2.64	5.59
fan,	bhp	3.45	5.10	6.65	7.70	5.80	9.00
1750 rpm	ME	0.630	0.776	0.828	0.700	0.421	0
	Noise	83.2	83.5	84.7	89.6	99.1	101.3
27-in	cfm	21,300	19,530	17,250	13,800	9,150	0
vaneaxial	SP	0	2.81	5.63	7.43	5.94	12.58
fan,	bhp	11.64	17.21	22.44	25.99	19.58	30.38
2625 rpm	ME	0.630	0.776	0.828	0.700	0.421	0
	Noise	92.0	92.3	93.5	98.4	107.9	110.1

mance points would move if the speed were varied gradually rather than by the ratio 1.5. We will say more on this later.

VARIATION IN FAN SIZE

Another important fan law concerns the conversion of fan performance if the fan size is varied. Most fan companies manufacture a certain fan design in various sizes and offer a complete line of geometrically similar fans. They do not have to run tests on all sizes of the line. They usually test only three sizes (say 12, 24, and 42 in) and compute the performance of the in-between sizes by using the fan laws for variations in size and speed.

The fan laws for size, however, can be used only if the two fans are in geometric proportion. Here is what *geometric proportionality* means:

1. Both fans have the same number of blades.
2. Both fans have the same blade angles and any other angles on the fan wheel and fan housing.
3. If the diameters of the two fan wheels are D_1 and D_2, for a size ratio D_2/D_1, all other corresponding dimensions of wheel and housing have the same ratio.

If all these requirements are fulfilled, we can proceed with the performance conversion using the following rules:

The air volume (cfm) varies as the cube of the size:

$$\frac{cfm_2}{cfm_1} = \left(\frac{D_2}{D_1}\right)^3 \tag{5.5}$$

The pressures vary as the square of the size:

$$\frac{SP_2}{SP_1} = \left(\frac{D_2}{D_1}\right)^2 \tag{5.6}$$

The brake horsepower varies as the fifth power of the size:

$$\frac{\text{bhp}_2}{\text{bhp}_1} = \left(\frac{D_2}{D_1}\right)^5 \tag{5.7}$$

The efficiency remains almost constant.

There is a minor increase in efficiency for larger sizes. For a size ratio of 1.5, the maximum efficiency will increase by less than 1 percent. This is called the *size effect*.

The noise level is increased (or decreased) by 50 times the logarithm (base 10) of the size ratio:

$$N_2 - N_1 = 50 \log_{10} \frac{D_2}{D_1} \tag{5.8}$$

Again, we note that the exponents add up: 3 (for air volume) plus 2 (for static pressure) equals 5 (for brake horsepower).

Let's again apply these rules in a specific example, the performance of our 27-in vaneaxial fan at 1750 rpm, as shown in Fig. 4.57. We replot the four performance curves in Fig. 5.2 in dashed lines. What will be the performance of this fan, converted in geometric proportion to, say, a 50 percent larger size, i.e., to 40½ in. Again, we take the performance data at six points and enter them in the top half of Table 5.2. Our conversion factors will be as follows:

$$\text{Air volume conversion factor} = \left(\frac{D_2}{D_1}\right)^3 = \left(\frac{40.5}{27}\right)^3 = 1.5^3 = 3.375$$

$$\text{Static pressure conversion factor} = \left(\frac{D_2}{D_1}\right)^2 = \left(\frac{40.5}{27}\right)^2 = 1.5^2 = 2.25$$

$$\text{Brake horsepower conversion factor} = \left(\frac{D_2}{D_1}\right)^5 = \left(\frac{40.5}{27}\right)^5 = 1.5^5 = 7.59$$

$$\text{Efficiency conversion factor} = 1$$

For the conversion of the noise level, we add an amount that can be calculated as

$$50 \log_{10} \frac{40.5}{27} = 50 \log_{10} 1.5$$

$$= 50 \times 0.17609 = 8.8 \text{ dB}$$

This is not quite accurate. It is approximate.

We apply these conversion factors to obtain the performance data for the 40½-in vaneaxial fan at 1750 rpm and enter these data in the bottom half of Table 5.2. We now plot these new data in solid lines in Fig. 5.2. We notice the following:

1. The geometrically similar, larger fan results in a flatter static pressure curve because the conversion factor for air volume is larger than the conversion factor for static pressure. In other words, the increased size boosts the air volume more than the static pressure. Note that we found the opposite condition for an increased speed.

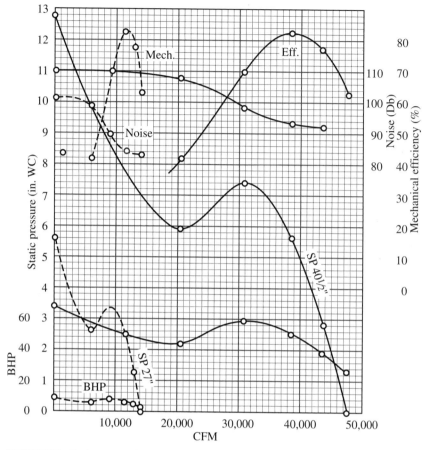

FIGURE 5.2 Performance of two geometrically similar vaneaxial fans, 27 and 40½ in, both running at 1750 rpm.

2. The power-consumption brake horsepower becomes very much higher because the conversion factor equals the fifth power of the size ratio. In the 27-in size a 7½-hp motor is adequate, but in the 40½-in size a 60-hp motor is needed. Note that only a 25-hp motor was needed for the increased speed.

3. The efficiency remains the same, as it did for the increased speed.

4. The noise level increases by about 8.8 dB, approximately the same as for the increased speed.

VARIATION IN BOTH FAN SIZE AND FAN SPEED

If both the fan size D and the fan speed (rpm) are varied, the two sets of rules discussed above can be applied consecutively, in either sequence. In fact, this is the

TABLE 5.2 Conversion of Fan Performance to a Geometrically Similar, Larger Fan Running at the Same Speed

		Free delivery	In-between point	Maximum efficiency	Maximum SP	Stalling dip	No delivery
27-in	cfm	14,200	13,020	11,500	9,200	6,100	0
vaneaxial	SP	0	1.25	2.50	3.30	2.64	5.59
fan,	bhp	3.45	5.10	6.65	7.70	5.80	9.00
1750 rpm	ME	0.630	0.776	0.828	0.700	0.421	0
	Noise	83.2	83.5	84.7	89.6	99.1	101.3
40½-in	cfm	47,925	43,943	38,813	31,050	20,588	0
vaneaxial	SP	0	2.81	5.63	7.43	5.94	12.58
fan,	bhp	26.20	38.73	50.50	58.47	44.04	68.34
1750 rpm	ME	0.630	0.776	0.828	0.700	0.421	0
	Noise	92.0	92.3	93.5	98.4	107.9	110.1

safest way to proceed. It most likely will avoid any possible errors. However, the two sets of rules also can be combined. The combined rules then will read as follows:

$$\frac{\text{cfm}_2}{\text{cfm}_1} = \left(\frac{D_2}{D_1}\right)^3 \times \frac{\text{rpm}_2}{\text{rpm}_1} \tag{5.9}$$

$$\frac{SP_2}{SP_1} = \left(\frac{D_2}{D_1}\right)^2 \times \left(\frac{\text{rpm}_2}{\text{rpm}_1}\right)^2 \tag{5.10}$$

$$\frac{\text{bhp}_2}{\text{bhp}_1} = \left(\frac{D_2}{D_1}\right)^5 \times \left(\frac{\text{rpm}_2}{\text{rpm}_1}\right)^3 \tag{5.11}$$

$$\frac{ME_2}{ME_1} = 1 \tag{5.12}$$

$$N_2 - N_1 = 50 \log_{10} \frac{D_2}{D_1} + 50 \log_{10} \frac{\text{rpm}_2}{\text{rpm}_1} \tag{5.13}$$

VARIATION IN SIZE AND SPEED WITH RECIPROCAL RATIOS

A special case of variation in size and speed occurs when the speed ratio is the reciprocal of the size ratio. This can be expressed as

$$\frac{\text{rpm}_2}{\text{rpm}_1} = \frac{D_1}{D_2} \tag{5.14}$$

This equation also can be written as

$$D_2 \times \text{rpm}_2 = D_1 \times \text{rpm}_1 \tag{5.15}$$

which means that the two fans will have the same tip speed. Our rules then will read as follows:

$$\frac{\text{cfm}_2}{\text{cfm}_1} = \left(\frac{D_2}{D_1}\right)^2 \qquad (5.16)$$

$$\frac{SP_2}{SP_1} = 1 \qquad (5.17)$$

$$\frac{\text{bhp}_2}{\text{bhp}_1} = \left(\frac{D_2}{D_1}\right)^2 \qquad (5.18)$$

$$\frac{\text{ME}_2}{\text{ME}_1} = 1 \qquad (5.19)$$

$$N_2 = N_1 \qquad (5.20)$$

Note that the exponents still add up: $2 + 0 = 2$.

Let's apply these rules to a specific example, the performance of our 27-in vaneaxial fan at 1750 rpm, as shown in Fig. 4.57. We replot the four performance curves in Fig. 5.3 in dashed lines. What will be the performance of this fan, converted in geometric proportion to 40½ in (a 50 percent larger size) and running at 1750/1.5 = 1167 rpm so that the higher speed is 50 percent larger than the lower speed. Again, we take the performance data at six points and enter them in the top half of Table 5.3. Our conversion factors now will be as follows:

$$\text{Air volume conversion factor} = \left(\frac{D_2}{D_1}\right)^2 = \left(\frac{40.5}{27}\right)^2 = 1.5^2 = 2.25$$

$$\text{Static pressure conversion factor} = 1$$

$$\text{Brake horsepower conversion factor} = \left(\frac{D_2}{D_1}\right)^2 = \left(\frac{40.5}{27}\right)^2 = 1.5^2 = 2.25$$

$$\text{Efficiency conversion factor} = 1$$

$$N_2 - N_1 = 0 \quad \text{(approximately)}$$

We apply these conversion factors to obtain the performance data for the 40½-in vaneaxial fan at 1167 rpm and enter these data in the bottom half of Table 5.3. We now plot these new data in solid lines in Fig. 5.3. We notice the following:

1. The larger fan again results in a flatter static pressure curve because the conversion factor for air volume is larger than the conversion factor for static pressure. In other words, the increased size boosts the air volume more than the static pressure. Even though the reduced speed works against this trend, the combination still boosts the air volume more than the static pressure.

2. The static pressure remains the same as a result of the same tip speed for both fans—an important observation.

3. The power-consumption brake horsepower becomes moderately higher. For the 27-in fan at 1750 rpm a 7½-hp motor is adequate, but the 40½-in fan at 1167 rpm requires a 15- or 20-hp motor.

4. The efficiency remains the same.

5. The noise level remains approximately the same.

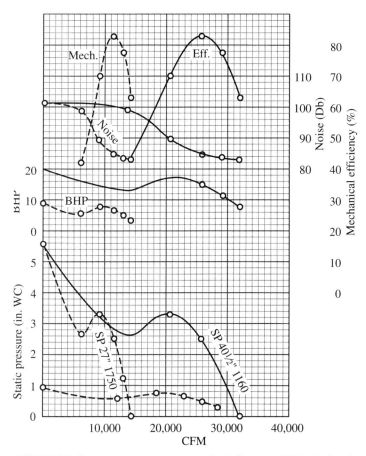

FIGURE 5.3 Performance of two geometrically similar vaneaxial fans having the same tip speed, 27 in at 1750 rpm and 40½ in at 1167 rpm.

VARIATION IN DENSITY

This fan law is used when the fan operates at high altitude where the air density is less (see Table 1.1), where the fan handles hot or cold air (the air density is inversely proportional to the absolute temperature), or where the fan handles a gas other than air, while the size and speed of the fan remain constant. Our conversion rules then will read as follows:

The air volume remains constant:

$$\frac{\text{cfm}_2}{\text{cfm}_1} = 1 \tag{5.21}$$

TABLE 5.3 Conversion of Fan Performance to a Geometrically Similar, Larger Fan Having the Same Tip Speed and Resulting in the Same Static Pressure

		Free delivery	In-between point	Maximum efficiency	Maximum SP	Stalling dip	No delivery
27-in	cfm	14,200	13,020	11,500	9,200	6,100	0
vaneaxial	SP	0	1.25	2.50	3.30	2.64	5.59
fan,	bhp	3.45	5.10	6.65	7.70	5.80	9.00
1750 rpm	ME	0.630	0.776	0.828	0.700	0.421	0
	Noise	83.2	83.5	84.7	89.6	99.1	101.3
40½-in	cfm	31,950	29,295	25,875	20,700	13,725	0
vaneaxial	SP	0	1.25	2.50	3.30	2.64	5.59
fan,	bhp	7.76	11.48	14.96	17.33	13.05	20.25
1167 rpm	ME	0.630	0.776	0.828	0.700	0.421	0
	Noise	83.2	83.5	84.7	89.6	99.1	101.3

The pressures vary directly as the density ρ:

$$\frac{SP_2}{SP_1} = \frac{\rho_2}{\rho_1} \tag{5.22}$$

The brake horsepower varies directly as the density ρ:

$$\frac{bhp_2}{bhp_1} = \frac{\rho_2}{\rho_1} \tag{5.23}$$

The efficiency remains constant.

The noise level remains constant.

Again, we note that the exponents add up: $0 + 1 = 1$.

Let's again apply these rules to a specific example, the performance of a 27-in vaneaxial fan at 1750 rpm, as shown in Fig. 4.57. We replot the four performance curves in Fig. 5.4 in dashed lines. What will be the performance of this fan if it handles air of 335°F instead of standard air at 70°F? The high temperature of 335°F corresponds to an absolute temperature of $460 + 335 = 795$ K. The standard air temperature of 70°F corresponds to an absolute temperature of $460 + 70 = 530$ K. The ratio of these two absolute temperatures is $795/530 = 1.5$. This ratio (or its reciprocal) also will be the ratio of the two air densities. Since the density of standard air is 0.075 lb/ft³, the density of the 335°F air will be $0.075/1.5 = 0.050$ lb/ft³.

Again, we take the performance data at six points in Fig. 4.57 and enter them in the top half of Table 5.4. Our conversion factors will be as follows:

$$\text{Air volume conversion factor} = 1$$

$$\text{Static pressure conversion factor} = \frac{\rho_2}{\rho_1} = \frac{0.050}{0.075} = \frac{1}{1.5} = 0.667$$

$$\text{Brake horsepower conversion factor} = \frac{\rho_2}{\rho_1} = \frac{0.050}{0.075} = \frac{1}{1.5} = 0.667$$

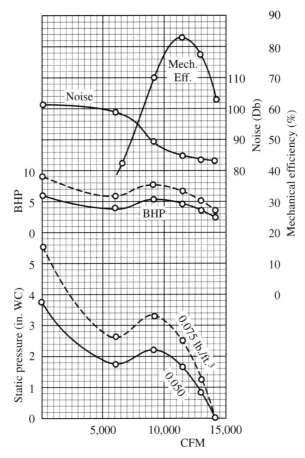

FIGURE 5.4 Performance of 27-in vaneaxial fan at 1750 rpm at two air densities: 0.075 and 0.050 lb/ft³.

$$\text{Efficiency conversion factor} = 1$$

$$\text{Noise difference} = N_2 - N_1 = 0$$

We apply these conversion factors to the data in the top half of Table 5.4 to obtain the performance data for the lower density and enter these new data in the bottom half of Table 5.4. We now plot these new data in solid lines in Fig. 5.4. We notice the following:

1. The smaller air density results in a flatter static pressure curve because the conversion factor for static pressure is smaller than the conversion factor for air volume.

2. The power-consumption brake horsepower becomes moderately lower because the conversion factor equals only the first power of the density ratio. While for

TABLE 5.4 Conversion of Fan Performance to a Different Density ρ

		Free delivery	In-between point	Maximum efficiency	Maximum SP	Stalling dip	No delivery
27-in	cfm	14,200	13,020	11,500	9,200	6,100	0
vaneaxial	SP	0	1.25	2.50	3.30	2.64	5.59
fan,	bhp	3.45	5.10	6.65	7.70	5.80	9.00
1750 rpm,	ME	0.630	0.776	0.828	0.700	0.421	0
ρ = 0.075	Noise	83.2	83.5	84.7	89.6	99.1	101.3
lb/ft³							
27-in	cfm	14,200	13,020	11,500	9,200	6,100	0
vaneaxial	SP	0	0.83	1.67	2.20	1.76	3.73
fan,	bhp	2.30	3.40	4.43	5.13	3.87	6.00
1750 rpm,	ME	0.630	0.776	0.828	0.700	0.421	0
ρ = 0.050	Noise	83.2	83.5	84.7	89.6	99.1	101.3
lb/ft³							

standard air density a 7½-hp motor is needed, for the lower air density a 5-hp motor might suffice. To be on the safe side, however, we probably will do the following two things:

a. Use a 7½-hp motor anyway, even for the lower air density, so that the motor will not be overloaded if the fan should ever be operated at standard air conditions.

b. Use belt drive rather than direct drive so that the motor will be located outside the housing and will not be heated up by the hot air stream passing through the fan housing.

3. The efficiency remains the same.
4. The noise level remains the same.

Again, this fan law for density can be combined with other fan laws whenever required in either sequence.

MACHINING DOWN THE FAN WHEEL OUTSIDE DIAMETER

Suppose that a company manufactures a line of turbo blowers. These are centrifugal fans for high static pressure and relatively small air volumes. They will be discussed in more detail in Chap. 7. Suppose that this line consists of the following eight sizes: 9, 12, 15, 18, 21, 24, 27, and 30 in and that the fan wheels are aluminum castings. The company has patterns for these eight sizes of fan wheels. They have narrow blades and run at high speeds, often at 3500 rpm for direct drive.

Figure 5.5 shows the static pressure and brake horsepower curves for two sizes of the line: a 15-in fan (dashes) and an 18-in fan (solid lines), both running at 3500 rpm. A customer wants one turbo blower to deliver 1600 cfm against a static pressure of 18 inWC. Figure 5.5 indicates that this is somewhat below the performance of the 18-in fan. Obviously, the company does not want to build a new pattern for just one unit with about a 17-in wheel diameter. Thus the company decides to use the 18-in fan wheel machined down to 17 in. The dotted lines in Fig. 5.5 show the performance

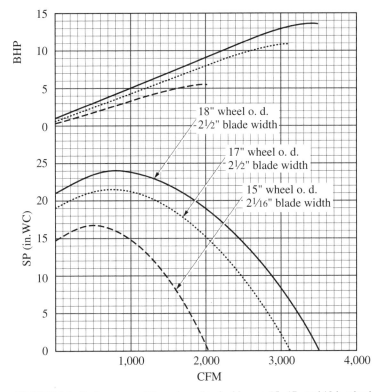

FIGURE 5.5 Performance of three sizes of turbo blowers, 15-, 17-, and 18-in wheel diameters, running at 3500 rpm.

we can expect from this 17-in fan. How did we obtain this predicted performance? Can we use Eqs. (5.5), (5.6), and (5.7) for variation in size, using the ratios $(17/18)^3$ for air volume, $(17/18)^2$ for static pressure, and $(17/18)^5$ for brake horsepower? Not really. Why not? Because these equations apply only if the units are in geometric proportion. By machining the wheel outside diameter down from 18 to 17 in, the wheel inlet diameter and the blade width will remain the same instead of being reduced in geometric proportion. In other words, the machined-down fan wheel will give us slightly more air volume than if we were to reduce all the dimensions in geometric proportion. We should modify our formulas as follows:

The air volume will vary as the square (not the cube) of the size:

$$\frac{\text{cfm}_2}{\text{cfm}_1} = \left(\frac{D_2}{D_1}\right)^2 \tag{5.24}$$

The static pressure will still vary as the square of the size:

$$\frac{SP_2}{SP_1} = \left(\frac{D_2}{D_1}\right)^2 \tag{5.25}$$

the same as in Eq. (5.6).

TABLE 5.5 Conversion of Fan Performance to a Machined-Down Fan Wheel Running at the Same Speed

18-in wheel outside diameter,	cfm	3490	2880	1850	950	0
2½-in blade width,	SP	0	10	20	24	21.2
3500 rpm	bhp	13.4	12.5	8.5	4.7	0.9
17-in wheel outside diameter,	cfm	3113	2569	1650	847	0
2½-in blade width,	SP	0	8.92	17.84	21.41	18.91
3500 rpm	bhp	10.66	9.95	6.76	3.74	0.72

The brake horsepower will vary as the fourth (not the fifth) power of the size:

$$\frac{\text{bhp}_2}{\text{bhp}_1} = \left(\frac{D_2}{D_1}\right)^4 \tag{5.26}$$

The efficiency will remain constant.

The noise level will be slightly decreased by 50 times the logarithm (base 10) of the size ratio:

$$N_2 - N_1 = 50 \log_{10} \frac{D_2}{D_1} \tag{5.27}$$

the same as in Eq. (5.23).

These formulas are not based on theory, as the other fan laws are, so they may not be 100 percent accurate, but they are close enough for practical purposes, and they help to solve a problem that is encountered often. Table 5.5 shows the data we obtained by using the preceding formulas.

This much for the fan wheel. What about the scroll housing? If it is fabricated (not a casting), we could reduce the scroll dimensions in geometric proportion, except we would still have to keep the housing width and the housing inlet diameter unchanged so that they will fit the machined-down fan wheel. Furthermore, the reductions in the scroll dimensions will be small, so we might as well use the same scroll housing as for the 18-in fan. This will be the simplest solution, and it will be close enough to be acceptable.

CHAPTER 6
SYSTEM RESISTANCE

AIRFLOW SYSTEMS

In Fig. 4.38 we showed a typical static pressure versus air volume curve for a vaneaxial fan. It showed the different static pressures a specific fan will produce when choked down to certain air volumes. This curve, then, is characteristic for this specific fan. It can be called the *fan characteristic.*

An airflow system consists of a fan (or several fans) and of various elements through which the airflow can pass. These may be ducts, elbows, expanding or converging transitions, heating and cooling coils, screens and guards, dampers, louvers and shutters, nozzles, bag houses and other filters, or bubble pools. Each component will offer some resistance to airflow, and the fan has to develop sufficient static pressure to overcome all these resistances. The total of all these resistances is called the *system resistance* or the *resistance pressure.* The static pressure produced by the fan has to be equal to the resistance pressure.

An airflow system also will have a characteristic curve of resistance pressure versus air volume. It will show the different static pressures that will be required to force certain air volumes through this specific system. This curve is called the *system characteristic.*

If we will plot the fan characteristic and the system characteristic on the same sheet of graph paper, usually there will be a point of intersection of the two curves. This point of intersection will be the only point that will satisfy both the fan characteristic and the system characteristic. It therefore will be the *point of operation.*

AIRFLOW THROUGH A POOL
OF STATIONARY LIQUID

Normally, the system resistance (i.e., the static pressure needed) will increase with the velocity and therefore with the volume of the air passing through the system. An exception is a pool of stationary liquid through which air or a gas is forced in bubbles, as in the aeration of sewage or of molten iron. Here the system resistance will be constant, regardless of the volume, because it is simply the hydrostatic pressure presented by the liquid. This hydrostatic pressure will be proportional to the depth of the pool and to the specific gravity of the liquid (so it will be very high for molten iron). However, the hydrostatic pressure obviously will not depend on the air volume forced through the liquid. (The air volume will depend only on the amount of air made available by the fan.) No air can bubble through if the maximum pressure produced by the fan is smaller than this hydrostatic pressure. This is shown in Fig. 6.1. Note that the system characteristic is a straight horizontal line. However, if the pressure produced by the fan is adequate, the bubbles will flow. From here on, no extra pressure (only more fan capacity) is needed to force more air through the liquid. Figure 6.2 shows the fan characteristic and the system characteristic for this case.

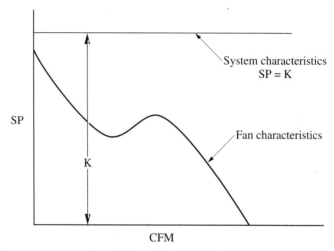

FIGURE 6.1 Fan characteristic and system characteristic for a bubble pool with no point of intersection. No bubbles can pass through the liquid.

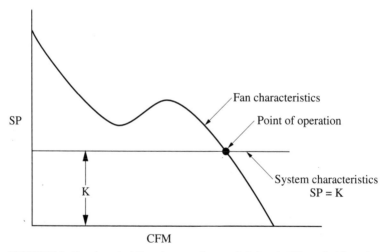

FIGURE 6.2 Fan characteristic and system characteristic for a bubble pool with a point of intersection. Gas bubbles will pass through the liquid.

The system characteristic is again a straight horizontal line, indicating that the static pressure needed to force bubbles through the liquid is constant, regardless of the air velocity, from 0 cfm to the point of operation and beyond. The formula for this system characteristic is

$$SP = K(\text{cfm})^0 = K \tag{6.1}$$

The constant K determines the height of the horizontal line above the air volume.

Apart from this exceptional case, the static pressure needed to blow or draw air through an airflow system is not constant, but it increases with the air volume or velocity, i.e., with the cubic feet per minute (cfm). The question now is: How fast will it increase? The answer is: It depends on the air velocity and on the resulting type of airflow (laminar or turbulent).

AIRFLOW THROUGH FILTER BAGS

The total area of filter bags in a bag house is made large in order to keep the flow resistance low, even when the bags start to get plugged up by the dust. As a result of the large area, the velocity of the air passing through the fabric is very low, about 3 to 4 ft/min (fpm), and the corresponding Reynolds number Re is small. For standard air, we can calculate the Reynolds number from Eq. (1.6) as

$$Re = 6375VR = 6375 \times \frac{3}{60} \times 1 = 319$$

This is far below the 2000 value where turbulent airflow might start. This means that the airflow through the filter bags is laminar. The system characteristic for laminar airflow can be calculated from the formula

$$SP = K \times \text{cfm} \tag{6.2}$$

This is a straight, inclined line through the origin, as shown in Fig. 6.3. The constant K determines the steepness of the straight line: $K = \tan \alpha$. As the bags get plugged up by dust, the filter efficiency will improve and the system resistance and the steepness angle α will increase, but it will still be a straight line.

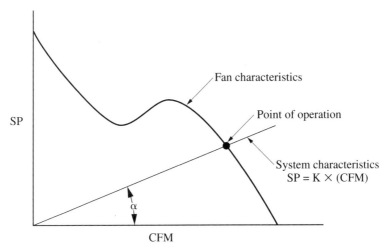

FIGURE 6.3 Fan characteristic and system characteristic for laminar airflow, such as air passing through filter bags.

AIRFLOW THROUGH A GRAIN BIN

Various grains, such as corn, soy beans, barley, and wheat, have to be dried after harvesting to prevent spoilage of the grain. For this purpose, they are stored in cylindrical grain bins that can be from 15 to 80 ft high. Vaneaxial fans or centrifugal fans are used to force heated air through the grain bin.

The static pressure needed to overcome the resistance of the system depends on the height of the bin and on the type of grain. It can be from 3 to 20 inWC. For the lower pressures, vaneaxial fans can be used, but for the higher pressures, centrifugal fans are needed. However, whatever the static pressure, the velocity of the air passing through the grain is about 20 fpm, approximately six times as large as the 3 to 4 fpm through filter bags.

The corresponding Reynolds number, then, is about 2100, the beginning of slightly turbulent airflow, and the formula for the system characteristic is

$$SP = K(\text{cfm})^{1.5} \tag{6.3}$$

This is a curve through the origin, as shown in Fig. 6.4. The constant K determines the steepness of the curve. For higher grain bins and for greater grain compaction (such as wheat), the curve gets steeper.

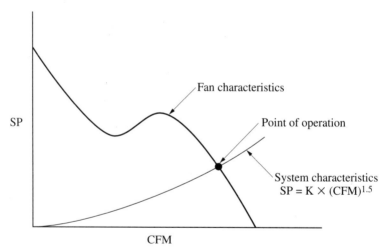

FIGURE 6.4   Fan characteristic and system characteristic for slightly turbulent airflow, such as air passing through grain bins.

AIRFLOW THROUGH A VENTILATING SYSTEM

In a conventional ventilating system, such as used in buildings, both the air velocities and the Reynolds number are considerably larger than in bag houses or grain bins. (Only in the filter section are they still low.) Let us check this statement. In our example on page 4.56 in Chap. 4, we discussed a 27-in vaneaxial fan delivering 11,500 cfm against a static pressure of 2.5 inWC. Our 27¼-in i.d. duct has an area of 4.050 ft²; thus our air velocity will be $V = 11500/4.050 = 2840$ fpm $= 47.3$ fps, and the velocity pressure will be $VP = (2840/4005)^2 = 0.503$ in WC. Let's assume that our system will consist of this $27\frac{1}{4}$-in i.d. duct plus some other equipment, resulting in a total resistance pressure of 2.5 in WC. Our Reynolds number will be $Re = 6375VR = 6375 \times 47.3 \times (27.25/24) = 342,400$. Since this Reynolds number is far above 2000, this is definitely turbulent airflow, as is normal in ventilating systems. The formula for the system characteristic now is

$$SP = K(\text{cfm})^2 \tag{6.4}$$

This is a parabola through the origin, as shown in Fig. 6.5. If one point of the system characteristic is known, the other points can be calculated, and the parabola can be plotted.

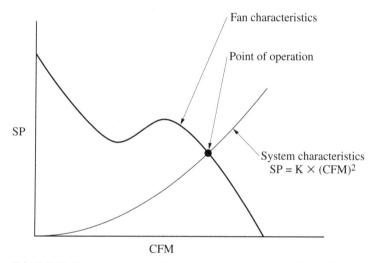

FIGURE 6.5 Fan characteristic and system characteristic for turbulent airflow prevailing in ventilating systems.

It might be interesting to note that on page 1.5 in Chap. 1 we showed in Eq. (1.4) that the friction loss f in a duct is proportional to $(\text{cfm})^2$. Now our Eq. (6.4) indicates the same thing for a ventilating system.

Table 6.1 summarizes our findings for four different airflow systems.

TABLE 6.1 Comparison of Flow Conditions for Four Airflow Systems

Type of system	Type of airflow	Air velocity (fpm)	Reynolds number	Formula for system characteristic
Bubbling pool	—			$SP = K(\text{cfm})^0$
Filter bags	Laminar	3	320	$SP = K(\text{cfm})$
Grain bins	Slightly turbulent	20	2,100	$SP = K(\text{cfm})^{1.5}$
Ventilating system	Turbulent	2840	342,000	$SP = K(\text{cfm})^2$

COMPARISON OF SYSTEM CHARACTERISTIC
CURVES AND CHANGING SPEED CURVES

Coming back to our 27-in vaneaxial fan running at 1750 rpm and delivering 11,500 cfm against a resistance pressure of 2.5 inWC, the performance of this fan was shown in Fig. 4.57 and again in Fig. 5.1. At 1750 rpm, this fan consumed a maximum of 7.7 bhp in the operating range.

Suppose that we want to boost the air volume passing through our system by 50 percent, from 11,500 to 17,250 cfm, as shown in Fig. 5.1. Our air volume ratio will be 17,250/11,500 = 1.5. How much static pressure will the system require so that 17,250 cfm will pass through it? According to Eq. (6.4) for the parabolic system characteristic, the static pressure will increase as the square of this ratio: $1.5^2 \times 2.5 = 5.63$ inWC. This will be the static pressure required to force 17,250 cfm through the system.

Here we note something interesting: The parabolic system characteristic curves are identical with the parabolic curves indicating how fan performance points move when the fan speed is changed. When we plotted the parabolic curves in Fig. 5.1, it was strictly to illustrate the fan laws for a variation in speed. The concept of system characteristics (how the resistance pressure of a system changes for a variation in air volume) did not enter our discussions. Now, in discussing system characteristics in ventilating systems, we find that the parabolic curves in Fig. 5.1 also represent system characteristics. Consequently, if we consider a fan-system combination, the point of operation (intersection) will remain at the same relative location on the pressure curve (corresponding points) if the fan speed should be changed. This is not a matter of course, it could be different, and in fact, it is different for low air velocities, as in grain bins or filter bags. However, in the range of the higher air velocities used in ventilating systems, the system characteristics are parabolic, and we are grateful for this because it simplifies our calculations in the application of fans. What a friendly gesture of nature!

On the other hand, since a system characteristic will always connect corresponding points on the fan performance curves, we cannot shift the performance to a point at a different relative location on the pressure curve by changing the speed. Suppose system resistance was underestimated, resulting in a vaneaxial fan, installed on a system, operating in the stalling range, which, of course, is highly undesirable (inadequate air volume, low efficiency, noisy operation). The customer suggests: Let's increase the speed, even if a larger motor will be required. This would not be a good solution. At the larger speed, the air volume, of course, would be increased, but the fan would still operate in the stalling range, at low efficiency and a high noise level.

SHIFTING THE OPERATING POINT OUT
OF THE STALLING RANGE

System characteristics are a useful tool. Skill in manipulating them can be of practical value in selecting the right fan for a system. Figure 6.6 illustrates this point. It shows a comparison of three fan curves and two system curves. The two solid lines are the same as in Fig. 4.42. They show the performance of two 29-in vaneaxial fans running at 1750 rpm with 7½-hp motors and 52 and 68 percent hub-tip ratios. The long dashes show a parabolic system characteristic that intersects the 52 percent fan

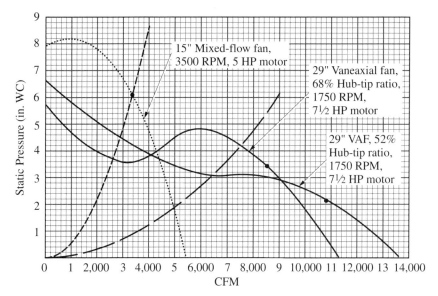

FIGURE 6.6 Comparison of static pressure curves for three fans: a 29-in vaneaxial fan with a 52 percent hub-tip ratio, 1750 rpm, 7½-hp motor; a 29-in vaneaxial fan with a 68 percent hub-tip ratio, 1750 rpm, 7½-hp motor; and a 15-in mixed-flow fan, 3500 rpm, 5-hp motor.

curve in the stalling range but the 68 percent fan curve in the good operating range. This illustrates a case where an increased hub diameter will solve the problem. The short dashes show a steeper parabolic system characteristic that intersects even the 68 percent fan curve in the stalling range. A mixed-flow fan, as shown by the dotted line, will solve this problem. It will deliver far less air volume in the low-pressure range, but it will outperform both vaneaxial fans in the high-pressure range, even though it requires only a 5-hp motor instead of a 7½-hp motor. All this is due to the following three features of the mixed-flow fan:

1. It has a higher pressure capability than vaneaxial fans.
2. It runs at 3500 rpm rather than at 1750 rpm.
3. At the steeper system characteristic, the mixed-flow fan operates in its efficient range, whereas the two vaneaxial fans operate in their inefficient stalling ranges.

Note that all three fans have approximately the same tip speeds.

PRESSURE LOSSES IN VENTILATING SYSTEMS

As mentioned earlier, a fan has to produce sufficient static pressure to overcome the friction in the various components of the ventilating system. Equation (1.4) showed that the friction loss f in a straight, round duct with constant diameter D and with smooth walls can be calculated as

$$f = 0.0195 \frac{L}{D} VP$$

This friction loss in a duct, therefore, is proportional to the velocity pressure VP. This is true not only for the duct friction but also for the friction loss in the other components (except for filters) in a ventilating system. The friction loss in all these components is proportional to the velocity pressure:

$$f = K \times VP \tag{6.5}$$

Only the factor K changes, depending on the type of equipment.

For example, for elbows, K is usually between 0.3 and 0.4, depending on the angle and the sharpness of the turn and on the cross section of the elbow (round, square, rectangular). In other words, elbow loss is approximately one-third the velocity pressure.

In the same manner, the friction losses in screens, transitions, coils, dampers, and so on are proportional to the velocity pressure; only the factor K will vary. Later we will discuss this in more detail.

Another loss to be considered is at the outlet of the system, where one velocity pressure is lost, whatever the velocity pressure at that point happens to be. This, too, has to be included in the static pressure required from the fan. We should attempt, therefore, to make the area at the outlet of our system not too small so that the velocity pressure and the resulting loss will not be too large. This can be accomplished with a gradually expanding outlet, whenever this is possible.

HOW THE STATIC PRESSURE VARIES ALONG A VENTILATING SYSTEM

Let's suppose our ventilating system consists simply of a fan blowing into a round duct that is of constant diameter and which contains some additional resistances such as screens, dampers, etc. Figure 6.7 shows how the static pressure will vary as we move along this ventilating system. We note the following:

1. The ambient static pressure near the fan inlet is zero.
2. The fan raises the static pressure to a maximum positive value.
3. The duct friction reduces the static pressure at a slow rate.
4. The resistance of certain components reduces the static pressure at a steeper rate.
5. The slow rate, due to duct friction, is resumed.
6. The velocity pressure at the outlet is a loss, as mentioned previously. This reduces the remaining static pressure at the outlet of the system to the zero ambient static pressure again.

Figure 6.8 shows how the static pressure will vary along the system if the fan is exhausting from rather than blowing into the system. We note the following:

1. The ambient static pressure near the inlet of the system is zero again.
2. As the air velocity increases from zero to a certain value in the duct, the static pressure decreases to a negative value equal to the velocity pressure in the duct

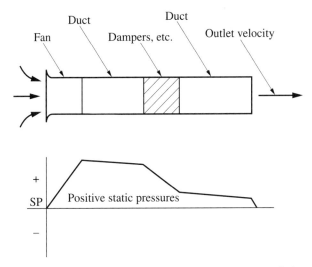

FIGURE 6.7 Variation in the static pressure along a ventilating system with the fan blowing into the system. All static pressures are positive.

plus a turbulence loss. This decrease would be smaller if the duct were equipped with a venturi inlet.

3. The duct friction reduces the static pressure at a slow rate to a more negative value.
4. The resistance of certain components reduces the static pressure at a steeper rate to an even more negative value.

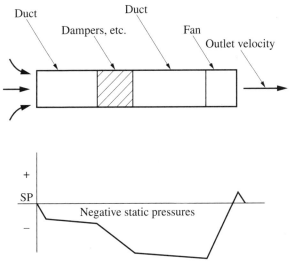

FIGURE 6.8 Variation in the static pressure along a ventilating system with the fan exhausting from the system. All static pressures are negative, except the velocity pressure loss at the end of the system.

5. The slow rate, due to duct friction, is resumed and reduces the static pressure to an even more negative value.

6. The fan raises the static pressure from its maximum negative value to a slightly positive static pressure.

7. The velocity pressure at the fan outlet is a loss that reduces the static pressure at the outlet back to the zero ambient static pressure.

Figure 6.9 shows how the static pressure will vary along the system if the fan is located somewhere in the middle. This means that the fan is exhausting from the left portion of the system and is blowing into the right portion of the system. We note the following:

1. The ambient static pressure near the inlet of the system is zero again.

2. As the air velocity increases from zero to the duct velocity, the static pressure decreases to a negative value, as pointed out in Fig. 6.8.

3. The duct friction reduces the static pressure at a slow rate to a more negative value.

4. The resistance of certain components reduces the static pressure at a steeper rate to a more negative value.

5. The slow rate, due to duct friction, is resumed and reduces the static pressure to an even more negative value.

6. The fan raises the static pressure from its maximum negative value to a maximum positive value.

7. The duct friction reduces the static pressure at a slow rate.

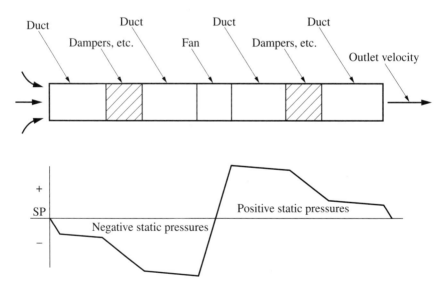

FIGURE 6.9   Variation in the static pressure along a ventilating system, with the fan located somewhere in the middle. The static pressures are negative in the exhaust section and positive in the blowing section.

8. The resistance of certain components reduces the static pressure at a steeper rate.

9. The slow rate, due to duct friction, is resumed.

10. The velocity pressure at the outlet is a loss that reduces the remaining static pressure at the outlet of the system to the zero ambient static pressure again.

So far we have assumed a constant duct diameter. Now let's consider the static pressure variation due to a converging conical transition, as shown in Fig. 6.10, with the fan blowing into the system. We note the following:

1. The ambient static pressure near the fan inlet is zero.

2. The fan raises the static pressure to a maximum positive value.

3. The duct friction reduces the static pressure at a slow rate.

4. The converging conical transition reduces the static pressure at a steeper rate not only because of the friction on the transition walls but also primarily because the increase in velocity pressure (kinetic energy) must be obtained at the expense of the static pressure (potential energy).

5. A somewhat slower rate of static pressure reduction, due to duct friction, is resumed, but not as slow as in item 3, because the smaller duct diameter results in an increased friction loss.

6. The velocity pressure at the outlet is a loss that is considerable here, due to the reduced duct diameter. This loss reduces the static pressure at the outlet back to the zero ambient static pressure.

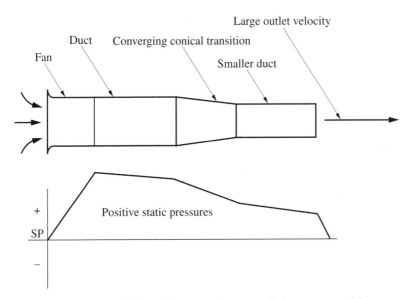

FIGURE 6.10 Variation in the static pressure along a ventilating system containing a converging conical transition, with the fan blowing into the system. All static pressures are positive.

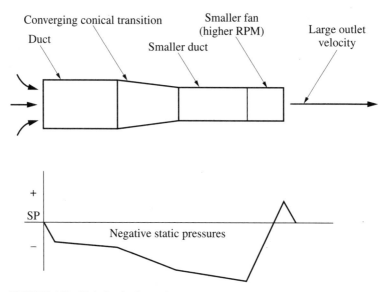

FIGURE 6.11   Variation in the static pressure along a ventilating system containing a converging conical transition, with the fan exhausting from the system. All static pressures are negative, except the velocity pressure loss at the end of the system, which is considerable, due to the smaller outlet diameter.

Figure 6.11 shows how the static pressure will vary along a system containing a converging conical transition if the fan is exhausting from the system rather than blowing into it. We note the following:

1. The ambient static pressure near the inlet of the system is zero again.

2. As the air velocity increases from zero to the duct velocity, the static pressure decreases to a negative value, as pointed out in Fig. 6.8.

3. The duct friction reduces the static pressure at a slow rate to a more negative value.

4. The converging conical transition reduces the static pressure at a steeper rate to a more negative value, mainly because the increase in velocity pressure must be accomplished at the expense of the static pressure and also somewhat because of friction on the transition walls.

5. A somewhat slower rate of static pressure reduction, due to duct friction, is resumed, but not as slow as in item 3, because the smaller duct diameter results in an increased friction loss.

6. The fan raises the static pressure from its maximum negative value to a positive static pressure.

7. The velocity pressure at the outlet is a loss that is considerable, due to the smaller duct diameter. This loss reduces the static pressure at the outlet back to the zero ambient static pressure.

Now let's consider the static pressure variation due to a diverging conical transition, as shown in Fig. 6.12, with the fan blowing into the system. We note the following:

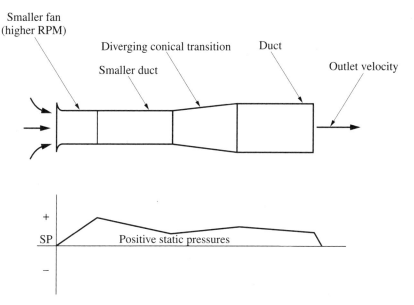

FIGURE 6.12 Variation in the static pressure along a ventilating system containing a diverging conical transition, with the fan blowing into the system. All static pressures are positive.

1. The ambient static pressure near the fan inlet is zero.
2. The fan raises the static pressure to a positive value, possibly the maximum.
3. The duct friction reduces the static pressure at a fairly fast rate due to the small duct diameter.
4. The diverging conical transition has two opposing effects: It slightly reduces the static pressure due to wall friction, but it increases the static pressure by static regain, as discussed in Chap. 1, and this will outweigh the slight friction loss. While air normally flows from higher static pressure to lower static pressure, a gradually diverging cone is an exception: Here air flows from lower to higher static pressure. The general rule, however, is that air will flow from points of higher total pressure to points of lower total pressure without exception.
5. The duct friction reduces the static pressure at a slow rate.
6. The velocity pressure at the outlet is a loss. It reduces the remaining static pressure at the outlet of the system to the zero ambient static pressure again.

Figure 6.13 shows how the static pressure will vary along a system containing a diverging conical transition, with the fan exhausting from the system. We note the following:

1. The ambient static pressure near the inlet of the system is zero.
2. As the air velocity increases from zero to the duct velocity, the static pressure decreases to a negative value, as pointed out in Fig. 6.8. Due to the smaller duct, this negative value is considerable.
3. The duct friction reduces the static pressure at a fairly fast rate, due to the smaller duct diameter, to a more negative value.

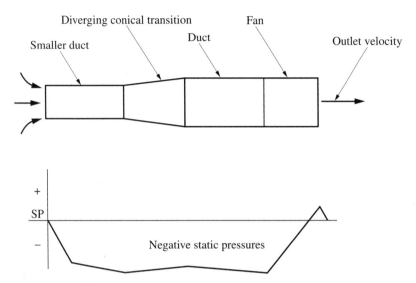

FIGURE 6.13 Variation in the static pressure along a ventilating system containing a diverging conical transition, with the fan exhausting from the system. All static pressures are negative, except the velocity pressure loss at the outlet end of the system.

4. The diverging conical transition will again have two opposing effects, as discussed in item 4 of Fig. 6.12, with the static regain outweighing the slight friction loss. The negative static pressure, therefore, will become slightly less negative.

5. A slow rate of static pressure reduction in the duct will make the static pressure slightly more negative.

6. The fan raises the static pressure from its negative value (it may or may not be a maximum) to a slightly positive static pressure.

7. The velocity pressure at the outlet is a loss. It reduces the static pressure at the outlet back to the zero ambient static pressure.

CHAPTER 7
CENTRIFUGAL FANS

FLOW PATTERN

Figure 3.17 showed a typical centrifugal fan wheel with backward-inclined blades. Figure 3.18 showed the same fan wheel in a scroll housing. The airflow enters the unit axially, the same as in an axial-flow fan, but then spreads out in a funnel-shaped pattern, turning 90° into various radially outward directions before meeting the blades. The blades then deflect these individual air streams into a spiral pattern to an almost circumferential direction. All these air streams are finally collected by the scroll housing and are reunited into a single air stream that leaves the unit at a right angle to the axis.

OPERATING PRINCIPLE

As discussed previously, the operating principle of axial-flow fans is simply deflection of the airflow by the fan blades from an axial direction into a helical flow pattern. In centrifugal fans, the operating principle is a combination of two effects: centrifugal force (this is why they are called *centrifugal* fans) and again deflection of the airflow by the blades, but here the deflection is from a radially outward direction into a spiral flow pattern, as can be seen from Figs. 3.17 and 3.18. As the fan wheel rotates, the air located between the blades and rotating along with them is, of course, subject to centrifugal force, and this is the main cause for the outward flow of the air. Simple outward deflection of the air by the blades is a contributing factor, but to a lesser degree in most types of centrifugal fans. Only in the case of forward-curved blades does the air deflection have a strong influence on the flow pattern and on the performance. In the other types of centrifugal fans, the centrifugal force is the predominant effect. This results in the following two differences between the performances of axial-flow and centrifugal fans:

1. Centrifugal fans normally produce more static pressure than axial-flow fans of the same wheel diameter and the same running speed because of the additional centrifugal force, which is missing in axial-flow fans.

2. Since in centrifugal fans the airfoil lift (if we can call it that) contributes only a small portion of the pressure produced (while most of it is produced by centrifugal action), the improvement due to airfoil blades (over single-thickness, backward-curved sheet metal blades) is not as pronounced in centrifugal fans as it is in axial-flow fans.

What did we mean by saying "if we can call it that"? Repeating a passage from Chap. 2: In a centrifugal fan wheel with backward-curved airfoil blades, the convex side of the airfoil is the pressure side. This is abnormal for an airfoil, even though the airfoil contours are superimposed on a spiral line. This type of blade may have the appearance of an airfoil, but it does not really have the function of an airfoil, since there is no lift force in the conventional sense. It is simply a backward-curved blade with a blunt leading edge, which helps broaden the range of good efficiencies and which improves the structural strength of the blade.

DRIVE ARRANGEMENTS

A centrifugal fan, as shown in Fig. 3.18, consists of a fan wheel and a scroll housing plus such accessories as the inlet cone, the cutoff, and the various supports for the housing and for the drive arrangement (motor, pulleys, bearings, and shaft). Most centrifugal fans use belt drive because it has the following three advantages:

1. Either 1750 or 3450-rpm motors can be used, and expensive slow-speed motors are avoided.
2. The exact fan speed for the required air volume and static pressure can be obtained.
3. The speed can still be adjusted in the field, if desired, by simply changing the pulley ratio.

On the other hand, direct drive is preferable whenever possible, particularly in small sizes, because it has the following three advantages:

1. It reduces the first cost whenever a 1750-rpm motor can be used, since extra supports, pulleys, bearings, and shaft are avoided.
2. It avoids a 5 to 10 percent loss in brake horsepower, consumed by the belt drive.
3. New belts usually stretch 10 to 15 percent in operation, requiring belt adjustments. Direct drive avoids this maintenance.

TYPES OF BLADES

According to their blade shapes, centrifugal fans can be subdivided into the following six categories: AF (airfoil), BC (backward-curved), BI (backward-inclined), RT (radial-tip), FC (forward-curved), and RB (radial blade). Figure 7.1 shows these six commonly used blade shapes. Each of them has its advantages and disadvantages. Accordingly, each is well suited for certain applications. Figure 7.1 also shows the approximate maximum efficiencies that usually can be attained with these blade shapes. Many fans, however, are built for low cost and have maximum efficiencies below those shown, and occasionally—as we will see—even higher efficiencies are

| AF | BC | BI | RT | FC | RB |
| 92% | 85% | 78% | 70% | 65% | 60% |

FIGURE 7.1 Six blade shapes commonly used in centrifugal fans. The approximate maximum efficiency attainable for each type is shown. *(From Bleier, F. P., Fans, in Handbook of Energy Systems Engineering. New York: Wiley, 1985, used with permission.)*

obtained with BI and RT blades. The highest efficiencies can be obtained with airfoil blades, the lowest with flat radial blades. Let us examine these six blade shapes and point out their features.

Centrifugal Fans with AF Blades

Airfoil centrifugal fans are the deluxe centrifugal fans. They have the following features:

1. They have the highest efficiencies of all centrifugal fans.
2. They have relatively low noise levels.
3. They have high structural strength so that they can run at high speeds and produce up to 30 inWC of static pressure.
4. They have stable performance, without surging or pulsation.
5. They are used primarily for clean air and gas applications and for general ventilation. They are good for systems with changing resistance because the performance curves have no stalling range. They can operate in parallel.
6. The passages between adjacent blades are gradually expanding for minimum turbulence.
7. The blades are usually made as hollow airfoils of steel, welded to the back plate and shroud (Fig. 7.2). In small sizes, the entire fan wheel is sometimes made as an aluminum casting (Fig. 7.3).

Figure 7.4 shows a schematic sketch of a typical airfoil centrifugal fan wheel. It shows the pattern of the airflow and the symbols for the principal dimensions. The

FIGURE 7.2 Angular view of 16-in centrifugal fan wheel (5 hp, 3450 rpm) with eight hollow airfoil blades, of welded steel, with flat back plate and spun shroud for tapered blades. *(Courtesy of Coppus Division, Tuthill Corporation, Millbury, Mass.)*

FIGURE 7.3 Angular view of a 10-in centrifugal fan wheel (1 hp, 3450 rpm) of cast aluminum with ten airfoil blades. *(Courtesy of Chicago Blower Corporation, Glendale Heights, Ill.)*

same symbols will be used in other types of centrifugal fans. In Fig. 7.4, notice the following:

1. The fan wheel has a back plate, a shroud, and ten backward-curved airfoil blades.

2. The principal dimensions, as shown, are the blade outside diameter d_2, blade inside diameter d_1, blade length l, blade width b, and the blade angles β_1 and β_2.

3. The leading edge of the blade is the blunt edge on the inside. It has a blade angle β_1, which is measured between the tangent to the circumference of the inner circle and a line bisecting the leading edge of the blade.

4. The trailing edge of the blade is the pointed edge at the blade tip. It has a blade angle β_2, which is measured between the tangent to the circumference of the outer circle and a line bisecting the trailing edge of the blade.

5. The figure also shows a spun inlet cone guiding the airflow into the wheel inlet and helping it turn radially outward with a minimum of turbulence.

6. The shroud shown in Fig. 7.4 has a curved portion at the inlet plus a flat portion extending outward. Some shrouds have a curved portion plus a conical (instead of a flat) portion so that the blade width is not constant but gets narrower toward the tip, making the area for the airflow constant. In other words, the blade width varies from b_1 at the leading edge to a smaller b_2 at the blade tip so that the areas $d_1 \pi b_1 = d_2 \pi b_2$. This used to be the generally accepted design because the constant area and the reduced angle for the airflow turning outward was thought to result in better efficiencies. This reasoning, however, overlooked the loss caused by the sudden expansion at the blade tip, from the blade width to the much larger housing width. A flat outer-shroud portion turned out to be just as efficient or even slightly more efficient because it results in less acceleration of the absolute air velocity as the air stream passes from the leading edge to the blade tip, and the sudden expansion at the blade tip, therefore, is less severe. As an added bonus, the airfoil blade is simpler and more economical in production if $b_1 = b_2$.

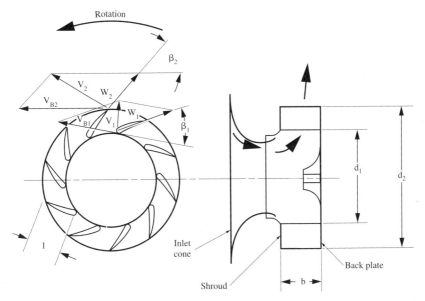

FIGURE 7.4 Schematic sketch of a typical centrifugal fan wheel with ten backward-curved airfoil blades having the maximum permissible width b. *Note:* d_1, blade i.d.; d_2, blade o.d.; b, blade width; l, blade length; β_1, blade angle at leading edge; β_2 blade angle at blade tip; V_{B1}, blade velocity (fpm) at leading edge of blade; V_{B2}, blade velocity (fpm) at blade tip; W_1, relative air velocity (fpm) at leading edge of blade; W_2, relative air velocity (fpm) at blade tip; V_1, resulting absolute air velocity (fpm) at leading edge of blade; V_2, resulting absolute air velocity (fpm) at blade tip.

Velocity Diagrams. Figure 7.4 also shows the velocity diagrams at the leading and trailing edges. Looking at these diagrams, we can make some observations and some calculations and we can derive certain dimensions from the requirements for speed, air volume, and static pressure as follows:

1. The vector sum of the relative air velocity W and the blade velocity V_B results in the absolute air velocity **V**. This is true at the leading edge as well as at the trailing edge.
2. At the leading edge, the following conditions exist:
   a. The cylindrical area A_1 through which the airflow will pass is

   $$A_1 = \frac{d_1 \pi b}{144} = 0.02182 d_1 b \qquad (7.1)$$

   where d_1 and b are in inches and A_1 is in square feet.
   b. The blade velocity V_{B1} at the leading edge can be calculated as

   $$V_{B1} = \frac{d_1 \pi}{12} \times \text{rpm} = 0.2618 d_1 \times \text{rpm} \qquad (7.2)$$

   where V_{B1} is in feet per minute (fpm).
   c. The absolute air velocity V_1 at the leading edge is radially outward and therefore perpendicular to the cylindrical area A_1. It can be calculated as

$$V_1 = \frac{\text{cfm}}{A_1} = 45.8 \frac{\text{cfm}}{d_1 b} \tag{7.3}$$

where V_1 is in feet per minute (fpm).

d. The blade angle β_1 at the leading edge can be calculated from

$$\tan \beta_1 = \frac{V_1}{V_{B1}} = 175 \frac{\text{cfm}}{d_1^2 b \times \text{rpm}} \tag{7.4}$$

It usually is between 10 and 30°.

e. The relative air velocity W_1 at the leading edge is about tangential to the leading edge of the blade.

3. As we move from the leading edge to the blade tip, the blade velocity V_B increases, the relative air velocity W decreases somewhat (due to the channel between adjacent blades becoming wider), and the absolute air velocity $\mathbf{V}$ (being the vector sum of the two) increases.

4. At the blade tip, the following conditions exist:

a. The cylindrical area A_2 is

$$A_2 = \frac{d_2 \pi b}{144} = 0.02182 d_2 b \tag{7.5}$$

where d_2 and b are in inches and A_2 is in square feet.

b. The blade velocity V_{B2} at the blade tip can be calculated as

$$V_{B2} = \frac{d_2 \pi}{12} \times \text{rpm} = 0.2618 d_2 \times \text{rpm} \tag{7.6}$$

where V_{B2} is in feet per minute (fpm).

c. The absolute air velocity V_2 at the blade tip is no longer radially outward or perpendicular to the cylindrical area A_2. By the time it reached the blade tip, the air stream got considerably deflected in the direction of fan rotation, from radially outward to an angle of only about 20° to 30° from the circumference.

d. The absolute air velocity V_2 can be resolved into two components: a radially outward component V_{2r} and a circumferential component V_{2c}. The radial component can be calculated easily as

$$V_{2r} = \frac{\text{cfm}}{A_2} = 45.8 \frac{\text{cfm}}{d_2 b} \tag{7.7}$$

The formula for the circumferential component V_{2c} is

$$V_{2c} = K \frac{SP}{\text{rpm} \times d_2} \tag{7.8}$$

similar to the corresponding equation (Eq. 4.6) for vaneaxial fans, but unfortunately, the constant K contains two correction factors (for hydraulic losses and for circulatory flow) that can only be estimated but cannot be calculated accurately. The preceding equation for V_{2c}, therefore, is only of theoretical interest. Whereas Eq. (4.6) gave us an accurate value for the circumferential component in vaneaxial fans, the preceding equation for centrifugal fans will not give us an accurate value for V_{2c}, the circumferential component for centrifugal fans.

e. The blade angle β_2 at the blade tip theoretically could be determined from

$$\tan \beta_2 = \frac{V_{2r}}{V_{B2} - V_{2c}}$$

but since V_{2c} cannot be determined accurately, β_2 cannot be determined accurately either.

As an indication of the general trend, it may be said that an increase of the blade angles β_1 and β_2 results in an increase in air volume and static pressure, but in a decrease in the fan efficiency. However, if β_1 and β_2 become too large, the passage between adjacent blades may become so wide and short that the airflow is no longer sufficiently guided and the circulatory flow will become excessive. In this case, a simultaneous increase in the number of blades will correct the situation by making the blade channel narrower.

Hydraulic Losses and Circulatory Flow. These two terms were mentioned earlier as the reason why we can not accurately calculate the circumferential component V_{2c} of the absolute air velocity V_2 at the blade tip. The reader, therefore, deserves an explanation of what these two terms mean. The hydraulic losses are the pressure losses due to friction as the air stream passes over the various surfaces. The circulatory flow is a peculiar phenomenon that takes place as the blade channel rotates around the axis of the fan wheel. Let's explain this in more detail.

The air particles occupying the space between the blades do not quite keep up with the rotation of the channel walls but—because of their inertia—lag behind. Relative to the channel, they therefore rotate slowly in the opposite direction. This relative vortex, called *circulatory flow,* which is superimposed on the main relative flow, is comparatively small, yet it is large enough to notably reduce the pressure produced by the fan. What is the mechanism of this pressure loss? The circumferential component V_{2c} is responsible for the pressure produced. Since the circulatory flow is an undesirable phenomenon that negatively affects fan performance, a correction factor has to be applied to the formula for V_{2c} to compensate for the pressure drop due to circulatory flow. This correction factor tends to increase V_{2c} and thereby the blade outlet angle β_2 The determination of this correction factor is difficult. The shape, width, and number of blades will affect it: The poorer the guidance of the air is while passing through the channel, the stronger will be the circulatory flow and the larger will have to be the correction factor.

Conclusion. In view of the preceding, the design calculation for centrifugal fans is not as accurate as the design calculation for vaneaxial fans. In other words, if the requirements for speed, air volume, and static pressure are given for a vaneaxial fan, it is entirely possible, without any previous experience, to go through the various formulas, as shown on page 4.63, to calculate the principal dimensions and to obtain a test sample (the first prototype) that will be satisfactory with respect to air volume, static pressure, and efficiency. For centrifugal fans, on the other hand, only an experienced fan designer who has some empirical data for this type of fan in his or her file will be able to meet the requirements with the first prototype.

Wheel Diameter d_2 and Blade Inside Diameter d_1 for an Airfoil Centrifugal Fan.
In Chap. 4 we pointed out that in designing a vaneaxial fan for a certain set of requirements (speed, air volume, static pressure), we can calculate the minimum hub diameter d_{min} by Eq. (4.1) and the minimum wheel diameter D_{min} by Eq. (4.3). Equa-

tion (4.1) indicated that the minimum hub diameter is only a function of speed and static pressure but not of air volume, which sounds reasonable. Equation (4.3) indicated that the minimum wheel diameter is a function of speed, air volume, and hub diameter, thus also of static pressure. The minimum wheel diameter obtained from Eq. (4.3) tells us whether the wheel diameter requested by the customer is acceptable.

In designing an airfoil centrifugal fan, the procedure is different. Again, we want to calculate the wheel diameter d_2 and the blade inside diameter d_1 from the requirements (speed, air volume, static pressure). To calculate the blade inside diameter d_1, we use the formula

$$d_{1,\min} = 10 \sqrt[3]{\frac{\text{cfm}}{\text{rpm}}} \qquad (7.9)$$

and note that the blade inside diameter depends only on speed and air volume but not on static pressure. This sounds reasonable. Obviously, a larger air volume and a smaller speed will result in a larger blade inside diameter. The static pressure will be produced after the blade inside diameter has been passed, by blades which start at d_1 and extend to d_2.

To calculate the wheel diameter d_2, we use the formula

$$d_{2,\min} = \frac{18,000}{\text{rpm}} \sqrt{SP} \qquad (7.10)$$

and note that the blade outside diameter depends only on speed and static pressure but not on air volume. At first glance, however, this may sound incomplete. Obviously, a larger static pressure and a smaller speed will result in a larger blade outside diameter, but what about the air volume? Won't air volume tend to increase the blade outside diameter, too? The answer is that the blade angles β_1 and β_2 will have to take care of the air volume. We have Eq. (7.4) to calculate β_1, and it contains air volume, as it should. The formula for β_2 also contains air volume, but it is inaccurate, as pointed out on page 7.7. The fact is that β_2 has a strong influence on both the air volume and the static pressure, as we will discuss in more detail in the section on forward-curve blades. An increased β_2 will result in a considerable increase in the air volume, but at the expense of the efficiency. Therefore, while a larger air volume could be obtained by increasing β_2, it sometimes is preferable to keep β_2 smaller, for better efficiency, and instead increase the wheel outside diameter beyond the value of $d_{2,\min}$ according to Eq. (7.10).

Blade Width b. Earlier we pointed out that the airflow enters a centrifugal fan axially and then turns 90° into various radially outward directions. In doing so, the airflow first passes through the circular area A_s of the shroud inlet and then—after the 90° turn—through the cylindrical area A_1 at the blade inside diameter, as can be seen in Fig. 7.4. These two areas can be calculated as follows:

$$A_s = \left(\frac{d_s}{24}\right)^2 \pi \qquad (7.11)$$

where d_s is the shroud inside diameter in inches and A_s is the circular shroud inlet area in square feet.

$$A_1 = \frac{d_1 \pi b}{144} \qquad (7.1)$$

where d_1 is the blade inside diameter in inches, b is the blade width in inches, and A_1 is the cylindrical area at the blade inside diameter in square feet.

Let's examine the three variables in these two formulas: d_1, d_s, and b. d_1 has been calculated from Eq. (7.9) as a function of speed and air volume. It has already been determined. d_s can be calculated as

$$d_s = 0.94d_1 \qquad (7.12)$$

The difference $d_1 - d_s$ is small, just large enough to allow for the curved portion of the shroud. d_s therefore has already been determined, too. b is the only variable that is still open for choice. How can we determine it? Obviously, the value of b will affect the air volume. As the blade width increases, so will the air volume, at least up to a point. When this point has been reached, a further increase in b will not result in a further increase in the air volume because the constant inlet cone inside diameter will act as a choke. It simply will not let any more air pass through. The question is how large can we make b for optimal performance, i.e., for maximum air volume, without impairing the fan efficiency. The answer to this question is: b_{max} should be such that

$$A_1 = 2.1A_s \qquad (7.13)$$

This is an empirical formula. It is the result of experimentation. It makes allowance for the inlet cone inside diameter being somewhat smaller than the shroud inside diameter d_s. Inserting the expressions of the Eqs. (7.1), (7.11), and (7.12) into Eq. (7.13), we get

$$\frac{d_1\pi b}{144} = 2.1\left(\frac{0.94d_1}{24}\right)^2\pi \qquad (7.14)$$

or

$$b = 0.46d_1 \qquad (7.15)$$

In other words, if $A_1 = 2.1A_s$ according to Eq. (7.13), which means a 5.2 percent deceleration of the airflow making the 90° turn from axial to radially outward, then the blade width b will be 46 percent of d_1, according to Eq. (7.15). This is the maximum recommended blade width b.

A right-angle turn with a 52 percent deceleration (or with any deceleration) is a ticklish proposition. The air stream tends to follow its inertia and to shoot across the blade width. In other words, instead of evenly filling the space between shroud and back plate, the airflow will crowd toward the back plate, and less air will flow near the shroud. Obviously, such an uneven condition is undesirable, even though the flow pattern is much improved by the smooth curves at the inlet of the shroud and at the outlet end of the inlet cone, as shown in Fig. 7.4.

While the blade width according to Eq. (7.15) is the maximum recommended, smaller blade widths are often used, whenever the requirements are for less air volume at the same static pressure. Such a reduced blade width, then, will result in less deceleration or even in some acceleration during the right-angle turn from axial to radially outward and, therefore, in a more even flow pattern. Acceleration will take place whenever

$$b \leq 0.221d_1 \qquad (7.16)$$

i.e., whenever the fan wheel has narrow blades (such as in turbo blowers with flat blades, which will be discussed later).

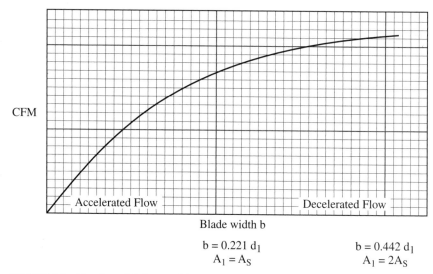

FIGURE 7.5   Air volume (cfm) versus blade width b for an airfoil centrifugal fan.

Figure 7.5 shows how the air volume will vary with the blade width b. It is not in direct proportion (which would be a straight line), as is sometimes assumed. As the blade width increases, the corresponding increase in air volume is (1) smaller than in proportion and (2) at a sacrifice of flow conditions. The line in Fig. 7.5 is somewhat curved so that at b_{max} (close to the point of saturation) the curve is almost horizontal. In the region of accelerated flow and narrow blades, the air stream will more evenly fill the space between back plate and shroud.

Summarizing, it may be said: Narrow blades have the advantage of being more compact; i.e., they deliver more air volume per inch of axial width. Wide blades have the advantage that the total air volume delivered is larger. Flow conditions are smoother in narrow wheels, but average air velocities are lower in wide wheels. Resulting efficiencies therefore can be comparable.

Number of Blades.   The decision regarding the number of blades is based on a compromise between two conflicting requirements. On the one hand, the channel between the blades should be narrow enough for good guidance of the air stream. On the other hand, it should be wide enough so that the resistance to airflow is not too great. Between 8 and 12 blades is a good selection for most designs, but occasionally, up to 16 blades are needed for designs having large diameter ratios d_1/d_2 or large inlet blade angles β_1.

It has been suggested that partial blades could be placed between the main blades, but only in the outer portion of the annular space, so that the channel width would be reduced where it is largest. However, tests on such a configuration have indicated that this will not improve the performance. The reason is that the partial blades result in double the number of blade edges, which—as mentioned previously—are a main source of turbulence.

Blade Length l.   The blade angle β_1 at the leading edge has been calculated from Eq. (7.4). The number of blades has been determined as a compromise, as explained earlier. The blade length *l*, then, will be found by simple layout of an airfoil shape, as shown in Figs. 7.2 through 7.4. The trailing edge of a blade should just slightly overlap or not overlap at all with the leading edge of the adjacent blade.

Scroll Housing. As mentioned earlier, the various air streams leaving the blade tips are collected in the scroll housing and are reunited into a single air stream that leaves the unit at right angles to the axis. The absolute air velocity V_2 at the blade tip (see the velocity diagram in Fig. 7.4) is considerably larger than the air velocity OV at the housing outlet. As the air stream gradually decelerates from V_2 to OV, some of the difference in velocity pressure is converted into static pressure, per Bernoulli (see pages 1.10 and 1.11). This is true for any type of blade. The effect is strongest for forward-curved blades, which have the largest air velocities V_2 at the blade tip.

Figure 7.6 shows a schematic sketch of a typical scroll housing for a 36½-in centrifugal fan with airfoil, backward-curved, or backward-inclined blades, for general ventilation. You will notice the following:

1. The spiral shape is approximated by three circular sections. Their radii are 71.2, 83.7, and 96.2 percent of the wheel diameter.
2. The centers of these three circular sections are located off the center lines by intervals of 6¼ percent of the wheel diameter.
3. For the maximum blade width (which is $0.46d_1$), the housing width is 75 percent of the wheel diameter, or 2.14 times the blade width. (Here is that sudden expansion mentioned on page 7.4.) If the blade width should be reduced by a certain amount, the housing width will be reduced by the same amount.
4. The height of the housing outlet is 112 percent of the wheel diameter.
5. The size of the spiral increases at an even rate. Some manufacturers deviate from this even rate and let the spiral increase faster at first and slower toward the outlet.

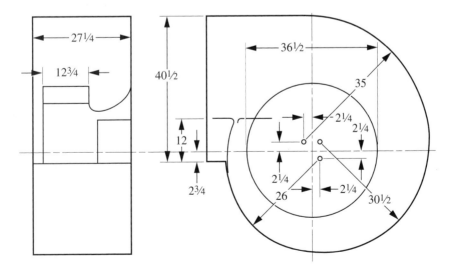

FIGURE 7.6 Schematic sketch of typical scroll housing assembly for a 36½-in centrifugal fan with airfoil, backward-curved, or backward-inclined blades for general ventilation (scale: ⅟₂₆).

6. At the housing outlet, you will notice a so-called cutoff, which looks like a continuation of the spiral and which protrudes into the housing outlet. Two types of cutoffs are in use. They can be described as follows:

a. One design uses a one-piece cutoff, as shown in Fig. 7.7, running all the way across the housing width. It continues the curvature of the scroll and is fastened to the scroll and to the two housing sides. It protrudes into the housing outlet by 20 to 30 percent of the outlet height. The top of the cutoff is the point closest to the blade tips. It leaves a cutoff clearance of 5 to 10 percent of the wheel diameter. The purpose of the cutoff is to minimize the recirculation of the bypassing air (which, of course, is a loss in air volume and efficiency) without producing excessive noise due to a small cutoff clearance. In other words, the 5 to 10 percent cutoff clearance is a compromise between prevention of recirculation and quiet operation.

b. The other design uses a two-piece cutoff placed only on the inlet side of the housing, opposite the inlet cone, where the fan wheel is not, as shown in Figs. 7.6 and 7.8. One cutoff piece is fastened to the scroll. The other piece extends all the way to the inlet cone and is fastened to the inlet cone. The fan wheel then discharges into the wide-open housing outlet, and the cutoff completely prevents recirculation on the inlet cone side but not at the wheel. For this type of cutoff, a better term is *recirculation shield,* particularly for the piece that is fastened to the inlet cone. The recirculation shield protrudes into the housing outlet by 30 to 35 percent of the outlet height.

Inlet Clearance. On pages 4.38 to 4.39, we discussed the tip clearance of vaneaxial fans, i.e., the gap between the rotating fan wheel and the stationary housing shell. We presented the performances we had obtained from a series of tests in which the tip clearance had been varied from very small to very large. We found that the smallest

FIGURE 7.7 Angular view of scroll housing showing drive side (shaft and bearings) and outlet side with cutoff running all the way across the housing width. *(Courtesy of Chicago Blower Corporation, Glendale Heights, Ill.)*

FIGURE 7.8 Angular view of scroll housing showing inlet side (inlet collar and inlet cone) and outlet side with recirculation shield opposite the inlet cone only. *(Courtesy of General Resource Corporation, Hopkins, Minn.)*

possible tip clearance results in the best performance in all respects (static pressure, efficiency, and noise level).

For centrifugal fans, there is a corresponding parameter. It is the small gap between the rotating fan wheel and the stationary inlet cone. It occurs at the inlet edge of the wheel shroud, where the inlet cone overlaps the shroud, as shown in Fig. 7.4. It is called the *inlet clearance*. Again, the smallest possible inlet clearance results in the best performance. For a 30-in wheel diameter, the inlet clearance should be not more than ⅛ in. Sometimes a labyrinth configuration is used instead of a simple overlap, to keep any possible back leakage to a minimum. However, such a labyrinth configuration requires some machined surfaces and therefore is used only in small sizes. In larger sizes, it would be too costly.

Inlet Boxes. For installations with inlet ducts, the inlet of a centrifugal fan can be equipped with an inlet box, bolted to the housing inlet, as shown in Fig. 7.9. The inlet box has two purposes:

FIGURE 7.9 Centrifugal fan with inlet box added, allowing connection to a vertical inlet duct. *(Courtesy of Chicago Blower Corporation, Glendale Heights, Ill.)*

1. If the configuration of the system requires an inlet duct that is perpendicular to the fan shaft, the inlet box will provide a tight, space-saving elbow guiding the air stream from the direction of the inlet duct into a 90° turn to the inlet of the scroll housing.

2. If the fan is wide in an axial direction so that a bearing on the inlet side is required, the inlet box will keep the inlet bearing out of the air stream.

The inlet box has a rectangular cross section. Its width (parallel to the housing side) usually is three times the diameter of the housing inlet. However, the depth of the inlet box (in the axial direction) is often only one-half the diameter of the housing inlet. This is what makes the inlet box space-saving and tight, but it results in considerable losses in air volume delivered (about 30 percent) and in static pressure produced by the fan. These losses can be reduced by the following two methods:

1. By an increase in the axial depth plus a taper in the lower portion of the inlet box, as shown in Fig. 7.9.

2. By providing one or two turning vanes inside the inlet box.

One inlet box is used on a single-inlet fan. Two inlet boxes are used on a double-inlet fan.

FIGURE 7.10 Two views of a double-inlet, double-width centrifugal fan wheel with 12 backward-curved, hollow-steel airfoil blades in staggered positions. *(Courtesy of Chicago Blower Corporation, Glendale Heights, Ill.)*

Double-Inlet, Double-Width Centrifugal Fans. The fans shown in Figs. 7.2 through 7.4 and 7.6 through 7.9 are of the single-inlet, single-width type. By combining a clockwise and a counterclockwise fan into one unit, a double-inlet, double-width centrifugal fan, as shown in Figs. 7.10 through 7.12, is obtained, delivering about 1.9 times the air volume against the same static pressure while consuming approximately double the brake horsepower. The blades of the two-in-one fan wheel usually are staggered for more even flow conditions. This construction is a practical and economical solution for many applications where the air volume requirements otherwise would be too large for the size and speed of the unit. In other words, a higher scroll housing is avoided; only the total width of the housing is increased.

Performance of Airfoil Centrifugal Fans. As mentioned previously, centrifugal fans of any type are for higher static pressure than vaneaxial fans. This means that for the same wheel diameter and for the same running speed, centrifugal fans produce a higher maximum static pressure. Figure 7.13 shows the performance curves for a typical 27-in airfoil centrifugal fan directly driven from a 5-hp, 1160-rpm motor. You will notice the following:

1. The maximum efficiency here is 88 percent, a high efficiency. For larger sizes, the maximum attainable efficiency can be even slightly higher. The optimal operating range is between 50 and 75 percent of the free-delivery air volume.

2. The maximum efficiency occurs at about 85 percent of the maximum static pressure.

3. The maximum static pressure is 3.7 inWC. If this fan were to run at 1750 rpm instead of 1160 rpm, the maximum static pressure (according to the fan laws) would be $(1750/1160)^2 \times 3.7 = 8.4$ inWC. Comparing this with the 3.3 inWC maximum static pressure produced by the 27-in vaneaxial fan at 1750 rpm, as shown in Fig. 4.21, the maximum static pressure produced by this airfoil centrifugal fan is 2.3 times higher.

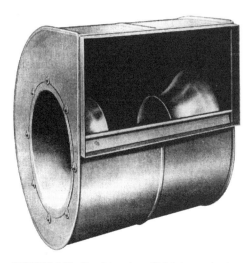

FIGURE 7.11 Scroll housing with inlet cones in place for a double-inlet, double-width centrifugal fan. Short cutoff permits full view into the inside, showing the two inlet cones. *(Courtesy of Chicago Blower Corporation, Glendale Heights, Ill.)*

FIGURE 7.12 Double-inlet, double-width centrifugal fan assembled with inlet cones, fan wheel, shaft, bearings, and braces in place. *(Courtesy of Chicago Blower Corporation, Glendale Heights, Ill.)*

4. The static pressure at the no-delivery point is slightly lower than the maximum static pressure, but there is no stalling dip such as usually occurs in the static pressure curves of forward-curved centrifugal fans and of vaneaxial fan units. There is no surging or pulsation. The performance is stable over the entire range. This makes airfoil centrifugal fans (and also backward-curved and backward-inclined

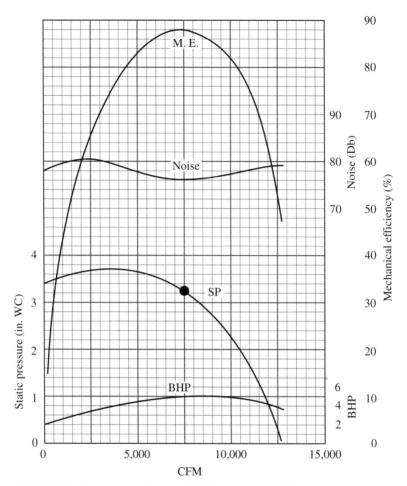

FIGURE 7.13 Performance of a typical airfoil centrifugal fan, 27-in wheel diameter, direct drive from a 5-hp, 1160-rpm motor.

centrifugal fans) suitable for fluctuating systems and for operation of two fans in parallel.

5. The maximum brake horsepower occurs at a static pressure of 2.8 inWC, just slightly below the 3.2 inWC point where the maximum efficiency occurs and far from the free-delivery point. In other words, the brake horsepower curve is nonoverloading. Regardless of where the point of operation is located on the performance curves, the motor can never be overloaded. By comparison, forward-curved centrifugal fans (to be discussed later) overload the motor at free delivery. Vaneaxial fans sometimes (when the blade tip is wider than the blade width at the hub) overload the motor at no delivery.

6. The minimum noise level occurs at the point of maximum efficiency.

Centrifugal Fans with BC and BI Blades

Centrifugal fans with backward-curved (BC) or with backward-inclined (BI) blades are similar in design and performance to those with backward-curved airfoil blades, except for a moderate decline in the maximum efficiency and in the structural strength. The BC centrifugal fan used to be the most efficient and popular type until the airfoil centrifugal fan took its place.

Most of our discussions about airfoil centrifugal fans apply to BC and BI centrifugal fans as well. In particular, the velocity diagrams are the same. The formulae for d_1, d_2, and b_{max} are the same. The discussions on number of blades, blade length, scroll housing, cutoff, inlet clearance, inlet boxes, double-inlet, double-width fans and nonoverloading brake horsepower curves all apply to BC and BI centrifugal fans as well. The advantages of BC and BI centrifugal fans are as follows:

1. They are less costly in production.
2. They can tolerate somewhat higher temperatures and slightly dust-laden gases.

Figure 7.14 shows a schematic sketch of a typical BC fan wheel. The outer portion of the shroud is flat, but it could be conical. The blade shape is a smooth curve. Often

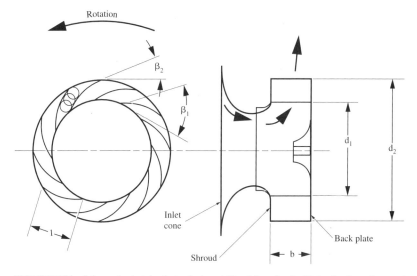

FIGURE 7.14   Schematic sketch of a typical centrifugal fan wheel with ten backward-curved single-thickness blades having the maximum permissible width b. (*Note:* d_1, blade inside diameter; d_2, blade outside diameter; b, blade width; l, blade length; β_1, blade angle at leading edge; β_2, blade angle at blade tip.)

it is just a simple circular arc, but sometimes it is a shape with more curvature near the leading edge, simulating an airfoil shape. The blade channel should be gradually expanding so that the relative airflow will decelerate gradually and at an even rate while passing through the blade channel. If the airflow is smooth and has a minimum of turbulence, a conversion from velocity pressure to static pressure can take place with a minimum of losses. To analyze the blade channel, the following procedure is employed: Circles are inscribed at different stations along the blade channel, as shown in Fig. 7.14, and the area of passage for the relative flow is determined at each station by multiplying the diameter of the inscribed circle by the blade width corresponding to the circle center. These areas of passage should increase at a slow and even rate. By calculating the radii of equivalent circles (having equal areas) and by plotting them against the center line of the channel, the shape of an equivalent cone can now be obtained. It should at all places have a cone angle of less than 7° with the center line; otherwise, the design should be revised.

For a centrifugal fan wheel with straight BI blades, there is a simple relationship between the blade angles β_1 and β_{2s}. After we have determined the ratio d_1/d_2 and the blade angle β_1 at the leading edge, we can calculate β_{2s} from

$$\cos \beta_{2s} = \frac{d_1}{d_2} \cos \beta_1 \qquad (7.17)$$

where d_1 = blade inside diameter in inches
$\quad d_2$ = blade outside diameter in inches
$\quad \beta_1$ = blade angle at the leading edge
$\quad \beta_{2s}$ = blade angle at the blade tip for straight blades

Suppose that we have found $d_1/d_2 = 0.75$ and $\beta_1 = 11°$, then we get $\cos \beta_{2s} = 0.75 \times 0.98163 = 0.73622$ and $\beta_{2s} = 42.6°$. $\beta_1 = 11°$ is a good blade angle for an efficient design, but larger angles up to 30° are acceptable. If Eq. (7.4) should result in a β_1 larger than 30° or 35° at the most, it will be advisable to increase d_1 (and probably d_2) so that Eq. (7.4) will give us a smaller β_1. Figure 7.15 gives a graphic representation

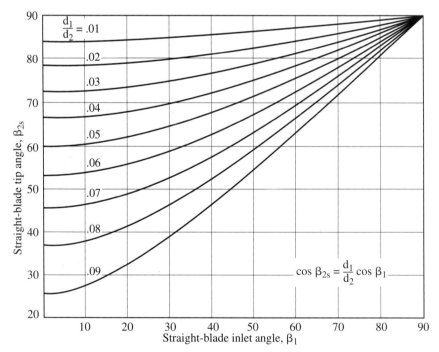

FIGURE 7.15 Tip angle β_{2s} as a function of the inlet blade angle β_1 and of the diameter ratio d_1/d_2 for straight blades.

of β_{2s} as a function of β_1 and of d_1/d_2 according to Eq. (7.17). If the blade were BC instead of BI, then β_2 would have to be somewhat smaller than β_{2s}.

Effect of Inlet Turbulence on the Sound Level of AF, BC, or BI Fans

The velocity diagrams in Fig. 7.4 indicated that the absolute air velocity increases from V_1 at the leading edge to V_2 at the blade tip as the airflow passes through the blade channel between adjacent blades but that the relative air velocity decreases from W_1 to W_2. This means that the highest relative air velocity occurs at the leading edge of the blade and therefore that correct design of this blade portion (tangential flow and the resulting conditions of minimum turbulence) is important.

As in the case of axial-flow fans, the sound level is affected by turbulence in the inlet air stream, but the sensitivity to inlet turbulence here is less than it is for axial-flow fans. The reason for this decreased sensitivity is found in the fact that the air stream meeting the leading edge of a centrifugal fan blade is turbulent already, from the 90° turn just performed. Some additional turbulence is therefore less noticeable than in the case of vaneaxial fans, where the inlet flow conditions are relatively smooth. Nevertheless, inlet turbulence produced by elbows ahead of the centrifugal fan inlet, such as inlet boxes, should be kept to a minimum. As mentioned earlier, this can be done by increasing the axial depth of the inlet box and by providing turning vanes inside the inlet box.

Types of Discharge

In most centrifugal fans, a scroll housing is provided, as shown in Figs. 7.6 through 7.8. Its main function is to collect the individual air streams from the blades and to reunite them into the single air stream discharging through the outlet opening. The location and direction of the discharging air stream relative to the incoming air stream will vary according to the demands of the installation. Sixteen different discharge arrangements (eight for each rotation) have been adopted as standards by the Air Movement and Control Association (AMCA). Additional in-between positions are also used occasionally. Most large centrifugal fans are designed in such a manner that the manufacturer can furnish them for any discharge arrangement required to suit the customer's need. Some designs are for universal discharge, meaning that even after complete assembly the discharge arrangement can still be changed, simply by removing a few bolts and moving the scroll housing to a different angular position.

A different type of discharge is sometimes used for applications in which air is exhausted from a space or from some equipment and delivered into a large space rather than into a duct. In such cases (roof ventilators, plug fans, old mine exhaust fans), the scroll housing is omitted, and the individual air streams leaving the blade channels simply diffuse outward into space. This arrangement is called *radial discharge* or *circumferential discharge.* The term *radial discharge* is used more often, even though the term *circumferential discharge* comes closer to the truth because the direction of the flow velocity is closer to circumferential than to radial. This type of discharge has the advantages of lower cost, greatly increased air volume (almost double) at low static pressure, and a considerable reduction in the sound level because the cutoff (a major source of noise) is eliminated. The maximum negative static pressure is reduced by about 15 percent. FC wheels cannot be used for this application because they require a scroll housing for the conversion of velocity pressure into static pressure, and without a scroll housing, their performance would be

FIGURE 7.16 BI centrifugal fan wheel with circumferential discharge.

poor, as indicated in Fig. 7.31. Figure 7.16 shows a BI centrifugal fan with circumferential discharge exhausting from a space.

Figure 7.17 illustrates another example of radial discharge, an RB centrifugal fan wheel drawing cooling air through the narrow passages of a truck generator. An axial-flow fan would not develop sufficient suction to overcome the resistance of the narrow passages; it would operate in the stalling range.

Centrifugal Fan Groups per AMCA

Figure 7.1 showed six types of centrifugal fans, classified according to their blade shapes. The AMCA (Air Movement and Control Association, Inc., Arlington

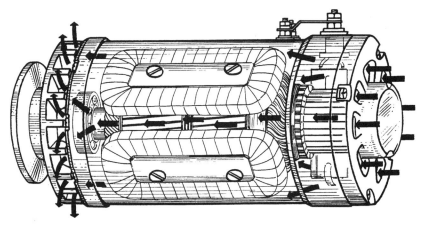

FIGURE 7.17 Radial-blade centrifugal fan wheel drawing cooling air through the narrow passages of a truck generator.

Heights, Ill.), in their publication 99-85 (*Standards Handbook*), shows two groups of centrifugal fans, classified according to their general application. AMCA calls these two groups (1) centrifugal fans and (2) industrial centrifugal fans. Let's describe these two groups in more detail.

Centrifugal Fans. The centrifugal fans we have discussed thus far (AF, BC, BI) belong in this group. They can be used for exhaust or for supply. With exception of the models with reduced blade width, these fans are for large air volumes and for moderately high static pressures. They have high efficiencies and nonoverloading brake horsepower curves. They are for general ventilation, forced or induced draft, boilers and bag houses, and clean or slightly dirty air. This means that they are for industrial applications, too, even though the heading does not indicate this. The AMCA standards show 25 standard sizes and specify the maximum wheel diameters and maximum outlet areas for each size. The maximum wheel diameters range from 12¼ to 132½ in, progressing in a ratio of 1.105. These fans can have wide blades (as discussed), large d_1/d_2 ratios (about 0.65 to 0.80), medium blade angles (10° to 30° at the leading edge, 35° to 50° at the tip), large scroll housings, and large inlet and outlet areas. They have an inlet ring with a diameter about 9 percent larger than the wheel diameter, for connection to an inlet duct, with a circular area about 15 percent larger than the reactangular outlet area. A spun inlet cone connects the large-area inlet ring with the much smaller wheel inlet, with a small overlap. These fans are available as single-inlet, single-width (SISW) and as double-inlet, double-width (DIDW) fans.

Industrial Centrifugal Fans. Most fan manufacturers offer these units with a choice of the following four different wheel types in each size:

Air-handling (AH) wheels (back plate, shroud, ten BI blades)

Material-handling (MH) wheels (back plate, shroud, six mostly radial blades)

Long-shavings (LS) wheels (same as MH, but no shroud)

Long-shavings open (LSO) wheels (no back plate, no shroud, six radial blades)

These fans used to be called *industrial exhausters* because the MH, LS, and LSO wheels are used predominantly for exhausting sawdust, shavings, and granular material. The AH unit, however, is used for supply as well as for exhaust. The AMCA standards show 16 standard sizes and specify the maximum wheel diameters, the maximum inlet diameters, and the maximum outlet area for each size. The maximum wheel diameters range from 19⅛ to 104¼ in, and the corresponding maximum housing inlet diameters range from 11 to 60 in. These inlet diameters are only 57.5 percent of the wheel diameters, and they match the wheel inlets. A simple cylindrical housing inlet leads to the wheel inlet, again with a small overlap whenever there is a shroud, such as in AH and MH wheels. No inlet cone spinning is needed as long as the unit is connected to an inlet duct. If the unit is used for blowing only (without an inlet duct), a venturi inlet is connected to the inlet ring, but this venturi inlet is not as large as the spun inlet cone used in group 1 fans. Compared with the group 1 fans, these group 2 fans, for equal size and speed, have a steeper pressure curve and deliver less air volume but produce a higher maximum static pressure.

Figure 7.18 shows an angular view of a scroll housing for an industrial centrifugal fan of group 2. It is smaller than the scroll housing for the group 1 fans, as shown in Figs. 7.7 and 7.8. The size of the scroll spiral is about 80 percent, the housing width is about 60 percent, and the outlet area is 32 percent. The housing has no extra cutoff, since the small outlet area brings the cutoff point up high enough. All four types of wheels use the same scroll housing; only the wheels are different. These fans are SISW only.

FIGURE 7.19   Angular view of a type AH industrial centrifugal fan wheel showing the BI blades and the conical shroud with a smooth inlet curve.

FIGURE 7.18   Angular view of scroll housing for an industrial centrifugal fan showing drive side (shaft and bearings) and outlet side with small outlet area.

Air-Handling (AH) Fans.   Figure 7.19 shows a typical AH wheel. It has a back plate and a conical shroud. Compared with the centrifugal fans of group 1, the AH wheel has a smaller diameter ratio d_1/d_2, narrower blades, and steeper blade angles. The leading edge may be slanted. The AH wheel is for handling air, gas, or fumes that are clean or only slightly dusty. It usually has ten BI blades, with blade angles of about 23° at the leading edge and about 62° at the blade tip. The blade width b_1 at the leading edge is again $0.46d_1$, according to Eq. (7.15), but since d_1/d_2 is smaller than for group 1 fans, b_1 is smaller, too. The blade width b_2 at the tip is even smaller, about 65 percent of the tip width for group 1 fans. Figure 7.20 shows in solid lines the performance of a 26⅛-in AH unit (26⅛ in is one of the AMCA standard sizes) at 1160 rpm. For comparison, we have plotted in dashed lines the performance of the group 1 27-in airfoil centrifugal fan at 1160 rpm (Fig. 7.13), converted according to the fan laws from 27 to 26⅛ in. We note the following:

1. The AH unit has a steeper static pressure curve than the airfoil centrifugal fan, resulting in less air volume but in a higher maximum static pressure.
2. The brake horsepower curve of the AH unit reaches its maximum at free delivery, but the brake horsepower curve is not steep, so any overload at free delivery will be moderate.
3. The maximum mechanical efficiency of the AH unit is 79 percent, lower than the 88 percent of the airfoil centrifugal fan, but still a very good efficiency.

Selection of an AH Centrifugal Fan from the Rating Tables.   In Chap. 4 on axial-flow fans we showed two rating tables for belt-driven vaneaxial fans and two possible selections for a fan delivering 20,600 cfm against a static pressure of 2 inWC. Let us now discuss in a similar way how we could select an AH centrifugal fan to deliver 7200 cfm against a static pressure of 5 inWC (less air volume but higher static pressure).

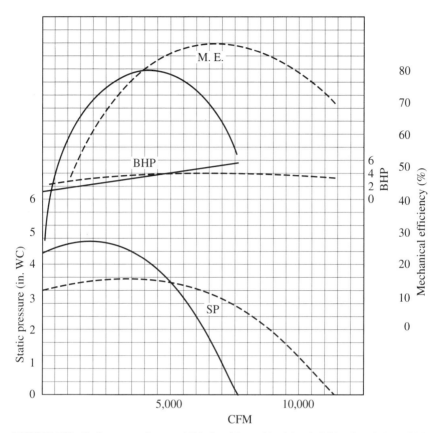

FIGURE 7.20 Performance of a type AH industrial centrifugal fan (*solid lines*) and of an airfoil centrifugal fan (*dashes*), both with 26⅛-in wheel diameters and 1160 rpm.

Table 7.1 is a rating table from a General Blower catalog for a 21-in AH unit. This means a 21-in-diameter housing inlet, corresponding to a 36½-in wheel diameter. The table again has an usual arrangement in which the performance of belt-driven fans is presented in catalogs, for convenient selection by the customer, with the static pressures shown on top, the volumes (cfm) and outlet velocities shown on the left side, and the speeds and brake horsepowers shown at the cross points. This particular table also shows the fan efficiencies (both the static and mechanical efficiencies) at the cross points. All these data were derived from the performance curves (solid lines) shown in Fig. 7.20. Later in this book we will explain how these data can be calculated from the performance curves.

Table 7.1 indicates that this fan, in order to deliver 7200 cfm at a static pressure of 5 inWC, will run at 888 rpm and will consume 8.07 bhp. A 10-hp, 1750-rpm motor will be used, with a pulley ratio of 888/1750 = 0.507. The table also claims that this fan will operate at a mechanical efficiency of 78 percent, which is a good efficiency for a BI centrifugal fan with flat blades. Let's check this efficiency value.

Since this fan has an outlet area of 2.40 ft^2, our outlet velocity will be 7206/2.40 = 3000 fpm, and the velocity pressure will be $(3000/4005)^2$ = 0.561 inWC. The total

TABLE 7.1 Performance Table for a 36½" AH Air-Handling Centrifugal Fan, Belt-Drive, from the Catalog of General Blower Co., Morton Grove, Illinois

Air-handling wheel

Wheel	Inlet	Outlet
36½ inch diameter	21.0 in. outside diameter	20 5/16 × 17 11/16 inch outside
9.56 ft. circumference	2.32 sq. ft. area inside	2.40 sq. ft. area inside

F.P.M. = Outlet velocity in feet per minute S.E. = Static efficiency M.E. = Mechanical efficiency

		½" SP				1" SP				2" SP				3" SP				4" SP			
cfm	fpm	rpm	bhp	SE	ME	rpm	bhp	SE	ME	rpm	bhp	SE	ME	rpm	bhp	SE	ME	rpm	bhp	SE	ME
2402	1000	284	.27	.69	.78	383	.53	.71	.76												
2882	1200	296	.34	.66	.78	390	.64	.71	.77	538	1.30	.70	.73								
3363	1400	311	.43	.62	.77	400	.76	.69	.78	542	1.49	.71	.76	658	2.30	.69	.72				
3843	1600	328	.53	.57	.75	412	.90	.67	.78	548	1.69	.71	.77	661	2.57	.71	.75	760	3.51	.69	.72
4324	1800	346	.65	.53	.74	427	1.06	.64	.77	556	1.92	.71	.78	666	2.86	.71	.76	762	3.86	.71	.74
4804	2000	364	.79	.49	.72	442	1.24	.61	.77	567	2.17	.69	.78	672	3.18	.71	.77	766	4.25	.71	.76
5284	2200	384	.94	.44	.71	459	1.44	.58	.76	579	2.45	.68	.78	681	3.52	.71	.78	772	4.66	.71	.77
5765	2400	403	1.12	.41	.70	476	1.66	.55	.75	593	2.75	.66	.78	692	3.90	.70	.78	780	5.10	.71	.77
6245	2600	424	1.32	.37	.68	494	1.91	.52	.74	607	3.08	.64	.77	703	4.30	.68	.78	790	5.57	.70	.78
6726	2800	446	1.56	.34	.67	513	2.18	.49	.73	623	3.44	.62	.77	716	4.73	.67	.78	800	6.07	.69	.78
7206	3000	468	1.82	.31	.66	532	2.48	.46	.72	639	3.83	.59	.76	730	5.20	.65	.77	812	6.61	.68	.78
7686	3200	490	2.11	.29	.65	551	2.81	.43	.71	656	4.25	.57	.75	745	5.70	.64	.77	825	7.19	.67	.78
8167	3400	512	2.43	.26	.65	571	3.18	.41	.70	674	4.71	.55	.75	761	6.24	.62	.77	839	7.80	.66	.78
8647	3600	535	2.79	.24	.64	591	3.58	.38	.69	691	5.20	.53	.74	777	6.81	.60	.76	853	8.46	.64	.77
9128	3800	558	3.18	.22	.63	613	4.02	.36	.68	710	5.72	.51	.73	793	7.43	.58	.76	868	9.15	.63	.77
9608	4000	581	3.61	.21	.63	635	4.50	.34	.67	729	6.29	.49	.72	811	8.08	.56	.75	884	9.88	.61	.77
10088	4200	604	4.08	.19	.62	657	5.02	.32	.66	748	6.89	.46	.72	828	8.77	.55	.75	900	10.7	.60	.76
10569	4400	627	4.59	.18	.62	679	5.59	.30	.66	767	7.53	.44	.71	846	9.51	.53	.74	917	11.5	.58	.76
11049	4600	651	5.14	.17	.62	701	6.19	.28	.65	787	8.22	.42	.70	864	10.3	.51	.73	934	12.4	.57	.75
11530	4800	675	5.74	.15	.61	723	6.85	.27	.65	807	8.96	.41	.70	883	11.1	.49	.73	952	13.3	.55	.75
12010	5000	699	6.39	.14	.61	746	7.55	.25	.64	827	9.74	.39	.69	902	12.0	.48	.72	970	14.2	.54	.74

cfm	fpm	5" SP				6" SP				7" SP				8" SP				9" SP			
		rpm	bhp	SE	ME	rpm	bhp	SE	ME	rpm	bhp	SE	ME	rpm	bhp	SE	ME	rpm	bhp	SE	ME
4324	1800	850	4.92	.69	.72																
4804	2000	852	5.37	.71	.74	931	6.55	.70	.72	1005	7.77	.68	.71								
5284	2200	856	5.84	.71	.76	934	7.09	.71	.74	1006	8.37	.70	.73	1075	9.70	.69	.71				
5765	2400	862	6.35	.71	.77	937	7.65	.71	.75	1009	9.01	.71	.74	1076	10.4	.70	.73	1140	11.8	.69	.72
6245	2600	869	6.89	.71	.77	943	8.26	.71	.76	1013	9.67	.71	.76	1079	11.1	.71	.75	1142	12.6	.70	.74
6726	2800	878	7.46	.71	.78	950	8.90	.71	.77	1018	10.4	.71	.76	1083	11.9	.71	.76	1145	13.5	.71	.75
7206	3000	888	8.07	.70	.78	958	9.57	.71	.78	1025	11.1	.71	.77	1089	12.7	.71	.76	1150	14.3	.71	.76
7686	3200	899	8.71	.69	.78	968	10.3	.70	.78	1033	11.9	.71	.77	1096	13.6	.71	.77	1155	15.2	.71	.76
8167	3400	911	9.40	.68	.78	979	11.0	.70	.78	1043	12.7	.70	.78	1104	14.4	.71	.77	1162	16.2	.71	.77
8647	3600	924	10.1	.67	.78	990	11.8	.69	.78	1053	13.6	.70	.78	1113	15.4	.71	.78	1170	17.2	.71	.77
9128	3800	938	10.9	.66	.78	1003	12.7	.68	.78	1064	14.5	.69	.78	1123	16.4	.70	.78	1180	18.3	.71	.77
9608	4000	952	11.7	.64	.77	1016	13.6	.67	.78	1076	15.5	.68	.78	1134	17.4	.69	.78	1190	19.4	.70	.78
10088	4200	967	12.6	.63	.77	1030	14.5	.65	.78	1089	16.5	.67	.78	1146	18.5	.68	.78	1200	20.5	.69	.78
10569	4400	983	13.5	.62	.77	1044	15.5	.64	.77	1103	17.5	.66	.78	1158	19.6	.68	.78	1212	21.7	.69	.78
11049	4600	999	14.4	.60	.76	1059	16.5	.63	.77	1117	18.6	.65	.77	1172	20.8	.67	.78	1224	23.0	.68	.78
11530	4800	1015	15.4	.59	.76	1075	17.6	.62	.77	1132	19.8	.64	.77	1186	22.0	.66	.78	1237	24.3	.67	.78
12010	5000	1032	16.5	.58	.75	1091	18.7	.61	.76	1147	21.0	.63	.77	1200	23.3	.65	.77	1251	25.6	.66	.78
12490	5200	1050	17.6	.56	.75	1107	19.9	.59	.76	1162	22.3	.62	.77	1215	24.7	.64	.77	1265	27.1	.65	.78
12971	5400	1067	18.8	.55	.75	1124	21.2	.58	.76	1178	23.6	.61	.76	1230	26.1	.63	.77	1280	28.5	.64	.77
13451	5600	1085	20.0	.53	.74	1141	22.5	.57	.75	1195	25.0	.59	.76	1246	27.5	.62	.77	1295	30.1	.63	.77
13932	5800	1103	21.2	.52	.74	1159	23.8	.56	.75	1212	26.4	.58	.76	1262	29.1	.60	.76	1310	31.7	.62	.77
14412	6000	1122	22.6	.51	.73	1176	25.3	.54	.74	1229	27.9	.57	.75	1278	30.6	.59	.76	1326	33.4	.61	.77
14892	6200	1140	24.0	.49	.73	1194	26.7	.53	.74	1246	29.5	.56	.75	1295	32.3	.58	.76	1342	35.1	.60	.76
15373	6400	1159	25.4	.48	.72	1213	28.3	.52	.74	1264	31.1	.55	.75	1312	34.0	.57	.75	1359	36.9	.59	.76
15853	6600	1178	26.9	.47	.72	1231	29.9	.51	.73	1281	32.8	.54	.74	1330	35.8	.56	.75	1376	38.8	.58	.76

7.25

TABLE 7.1 Performance Table for a 36½" AH Air-Handling Centrifugal Fan, Belt-Drive, from the Catalog of General Blower Co., Morton Grove, Illinois. (*Continued*)

cfm	fpm	10" SP				12" SP				14" SP				16" SP				18" SP			
		rpm	bhp	SE	ME	rpm	bhp	SE	ME	rpm	bhp	SE	ME	rpm	bhp	SE	ME	rpm	bhp	SE	ME
5765	2400	1202	13.3	.68	.71																
6245	2600	1203	14.2	.70	.73	1316	17.4	.68	.71												
6726	2800	1205	15.1	.71	.74	1317	18.4	.69	.72												
7206	3000	1208	16.0	.71	.75	1319	19.4	.70	.74	1422	23.0	.69	.72	1520	26.7	.68	.71				
7686	3200	1213	17.0	.71	.76	1322	20.5	.71	.75	1424	24.2	.70	.73	1521	28.1	.69	.72	1612	32.0	.68	.71
8167	3400	1219	18.0	.71	.77	1326	21.7	.71	.76	1427	25.5	.71	.74	1522	29.5	.70	.73	1613	33.5	.69	.72
8647	3600	1226	19.1	.71	.77	1331	22.9	.71	.76	1431	26.8	.71	.75	1525	30.9	.71	.74	1614	35.1	.70	.73
9128	3800	1234	20.2	.71	.77	1338	24.1	.71	.77	1435	28.2	.71	.76	1528	32.4	.71	.75	1617	36.7	.71	.74
9608	4000	1243	21.3	.71	.78	1345	25.4	.71	.77	1441	29.6	.71	.76	1533	34.0	.71	.76	1620	38.4	.71	.75
10088	4200	1253	22.6	.70	.78	1353	26.8	.71	.77	1448	31.1	.71	.77	1538	35.6	.71	.76	1625	40.2	.71	.76
10569	4400	1264	23.8	.69	.78	1363	28.2	.71	.78	1456	32.7	.71	.77	1545	37.2	.71	.77	1630	42.0	.71	.76
11049	4600	1275	25.2	.69	.78	1373	29.7	.70	.78	1465	34.3	.71	.78	1552	39.0	.71	.77	1636	43.8	.71	.77
11530	4800	1288	26.5	.68	.78	1383	31.2	.70	.78	1474	35.9	.70	.78	1561	40.8	.71	.77	1644	45.7	.71	.77
12010	5000	1300	28.0	.67	.78	1395	32.8	.69	.78	1484	37.6	.70	.78	1570	42.6	.71	.78	1651	47.7	.71	.77
12490	5200	1314	29.5	.66	.78	1407	34.4	.68	.78	1495	39.4	.69	.78	1579	44.5	.70	.78	1660	49.8	.71	.78
12971	5400	1328	31.0	.66	.78	1420	36.1	.68	.78	1507	41.3	.69	.78	1590	46.5	.70	.78	1669	51.9	.71	.78
13451	5600	1342	32.7	.65	.77	1433	37.9	.67	.78	1519	43.2	.68	.78	1601	48.6	.69	.78	1679	54.1	.70	.78
13932	5800	1357	34.3	.64	.77	1446	39.7	.66	.78	1531	45.2	.68	.78	1612	50.7	.69	.78	1690	56.3	.70	.78
14412	6000	1372	36.1	.63	.77	1461	41.6	.65	.77	1544	47.2	.67	.78	1624	52.9	.68	.78	1701	58.7	.69	.78
14892	6200	1388	37.9	.62	.77	1475	43.6	.64	.77	1558	49.3	.66	.78	1637	55.2	.68	.78	1713	61.1	.69	.78
15373	6400	1404	39.8	.61	.76	1490	45.6	.64	.77	1572	51.5	.66	.78	1650	57.5	.67	.78	1725	63.6	.68	.78

pressure, then, will be 5 + 0.561 = 5.561 inWC, the air horsepower will be (7206 × 5.561)/6356 = 6.30, and the mechanical efficiency will be 6.30/8.07 = 0.781 = 78.1 percent, as claimed in the rating table.

The table also shows that for 884 rpm and a static pressure of 4 inWC (instead of 5 inWC), the fan will consume 9.88 bhp, so the 10-hp motor will still be safe, even if the static pressure should be somewhat lower than anticipated. However, if the static pressure should be as low as 3 inWC, this fan at 888 rpm would consume about 11.3 bhp, which might be too much overload for the motor. For larger pressures, on the other hand, the brake horsepower will be lower, and the 10-hp motor will be safe.

Suppose the customer says that a 36½-in wheel diameter seems kind of large for a 7200-cfm requirement and asks, "Couldn't we use a smaller size?" A good question. Let's look at Table 7.2, which is the rating table for the next smaller size, the 19-in AH unit, having a 19-in diameter inlet ring, corresponding to a 33-in wheel diameter. This table indicates that this fan, running at a somewhat higher speed of 1047 rpm, will deliver 7458 cfm against a static pressure of 5 inWC, thus consuming 8.78 bhp. The table also shows that for 1041 rpm and a static pressure of 4 inWC (instead of 5 inWC), the fan will consume 9.89 bhp, so the 10-hp motor will still be safe, even if the static pressure should be somewhat lower than anticipated. We will tell the customer, yes, we can use the next smaller size with the same 10-hp motor, but the 33-in fan will run at 1047 rpm (instead of 888 rpm), the brake horsepower will be 9 percent larger (8.78 instead of 8.07 bhp), and the tip speed will be 7 percent higher for a slightly higher noise level. These are minor disadvantages that probably will be acceptable.

Material-Handling (MH) Fans. Figure 7.21 shows an MH wheel. It still has the same back plate and the same conical shroud as the AH wheel, but it has only six blades that are mostly radial; only the inner portion of the blade is somewhat backward inclined, similar to a radial-tip blade. The MH wheel can handle air or gases containing small-particle dust and granular materials from wood or metal working

FIGURE 7.21   Angular view of a type MH industrial centrifugal fan wheel showing the RT blades and the conical shroud with a smooth inlet curve.

TABLE 7.2 Performance Table for a 33-in Air-Handling Centrifugal Fan, Belt Drive, from the Catalog of General Blower Co., Morton Grove, Illinois.

Air-handling wheel

Wheel	Inlet	Outlet
33 inch diameter	21.0 in. outside diameter	18⅜ × 15⅞ inch outside
8.64 ft. circumference	1.89 sq. ft. area inside	1.96 sq. ft. area inside

F.P.M. = Outlet velocity in feet per minute S.E. = Static efficiency M.E. = Mechanical efficiency

cfm	fpm	½″ SP				1″ SP				2″ SP				3″ SP				4″ SP			
		rpm	bhp	SE	ME	rpm	bhp	SE	ME	rpm	bhp	SE	ME	rpm	bhp	SE	ME	rpm	bhp	SE	ME
1963	1000	317	.22	.70	.78	427	.43	.71	.76	599	.93	.67	.69								
2355	1200	331	.27	.67	.79	435	.53	.71	.77	600	1.07	.70	.73								
2748	1400	348	.35	.63	.78	446	.62	.70	.78	604	1.21	.71	.76	735	1.88	.70	.72				
3140	1600	366	.43	.59	.77	461	.72	.68	.79	611	1.39	.71	.77	738	2.11	.71	.75	848	2.87	.69	.72
3533	1800	386	.53	.54	.76	476	.85	.65	.78	621	1.56	.71	.78	743	2.35	.71	.76	850	3.18	.71	.74
3925	2000	405	.62	.50	.74	493	.99	.63	.78	633	1.76	.70	.78	750	2.60	.71	.77	855	3.48	.71	.76
4318	2200	427	.75	.46	.73	511	1.15	.60	.77	647	1.98	.68	.79	761	2.87	.71	.78	862	3.81	.71	.77
4710	2400	450	.89	.42	.71	530	1.32	.57	.77	662	2.22	.67	.79	772	3.16	.70	.78	871	4.15	.71	.77
5103	2600	473	1.07	.38	.70	550	1.53	.53	.76	678	2.47	.65	.78	785	3.48	.69	.78	881	4.53	.70	.78
5495	2800	497	1.24	.34	.68	571	1.76	.50	.74	695	2.76	.63	.78	800	3.81	.67	.79	894	4.91	.70	.78
5888	3000	522	1.47	.32	.67	592	1.99	.47	.73	712	3.06	.61	.78	815	4.18	.66	.79	906	5.35	.69	.78
6280	3200	549	1.71	.29	.66	614	2.27	.44	.72	731	3.40	.59	.77	831	4.58	.65	.78	921	5.79	.68	.79
6673	3400	574	1.96	.27	.65	636	2.55	.41	.71	750	3.77	.56	.77	849	5.01	.63	.78	936	6.27	.66	.79
7065	3600	601	2.25	.24	.64	659	2.87	.39	.70	770	4.16	.54	.76	866	5.46	.61	.78	952	6.80	.65	.78
7458	3800	627	2.57	.23	.63	683	3.22	.36	.69	791	4.58	.52	.75	884	5.95	.60	.77	969	7.34	.64	.78
7850	4000	654	2.92	.21	.63	708	3.62	.34	.68	811	5.04	.50	.74	903	6.46	.58	.77	986	7.93	.63	.78
8243	4200	681	3.30	.19	.62	734	4.05	.32	.67	833	5.52	.48	.74	923	7.02	.56	.76	1004	8.54	.61	.78
8635	4400	708	3.72	.18	.61	759	4.52	.30	.66	854	6.05	.46	.73	943	7.61	.54	.76	1022	9.19	.60	.77
9028	4600	735	4.16	.16	.61	785	5.01	.28	.66	876	6.61	.44	.72	963	8.25	.53	.75	1041	9.89	.58	.77
9420	4800	762	4.64	.15	.60	811	5.54	.27	.65	899	7.20	.42	.71	983	8.90	.51	.75	1061	10.6	.57	.77
9813	5000	790	5.15	.14	.60	838	6.11	.25	.64	922	7.83	.40	.70	1005	9.62	.49	.74	1081	11.4	.55	.76

7.28

cfm	fpm	5″ SP rpm	bhp	SE	ME	6″ SP rpm	bhp	SE	ME	7″ SP rpm	bhp	SE	ME	8″ SP rpm	bhp	SE	ME	9″ SP rpm	bhp	SE	ME
3533	1800	948	4.04	.69	.72																
3925	2000	951	4.40	.71	.74	1039	5.38	.70	.72	1122	6.37	.69	.71								
4318	2200	956	4.79	.71	.75	1042	5.81	.71	.74	1123	6.88	.70	.73	1199	7.96	.69	.72				
4710	2400	962	5.20	.71	.77	1047	6.27	.71	.75	1126	7.39	.71	.74	1201	8.54	.70	.73	1272	9.70	.69	.72
5103	2600	970	5.62	.71	.77	1052	6.77	.71	.76	1130	7.93	.71	.75	1204	9.13	.71	.75	1275	10.4	.70	.73
5495	2800	980	6.08	.71	.78	1060	7.26	.71	.77	1137	8.49	.71	.76	1209	9.75	.71	.76	1278	11.0	.71	.75
5888	3000	991	6.56	.70	.78	1070	7.80	.71	.78	1144	9.08	.71	.77	1215	10.4	.71	.76	1283	11.7	.71	.75
6280	3200	1003	7.05	.69	.78	1081	8.36	.70	.78	1153	9.70	.71	.78	1223	11.1	.71	.77	1290	12.5	.71	.76
6673	3400	1016	7.59	.69	.79	1092	8.95	.70	.78	1164	10.3	.71	.78	1232	11.8	.71	.77	1298	13.2	.71	.76
7065	3600	1032	8.17	.68	.79	1105	9.57	.69	.78	1176	11.0	.70	.78	1242	12.5	.71	.78	1306	14.0	.71	.77
7458	3800	1047	8.78	.66	.79	1119	10.2	.68	.79	1188	11.7	.69	.78	1254	13.3	.70	.78	1317	14.9	.71	.77
7850	4000	1062	9.41	.65	.78	1134	10.9	.67	.79	1202	12.5	.69	.78	1266	14.1	.70	.78	1328	15.7	.70	.78
8243	4200	1079	10.1	.64	.78	1149	11.7	.66	.79	1216	13.3	.68	.79	1279	14.9	.69	.78	1340	16.6	.69	.78
8635	4400	1096	10.8	.63	.78	1165	12.4	.65	.78	1231	14.1	.67	.79	1293	15.8	.68	.79	1353	17.6	.69	.78
9028	4600	1114	11.6	.62	.78	1182	13.3	.64	.78	1246	15.0	.66	.78	1308	16.8	.67	.79	1367	18.5	.68	.78
9420	4800	1132	12.4	.60	.78	1199	14.1	.63	.78	1263	15.9	.65	.78	1323	17.7	.67	.79	1381	19.6	.68	.79
9813	5000	1150	13.2	.59	.77	1217	15.0	.62	.78	1279	16.9	.64	.78	1339	18.7	.66	.78	1396	20.6	.67	.79
10205	5200	1169	14.1	.58	.77	1235	16.0	.61	.78	1297	17.9	.63	.78	1355	19.8	.65	.78	1412	21.7	.66	.78
10598	5400	1189	15.0	.56	.76	1253	16.9	.60	.77	1314	19.0	.62	.77	1372	20.9	.64	.78	1428	23.0	.65	.78
10990	5600	1209	16.0	.55	.76	1272	18.0	.58	.77	1332	20.1	.61	.77	1389	22.0	.63	.78	1445	24.1	.64	.78
11383	5800	1229	17.1	.54	.76	1291	19.1	.57	.77	1351	21.2	.60	.77	1408	23.3	.62	.77	1461	25.4	.63	.78
11775	6000	1249	18.0	.52	.75	1311	20.3	.56	.76	1370	22.3	.59	.76	1425	24.6	.61	.78	1479	26.8	.63	.78
12168	6200	1270	19.1	.51	.75	1331	21.4	.55	.76	1389	23.6	.57	.77	1444	25.8	.60	.77	1497	28.1	.62	.78
12560	6400	1291	20.4	.49	.74	1351	22.7	.53	.75	1408	24.9	.56	.76	1462	27.3	.59	.77	1515	29.5	.61	.78
12953	6600	1312	21.5	.48	.74	1371	23.9	.52	.75	1427	26.3	.55	.76	1482	28.7	.58	.77	1533	31.1	.60	.77

TABLE 7.2 Performance Table for a 33-in Air-Handling Centrifugal Fan, Belt Drive, from the Catalog of General Blower Co., Morton Grove, Illinois. (*Continued*)

cfm	fpm	10" SP				12" SP				14" SP				16" SP				18" SP			
		rpm	bhp	SE	ME	rpm	bhp	SE	ME	rpm	bhp	SE	ME	rpm	bhp	SE	ME	rpm	bhp	SE	ME
4710	2400	1340	10.9	.69	.71																
5103	2600	1342	11.6	.70	.73																
5495	2800	1344	12.4	.71	.74	1469	15.1	.70	.72												
5888	3000	1348	13.1	.71	.75	1472	15.9	.70	.74	1586	18.8	.69	.72	1696	21.9	.68	.71				
6280	3200	1354	13.9	.71	.76	1475	16.9	.71	.75	1589	19.9	.70	.73	1697	23.0	.69	.72				
6673	3400	1360	14.7	.71	.77	1480	17.7	.71	.75	1592	20.9	.71	.74	1698	24.3	.70	.73	1799	27.4	.69	.72
7065	3600	1368	15.6	.71	.77	1486	18.8	.71	.76	1596	22.0	.71	.75	1701	25.4	.71	.74	1801	28.9	.70	.73
7458	3800	1377	16.4	.71	.77	1493	19.8	.71	.77	1602	23.1	.71	.76	1705	26.6	.71	.75	1804	30.2	.71	.74
7850	4000	1388	17.4	.71	.78	1501	20.7	.71	.77	1609	24.3	.71	.76	1711	27.9	.71	.76	1808	31.6	.71	.75
8243	4200	1399	18.3	.70	.78	1510	21.9	.71	.78	1617	25.4	.71	.77	1717	29.2	.71	.76	1813	32.9	.71	.76
8635	4400	1411	19.3	.70	.78	1521	23.0	.71	.78	1625	26.6	.71	.78	1724	30.5	.71	.77	1819	34.3	.71	.76
9028	4600	1423	20.4	.69	.78	1532	24.1	.70	.78	1635	27.9	.71	.78	1733	31.9	.71	.77	1826	35.9	.71	.77
9420	4800	1438	21.4	.69	.79	1545	25.2	.70	.78	1646	29.2	.71	.78	1742	33.2	.71	.77	1834	37.3	.71	.77
9813	5000	1452	22.5	.68	.79	1557	26.5	.69	.78	1657	30.6	.70	.78	1752	34.8	.71	.78	1844	38.9	.71	.77
10205	5200	1466	23.8	.67	.79	1571	27.8	.69	.78	1669	31.9	.70	.78	1763	36.2	.70	.78	1853	40.5	.71	.77
10598	5400	1482	25.1	.66	.79	1585	29.2	.68	.79	1681	33.3	.69	.79	1775	37.8	.70	.78	1864	42.3	.71	.78
10990	5600	1498	26.3	.66	.78	1599	30.5	.67	.78	1695	34.9	.69	.78	1787	39.4	.70	.78	1875	43.9	.71	.78
11383	5800	1514	27.6	.65	.78	1614	31.9	.67	.79	1709	36.5	.68	.79	1799	41.0	.69	.78	1887	45.8	.71	.78
11775	6000	1531	29.0	.64	.78	1630	33.5	.66	.79	1723	38.1	.68	.79	1813	42.8	.69	.78	1899	47.5	.71	.78
12168	6200	1548	30.5	.63	.78	1646	35.1	.65	.78	1738	39.7	.67	.79	1827	44.5	.68	.79	1912	49.5	.69	.79
12560	6400	1566	31.9	.62	.78	1662	36.7	.65	.78	1754	41.5	.66	.78	1841	46.4	.68	.78	1925	51.4	.69	.78

operations without plugging up the blade passages. The blades are wider than for an AH wheel but in such a way that the same shroud can be used: At the leading edge they are wider by 22 percent and at the blade tip by 30 percent. The blades are still not as wide as the maximum-width blades of group 1 wheels.

Long Shavings (LS) Fans. Figure 7.22 shows an LS wheel. Back plate and blades are the same as for the MH wheel, but the shroud has been omitted, so the risk of plugging up has been further reduced and even long shavings and abrasive materials can be handled. Temperatures up 1600°F can be tolerated.

Long Shavings Open (LSO) Fans. Figure 7.23 shows an LSO wheel. It has no back plate and no shroud and only six radial blades welded to a heavy spider and reinforced by heavy ribs. It is a rugged wheel that can handle not only long shavings but extremely abrasive and corrosive materials and can tolerate high temperatures. The pressure side of the blades can be protected by wearplates that can be replaced when worn.

Performance of Industrial Centrifugal Wheels. Figure 7.24 shows a comparison of the performances for the four types of industrial centrifugal fans of group 2. We note the following:

1. The static pressure curves of the four types are only slightly different. Particularly in the important operating range from 2 to 5 inWC of static pressure, the differences are minor.

2. For all four wheels, the brake horsepower curves reach their maximum at free delivery. Any overload at free delivery is moderate for the AH wheel but considerable for the three other wheels as a result of their radial blades.

3. The efficiencies decrease in the sequence AH, MH, LS, and LSO. The maximum efficiencies are 79 percent for the AH wheel, 71 percent for the MH wheel, 67 percent for the LS wheel, and 63 percent for the LSO wheel.

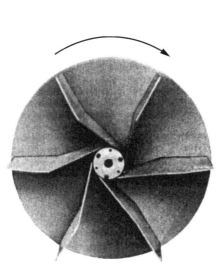

FIGURE 7.22 Type LS industrial centrifugal fan wheel showing the back plate and RT blades.

FIGURE 7.23 Type LSO industrial centrifugal fan wheel showing the radial blades and the reinforcement ribs.

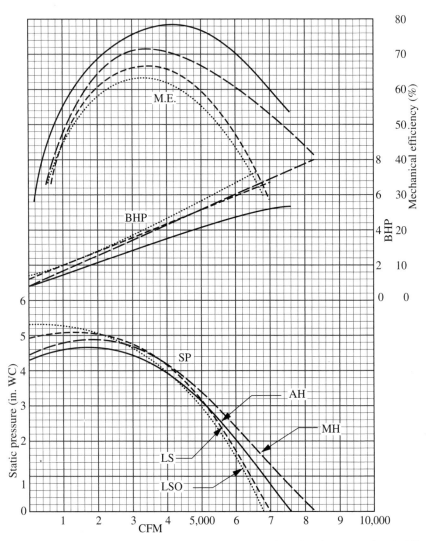

FIGURE 7.24 Performance of the four types of industrial centrifugal fans (AH, MH, LS, and LSO) for a 26⅛-in wheel diameter and 1160 rpm.

Conclusion

1. Shrouds improve the smooth inlet flow pattern and thereby increase the fan efficiency.

2. Radial blades improve ruggedness, but they increase the brake horsepower at free delivery and reduce the efficiency, mainly because of the poor flow conditions at the leading edge, where the relative air velocity hits a radial blade at an angle of 60° to 70° (thereby producing considerable turbulence), instead of being nearly tangential. Radial blades handle the air stream with brute force. They are used mainly when the air stream is contaminated with granular material that cannot be tolerated by other fan types.

Centrifugal Fans with RT Blades

Figures 7.20 and 7.24 indicated that the industrial centrifugal fans of group 2 produce higher static pressure but deliver less air volume (steeper pressure curve) than the AF, BC, and BI fans of group 1. We also indicated that the AH and MH wheels of group 2 have fairly good efficiencies (79 percent for the AH wheel, 71 percent for the MH wheel) and that the AH wheel can handle only clean or slightly dusty air, while the MH wheel can handle air carrying granular material without material buildup on the blades.

Another type of centrifugal fan that is somewhat similar to the MH centrifugal fan is the radial-tip (RT) centrifugal fan. It is used mainly in large sizes for forced and induced draft, for process exhaust, for industrial applications involving hot gases up to 800°F, and in combination with bag houses for air pollution control. The shape of the RT blade results in a self-cleaning action, an important feature in the presence of dust, fly ash, and granular material. Again, material buildup on the blades (which would cause unbalance and vibration) is avoided.

Inlet boxes are often provided on RT units. Renewable wearing plates, fastened to the pressure side of the blades, and interchangeable scroll linings are sometimes provided. The bearings are often water cooled. Instead of RT blades, BI blades with steep blade angles are sometimes used advantageously to handle dust-laden air. They still have the self-cleaning feature, if the blade angles are large enough, and the efficiencies are slightly higher, the brake horsepower curve is less overloading, and the manufacturing cost is lower.

We mentioned earlier that the RT fan is somewhat similar to the MH fan. The similarity between these two types is in the following four respects:

1. They both can handle granular material.

2. At the same size and speed, they will produce the same maximum static pressure, which is higher than for AF, BC, and BI fans.

3. They both have a maximum efficiency of about 71 percent, not quite as high as the AF, BC, BI, and AH types but higher than the FC and RB types.

4. They both have steadily rising pressure curves and therefore stability of performance for systems with fluctuating resistance and for parallel operation, which is frequently encountered in mechanical draft (this feature is also found in AF, BC, and BI fans but not in FC fans). They both have steeper pressure curves than FC fans, which is desirable in case of fluctuating operating conditions (air deliveries will remain more nearly constant despite pressure fluctuations).

However, the RT and MH fans are different in the following three respects:

1. At the same size and speed, the RT centrifugal fan will deliver almost twice as much air volume as the MH fan (see Fig. 7.27).

2. The RT fan is of more rugged construction; therefore, it can run at higher speeds and can produce up to 40 inWC of static pressure compared to about 18 inWC for the MS fan.

3. The RT fan can be built both SISW and DIDW, while the MH fan is SISW only.

In some respects, the RT fan takes a place between the BI and the FC fan, but it is closer to the BI fan. This holds true not only for the shape of the blade and other design features but also for the resulting performance (see Fig. 7.30).

Figure 7.25 shows a typical RT wheel. Different designs are available from various manufacturers. The wheel diameters range in size from 26 to 110 in. The inlet, again, is equipped with a converging inlet cone, providing a small inlet clearance,

FIGURE 7.25 Centrifugal fan wheel with 12 radial-tip blades, diameter ratio of 0.70. Blades are spaced far enough apart to prevent material buildup. Welded of abrasion-resistant steel. *(Courtesy of Chicago Blower Corporation, Glendale Heights, Ill.)*

the same as for BI blades. The diameter ratio d_1/d_2 may vary from 0.5 to 0.8. This is a wide range, wider than for other types of centrifugal fans. The number of blades usually is between 12 and 16 but occasionally up to 24. More blades are desirable for large diameter ratios. The shroud is conical rather than flat. The maximum blade width at the tip is $0.46d_1$, but at the leading edge it may be as large as $0.6d_1$. The blade angles are $25°$ to $40°$ at the leading edge and $80°$ to $90°$ at the blade tip. An $80°$ tip angle is, of course, not quite radial, but it will result in a better fan efficiency, at the expense of slightly less air volume and static pressure. The scroll size is about the same as for group 1 fans, but the maximum housing width is only about 75 percent of that for group 1 fans. At the housing outlet, the bottom is often extended like an outlet diffuser, simulating a long cutoff. This results in an outlet area almost as large as for group 1 fans.

A special type of RT centrifugal fan was designed and tested by me in 1970 for the Coppus Division of the Tuthill Corporation (Millbury, Mass.). It was a small fan for welding fume exhaust. The wheel outside diameter was 8¾ in. The fan was designed for direct drive from a ½-hp, 3450-rpm motor. Figure 7.26 shows the configuration of the fan wheel. It had 12 airfoil blades with radial tips, a design not commonly used by manufacturers. Both the fan wheel and the scroll housing were cast aluminum. Please note that here the concave side of the airfoil is the pressure side, the way it is in a normal airfoil. The housing inlet overlapped with the wheel shroud in a labyrinth-style arrangement for minimum leakage through the inlet clearance. This unit had a surprising 84 percent maximum efficiency, a high efficiency considering the small size of the unit and the radial-tip blades.

Performance of RT Centrifugal Fans. Figure 7.27 shows a comparison of the performance curves for the three types of centrifugal fans we just discussed, all converted by the fan laws to the same size and speed. Table 7.3 lists the principal design data for the three fans, again converted to the same size. The solid lines are for a typical RT centrifugal fan. The dashes are for an MH centrifugal fan. The dots are for the special design with airfoil RT blades. You will note the following:

1. The RT fan delivers the largest air volume, even though it has the narrowest blades. This large air volume is a result of larger inlet and outlet areas, a larger scroll size, and more blades, resulting in better airflow guidance. The airfoil RT fan delivers the smallest air volume, even though it has the widest blades. This small air volume is a result of small inlet and outlet areas and a tight scroll.

2. The RT and MH fans produce the same maximum static pressure, but the airfoil RT fan produces a 15 percent higher static pressure as a result of the efficient airfoil blades.

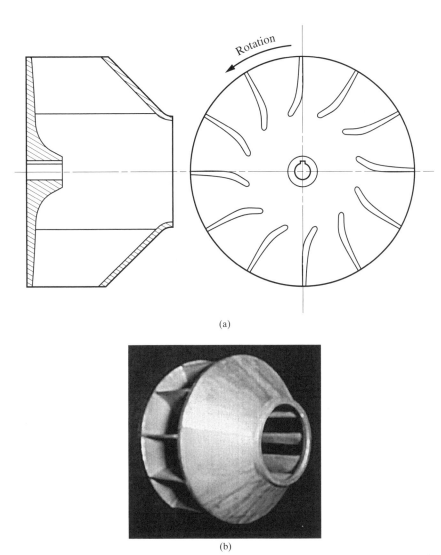

(a)

(b)

FIGURE 7.26 Cast-aluminum centrifugal fan wheel with 12 radial-tip airfoil blades and with a conical shroud. (*a*) Sketch (shroud omitted in right-hand view). (*b*) Photograph. *(Courtesy of Coppus Division of Tuthill Corporation, Millbury, Mass.)*

3. The static pressure curves are still stable (no dip).

4. All three fans consume approximately the same brake horsepower and are somewhat overloading at free delivery.

5. The RT and MH fans have the same 71 percent maximum efficiency, but the airfoil RT fan has a much higher 84 percent maximum efficiency.

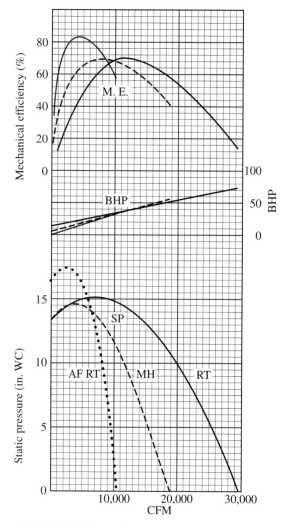

FIGURE 7.27 Comparison of performance for three centrifugal fans, all converted to 30 in, 1750 rpm.

Centrifugal Fans with Forward-Curved (FC) Blades

If we take a flat BI blade and curve the outer portion in the forward direction of rotation until the blade tip is radial, we will obtain a radial-tip (RT) blade. If we now continue curving the blade even more in the forward direction, we will obtain a forward-curved (FC) blade.

Forward-curved centrifugal fans are special in that they deliver considerably more air volume and also produce higher static pressures than AF, BC, or BI centrifugal fans of the same size and speed (but at the expense of lower efficiencies).

TABLE 7.3 Principal Design Data for the Three Fans Compared in Fig. 7.27, 30-in, 1750 rpm.

	RT	MH	AF RT
Diameter ratio d_1/d_2	0.50	0.59	0.50
Number of blades z	16	6	12
Blade width at			
Leading edge (in)	9.3	11.2	16.5
Blade tip (in)	6.5	8.6	9.4
Blade angles			
At leading edge	30°	70°	48°
At blade tip	80°	90°	90°
Shroud i.d. (in)	20	17.7	12.7
Scroll outlet height (in)	32.5	26.4	23.5
Scroll housing width (in)	16.8	14.2	12.3
Outlet area (ft²)	4.9	1.62	0.79
Brake horsepower at free delivery	74	55	32

Conversely, FC fans can run at about half the speed needed to operate in a comparable range of air volume and static pressure.

FC centrifugal fans are often used for furnaces, for various heating, ventilating, and air-conditioning applications, and for the cooling of electronic and other equipment where low operating speeds are desirable to prevent vibration. They are used mainly in small and medium sizes (2- to 36-in wheel diameter) where their lower efficiency is less objectionable but occasionally in sizes up to 73-in wheel diameter. To accommodate the airflow, which is large relative to size and speed, the diameter ratio d_1/d_2 is also large, from 0.75 for small sizes to 0.90 for large sizes. This results in the typical appearance of FC fan wheels, as seen in Fig. 7.28. They always have a large blade inside diameter d_1 and a narrow annular space left for the blades. This narrow annulus calls for a greater number of blades, usually between 24 (in the small sizes) and 64 (in the larger sizes). In other words, the passages between adjacent blades are short and are made narrower (by using more blades) for better guidance of the airflow.

The shroud is a flat ring, so $b_1 = b_2$. The shroud inside diameter is often slightly larger than the blade inside diameter so that portions of the blades protrude inward beyond the shroud inside diameter. This will somewhat improve the flow conditions by leaving more room for the right-angle turn from axial to radially outward. The inlet clearance is made larger than for AF, BC, BI, or RT fans. The maximum blade width is large, often as much as 65 percent of the blade inside diameter d_1.

FIGURE 7.28 Typical centrifugal fan wheel with 52 forward-curved blades and with a flat shroud ring, shown with a spun inlet bell for guiding the airflow into the wheel inlet. Because of its appearance, this type of fan wheel is sometimes called a *squirrel cage wheel*.

The blade angles are very large to obtain the large air volume. At the leading edge, the blade angle β_1 is usually between 80° and 120°, so the relative airflow hits the leading edge of the blade at a large, unfavorable angle, far from any tangential condition. At the blade tip, the blade angle β_2 is even larger, between 120° and 160°. This results in a large and almost circumferential (about 20° from circumferential) absolute air velocity V_2 at the blade tip. V_2 is larger than the tip speed, i.e., the velocity of the blade tip itself. This is a result of the scooping action of the blades. The scroll housing is the same size and shape as for AF, BC, and BI fans, but the cutoff protrudes higher into the outlet. The large air velocity V_2 (kinetic energy) is gradually slowed down in the scroll housing and partially converted into static pressure (potential energy). A good portion of the static pressure is produced in the scroll housing as a result of this conversion from velocity pressure into static pressure. For this reason, FC centrifugal fans can function properly only in a scroll housing. For this reason, FC centrifugal fans can function properly only in a scroll housing. For radial discharge, as in plug fans or in roof ventilators, AF, BC, and BI centrifugal fans can be used, but not FC fans (see Fig. 7.31).

Figure 7.29 shows a water jet hitting an egg cup (a forward-curved surface), which reverses its flow by almost 180°. A similar action can be observed when children play

FIGURE 7.29 View of a water jet being reversed when hitting a forward-curved surface. *(From Bleier, F.P., Fans, in Handbook of Energy Systems Engineering. New York: Wiley, 1985. Used with permission.)*

in a pool. By curving their fingers forward and scooping the water at the surface, they can produce water jets of high velocities, higher than the velocity of their moving hand.

The main reason for the lower efficiency of FC fans is that the airflow through the blade channels of FC fans has to change its direction by almost 180°, i.e. more than in other centrifugal fan types. The air stream can hardly follow the strong curvature of the blades, tangential conditions are no longer prevailing, and the flow is far from being smooth. It is more turbulent than in AF, BC, BI, or RT fans. Because the fan efficiency is comparatively low anyway, slight manufacturing inaccuracies will not reduce it much further and therefore will be less objectionable than with BI blades. Aerodynamic conditions are often secondary in the design of FC fans. Refinements such as overlapping at the inlet, sloping of the shroud, and so on can be left out.

The shape of FC blades is a smooth curve: In small sizes, it is a simple circular arc; in larger sizes, a shape with more curvature near the leading edge is of advantage, since it results in a gradual expansion of the blade channel at a more even rate.

Performance of FC Centrifugal Fans.   Figure 7.30 shows a comparison of four static pressure curves, all for 27-in wheel diameters. You will note the following:

1. At 1140 rpm, the RT fan delivers slightly more air volume and produces slightly more static pressure than the BI fan, but the FC fan delivers considerably more air volume (about 2.5 times as much) and produces a much higher maximum static pressure (about double) than the BI fan. This, as mentioned previously, is at the expense of a lower efficiency for the FC fan.

2. If the FC fan runs at half the speed, the static pressure curve covers a range comparable with that of the BI fan. To be more specific, the FC fan at 570 rpm still delivers about 28 percent more air volume at free delivery, and the maximum static pressure is about one-half. It should be mentioned that the FC fan has a much higher noise level than a BI fan of the same size and speed due to the highly turbulent airflow. The noise levels of the two fans are only comparable if the FC fan runs at half the speed.

3. The static pressure curve of the FC fan has a dip that in some installations may cause unstable operation. Precaution, therefore, should be taken so that FC fans are not used for applications such as fluctuating systems or parallel operation and

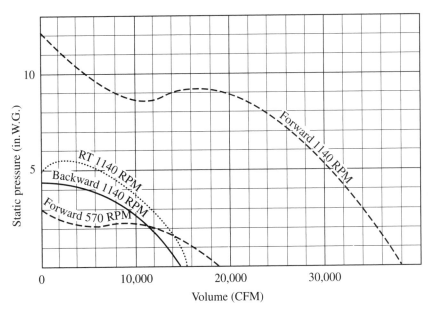

FIGURE 7.30   Comparison of static pressure curves for BI, RT, and FC centrifugal fans, 27-in wheel diameter.

that even in other installations the unit will not operate in the unstable range of the static pressure curve.

Figure 7.31 shows the complete performance (static pressure, brake horsepower, efficiency, and sound level versus air volume) for a 27-in FC fan at 570 rpm. Note that the air volume scale here is double that in Fig. 7.30. You will note the following:

1. The brake horsepower curve is overloading in the low-pressure range. At free delivery, the brake horsepower is more than twice the brake horsepower in the range of maximum efficiency. This shape of the brake horsepower curve results in power requirements outside the operating range that are higher by far than those within the operating range. While centrifugal fans in general are not built for operation at or near free delivery (propeller fans or tubeaxial fans perform this function with greater efficiency and at lower cost), the overloading brake horsepower curve is a serious disadvantage of FC fans. In small sizes, an oversize motor with a horsepower rating equal to the maximum brake horsepower at free delivery is normally selected so that operation at any condition will be safe. For larger units, however, the increased price of the oversized motor would be prohibitive, and the motor horsepower is selected only slightly larger than the brake horsepower for the prospective operating condition. Precaution must be taken so that the unit, when installed in the field, will not operate against too low a static pressure because this would result in an overload for the motor. The permissible operating range of the FC fan, being limited to the left by the dip in the static pressure curve and to the right by the rising brake horsepower curve, therefore, is narrower than that of AF, BC, or BI fans.

2. The sound-level curve has its minimum in the range of best efficiencies.

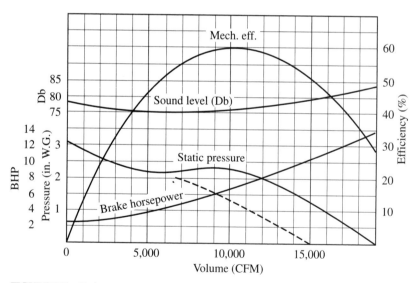

FIGURE 7.31 Performance curves for a typical FC centrifugal fan, 10 hp, 27-in wheel diameter, 570 rpm. The dashes show the performance for radial discharge (no scroll housing). *(From Bleier, F.P., Fans, in Handbook of Energy Systems Engineering. New York: Wiley, 1985. Used with permission.)*

3. Note the dashed line, indicating a poor performance, if an FC fan were used for circumferential discharge, i.e., without a scroll housing, as in plug fans or roof ventilators. Without a scroll housing, AF, BC, and BI fans will deliver larger air volumes, but FC fans would deliver smaller air volumes because they need the scroll housing to convert some of the high air velocity at the blade tips into additional static pressure. Without this conversion, the FC fan will have a poor performance, as shown by the dashes in Fig. 7.31.

Summarizing, we can say that BC and BI blades have the following advantages over FC blades:

1. Stable static pressure curve (no dip)

2. Higher efficiencies, resulting in lower operating costs

3. Nonoverloading brake horsepower characteristic

4. Higher operating speeds, which for direct drive may result in less expensive motors

FC blades, on the other hand, have the following advantages over BC and BI blades:

1. Compactness

2. Lower running speeds, resulting in easier balancing

3. Lower first cost, particularly in small sizes

Conclusion

From the preceding it appears that either type (BI or FC) has its advantages for certain applications. In small sizes, the advantages of FC fans will outweigh their disadvantages. In larger sizes, however, the BI fan will be preferable. The RT fan takes the place between BI and FC fans, but—as indicated in Fig. 7.30—it is closer to the BI fan. This intermediate condition is true for performance, brake horsepower, efficiency, and sound level. The RT fan, however, is a more rugged unit than either the BI or the FC fan and therefore can tolerate higher running speeds (resulting in high static pressures), higher temperatures, and severe service conditions. RT fans are often employed for conveying materials, such as grinding dust, saw dust, cotton, grains, and shavings, if the blades are spaced far enough from each other that narrow passages are avoided, which would tend to become plugged up by deposits of dirt or of the material conveyed.

Centrifugal Fans with Radial Blades

Of the six blade shapes shown in Fig. 7.1, we have discussed the first five types in detail. With regard to the sixth type, radial blades, we have already discussed the LSO (long shavings open) wheel, as shown in Fig. 7.23. This LSO wheel has no back plate or shroud and has only six radial blades welded to a heavy spider. It is a rugged wheel that can tolerate high temperatures and corrosive and abrasive materials, but it has a low efficiency (60 to 63 percent maximum), as shown in Fig. 7.24. These comparatively low efficiencies are due to the lack of back plate and shroud and the poor flow conditions at the leading edge, as pointed out earlier.

PRESSURE BLOWERS, TURBO BLOWERS

Another type of centrifugal fan where radial blades are often used is the pressure blower or turbo blower. These units are used for high pressures and relatively small air volumes. Their fan wheels have a back plate and a shroud, resulting in maximum efficiencies up to about 70 percent. The blades are narrow, usually welded or riveted to back plate and shroud. In small sizes, back plate, shroud, and blades are sometimes one-piece aluminum castings. The blades extend far inward so that the diameter ratio d_1/d_2 is small, usually between 0.4 and 0.55. Because of the narrow blades, these fan wheels are structurally strong, can run at high speeds, and produce high pressures. These units can be used for blowing (positive pressures) or for exhausting (negative pressures).

Applications of High-Pressure Blowers and Exhausters

Here are some of the applications for these high-pressure blowers and exhausters:

Pneumatic conveying of agricultural grains (corn, oats, barley, wheat, rice) and other materials

Aeration of waste water, molten iron, and other fluids

Central vacuum cleaning systems

Gas boosting

Combustion air

Air flotation systems

Various industrial processes

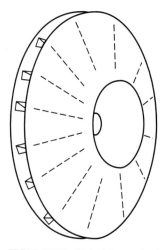

FIGURE 7.32 Turbo blower wheel with a diameter ratio of 0.50 and with 14 radial blades 1¾ to 1 in wide. Each blade has a bent-over strip for riveting to the back plate and little tabs that fit into slots in the shroud for fastening.

Pressure Blowers with Radial Blades

Let's discuss a few examples of such pressure blowers and exhausters. Figure 7.32 shows a typical turbo blower wheel that was designed for pneumatic conveying. It has a *26-in blade outside diameter,* a back plate, a spun shroud and *14 radial blades,* so the blade angles are 90° at the leading edge as well as at the blade tip. The blade width varies from 1¾ in at the leading edge to ⅞ in at the blade tip. The diameter ratio is 0.5.

Pressure Blowers with Steep BI Blades

As mentioned previously, a 90° blade angle at the leading edge results in poor flow conditions. The efficiency of pressure blowers, therefore, can be further improved if the blades are not strictly radial but somewhat backwardly inclined so that the blade angle at the leading edge is more favorable. These are still inexpensive flat blades, but the blade angle at the tip is 65° to 75° instead of 90°.

The blade is extended inward until the blade angle at the leading edge is about 30°
for more nearly tangential flow conditions at the leading edge. This will increase the
maximum efficiency from 70 or 71 percent to about 79 percent, without a loss in the
maximum static pressure produced.

Figure 7.33 illustrates this improvement. It shows a comparison of the performance
for two *18-in turbo blowers,* both having diameter ratios of 0.45 and blade widths varying from 3 to 2 in. Both blowers use the same scroll housing and both are directly
driven by a 10-hp, 3500-rpm motor, resulting in a tip speed of 16,500 fpm. One blower
has *twelve radial blades* (dashes); the other has *ten steep BI blades,* with the blade
angles varying from 30° to 70° (solid lines). You will note that the BI blades result in a
2 percent increase in the maximum static pressure, in a 10 to 18 percent increase in the
air volume, in an 8 percent reduction in the brake horsepower, and in an increase in the
efficiency of 11 percent at the maximum efficiency point but even more at lower pressures. The maximum pressure for the BI blades is 23.9 inWC. Figure 7.33 shows the
considerable improvement in performance obtained with BI blades over radial blades.

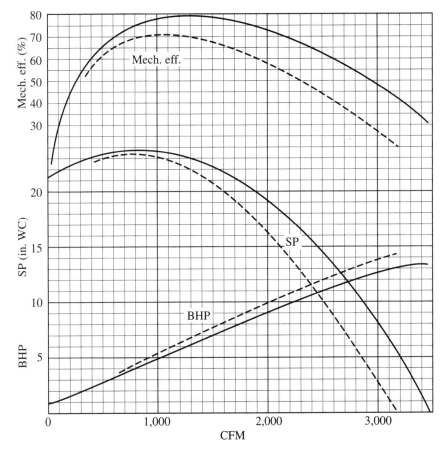

FIGURE 7.33 Performance of two 18-in turbo blowers, direct drive from a 10-hp, 3500-rpm motor.
Diameter ratio = 0.44, blades 3 to 2 in wide, blade angles 30° to 70° (*solid lines*) and 90° to 90°
(*dashes*).

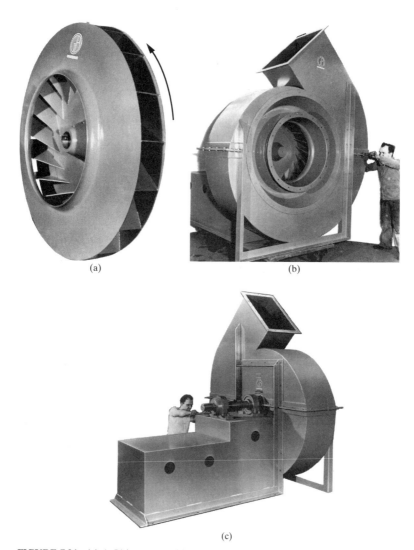

(a)

(b)

(c)

FIGURE 7.34 (*a*) A 54-in pressure blower wheel with spun shroud, 0.54 diameter ratio, 18 BI blades, 6½ to 4⅞ in wide, blade angles 31° to 64°. *(Courtesy of General Blower Company, Morton Grove, Ill.)* (*b*) A 54-in pressure blower showing scroll housing with angular discharge and housing inlet and outlet. *(Courtesy of General Blower Company, Morton Grove, Ill.)* (*c*) A 54-in pressure blower showing scroll housing with angular discharge, housing outlet, and drive side with shaft and bearings. *(Courtesy of General Blower Company, Morton Grove, Ill.)*

Another example of a pressure blower is shown in Fig. 7.34. This unit has a 54-in wheel diameter, which is three times as large as the 18-in blower wheel we just discussed. It is directly driven by a 150-hp, 1140-rpm motor, which is about one-third the 3500 rpm used in the 18-in unit. This means that the tip speed is about the same for the two units. The 54-in blower wheel has a diameter ratio of 0.54, *18 flat BI blades,* 6½ to 4⅞ in wide, with blade angles varying from 31° to 64°. Since the 54-in unit has about the same tip speed as the 18-in unit, the two units—according to the fan laws—will produce the same maximum static pressure if the two blowers are in geometric proportion. Actually, the 54-in pressure blower produces a maximum static pressure of only 20 inWC (even though it has 18 BI blades, while the 18-in

(a)

(b)

FIGURE 7.35 (*a*) A 25-in turbo blower, DIDW, 50 hp, 3500 rpm, diameter ratio of 0.55, 15 radial blades, 3 in wide on each side, viewed from the motor side. *(Courtesy of General Blower Company, Morton Grove, Ill.)* (*b*) A 25-in turbo blower, DIDW, 50 hp, 3500 rpm, diameter ratio of 0.55, 15 radial blades, 3 in wide on each side, viewed from the side opposite the motor. *(Courtesy of General Blower Company, Morton Grove, Ill.)*

turbo blower has only 10 BI blades), compared with the 23.9 inWC produced by the 18-in unit. This 16 percent deficiency in the maximum static pressure is caused by some deviations from geometric proportionality, such as a larger diameter ratio (shorter blades) and lower blade angle at the tip (64° instead of 70°).

Still another example of a high-pressure turbo blower is shown in Fig. 7.35. This figure shows two views of a unit with a *25-in wheel diameter,* DIDW, directly driven by a 50-hp, 3500-rpm motor. The blower wheel has a diameter ratio of 0.55 and *15 radial blades,* 3 to 3 in wide, on each side. This unit delivers 4000 cfm against 24 oz/in^2 = 41.5 inWC (1 oz/in^2 = 1.73 inWC). It has a tip speed of 22,910 fpm. Let's try to compare the maximum static pressures of this unit and of the 54-in pressure blower we just discussed, even though this 25-in wheel has 15 radial blades, while the 54-in wheel has 18 BI blades. Applying the square ratio of the two tip speeds to the maximum static pressure of the 54-in blower, we get

$$\left(\frac{22{,}910}{16{,}120}\right)^2 \times 20 = 40.4 \text{ inWC}$$

which is close to the 41.5 inWC actually produced by this 25-in blower.

TURBO COMPRESSORS

Pressure blowers and turbo blowers, as discussed earlier, will produce static pressures up to about 50 inWC. Even higher pressures are required for some applications. There are three ways to produce these higher pressures: increased speed, two separate blowers connected in series, and multistage turbo blowers. Let's discuss these three methods.

Increased speed, resulting in higher tip speeds and in higher pressures, can be obtained with belt drive or—for even higher speeds—with gear drive, using speed ratios up to 4.5. Figure 7.36 shows an example of a gear-driven turbo blower/exhauster. The 17⅜-in o.d. impeller has ten backward-curved blades, as shown in Fig. 7.36a. The scroll housing is mounted to a gear box, as shown in Fig. 7.36b. Different gear boxes are used to produce different blower speeds, varying from 6000 to 11,000 rpm. With a motor speed of 3450 rpm, the corresponding gear ratios vary from 1.8 to 3.2. The units consume up to 100 hp.

Figure 7.37 shows the performance as a blower; Fig. 7.38 shows the performance as an exhauster. The blower produces pressures up to 8.3 psig = 230 inWC (1 psi = 27.7 inWC). The exhauster produces vacuums up to 10.9 inHg = 148 inWC (1 inHg = 13.6 inWC). The reason for the lower vacuum will be explained later. It is customary to measure high pressures in pounds per square inch (psi) and high vacuums in inches of mercury (inHg).

Eight pounds per square inch (8 psi) really is a high pressure, considering that the atmospheric pressure at the surface of the earth equals 14.7 psi. A blower producing 8 psi, therefore, adds 54 percent to the atmospheric pressure. These very high pressure units, often called *turbo compressors,* have good efficiencies (80 to 85 percent), better than multistage units. Due to the high tip speeds, the noise levels are high, but they are reduced by silencers. The air stream leaving the unit is hot, due to adiabatic compression, with temperature rises of 60 to 80°F.

In order to prevent excessive running speeds and noise levels and still produce such high pressures, one can use *two turbo blowers in series,* as shown in Fig. 7.39. By connecting the outlet of the first blower to the inlet of the second blower, the two

(a)

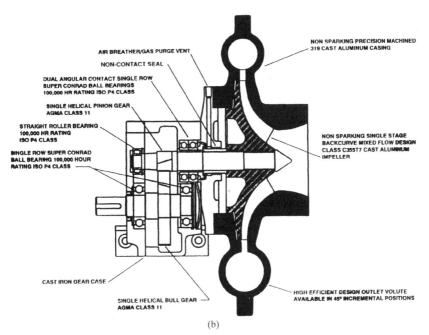

(b)

FIGURE 7.36 High-pressure (vacuum) turbo blower (exhauster), 6000 to 11,000 rpm. (*a*) Impeller, 17⅜-in o.d. (*b*) Assembly of housing, impeller, and gear box. *(Courtesy of Invincible Air Flow Systems, Baltic, Ohio.)*

static pressures (or vacuums) add up, and the total produced will be about doubled. (As mentioned previously, two fans in parallel will double the air volume, while two fans in series will about double the pressure.)

Another method to produce higher static pressures (without excessive speed and noise) is the *multistage turbo blower*, as shown in Figs. 7.40 through 7.42. In these units, as can be seen, the air stream flows radially out and in again, thereby making

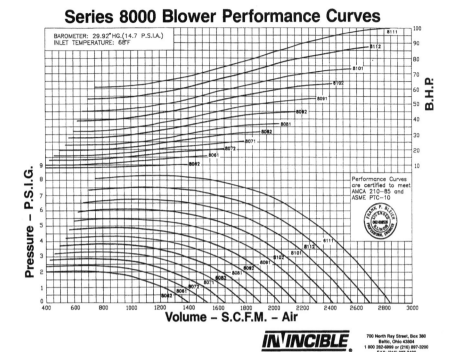

Series 8000 Blower Performance Curves

FIGURE 7.37 Performance of high-pressure turbo blower at 12 different speeds. *(Courtesy of Invincible Air Flow Systems, Baltic, Ohio.)*

several 180° turns. This results in lower efficiencies, but the pressure produced can always be increased simply by adding more stages. A typical unit is shown in Fig. 7.40. It has six stages and 8-in o.d. blower wheels, with 13 radial blades, $\frac{7}{16}$ to $\frac{7}{16}$ in wide. The diameter ratio is 0.4. A stator is located between each two blower wheels. The stator has 10 curved vanes, removing the air spin past the wheel before the airflow enters the next wheel. The unit is driven by a shaded-pole motor, $\frac{1}{4}$-hp, 3300 rpm. It delivers 30 cfm against 13 inWC, but the maximum static pressure produced is 15 inWC, or an average of 2.5 inWC per stage (the early stages produce less, the later stages more). The tip speed is 6900 fpm. An important figure is the ratio of housing inside diameter to blade outside diameter. It is called the *diffuser ratio* because it indicates the radial space available beyond the blade tip for the airflow to diffuse before making the 180° turn inward. This 8-in turbo blower, shown in Fig. 7.40, was designed for compactness and low cost. It has a diffuser ratio of only 1.2, which results in low efficiencies. In larger units, consuming up to 10,000 hp, good efficiencies are important, and long diffuser passages are provided, resulting in diffuser ratios of 1.5 to 2.0.

As mentioned previously, the maximum pressure produced per stage by this 8-in turbo blower is 2.5 inWC and the tip speed is 6900 fpm. Let's try to compare this with the 18-in turbo blower with radial blades that has a tip speed of 16,500 fpm and produces a maximum pressure of 23.3 inWC. Applying the square ratio of the two tip speeds to the maximum static pressure produced by the 18-in blower, we get

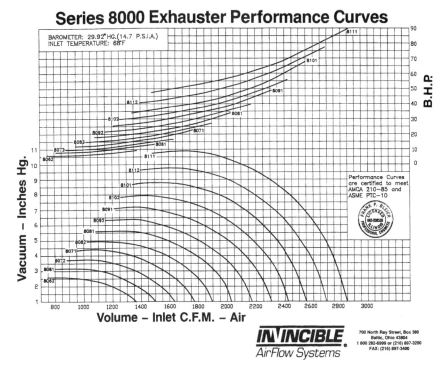

FIGURE 7.38 Performance of high-vacuum turbo exhauster at 12 different speeds. *(Courtesy of Invincible Air Flow Systems, Baltic, Ohio.)*

FIGURE 7.39 Two turbo blowers, directly driven from a double-shaft extension motor, connected in series by external ducting. *(Courtesy of Andritz Sprout-Bauer, Muncy, Pa.)*

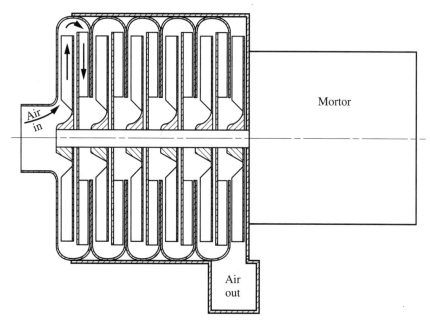

FIGURE 7.40 Six-stage turbo blower with direct drive from a shaded-pole motor, ¼ hp, 3300 rpm, 8-in wheel diameter, diameter ratio of 0.4, 13 radial blades, ⁷⁄₁₆ in wide, diffuser ratio of 1.2.

FIGURE 7.41 Four-stage turbo blower, belt drive from 300-hp engine, 26-in wheel diameters, 4800 rpm, for grain conveying. *(Courtesy of Christianson Systems, Inc., Blomkest, Minn.)*

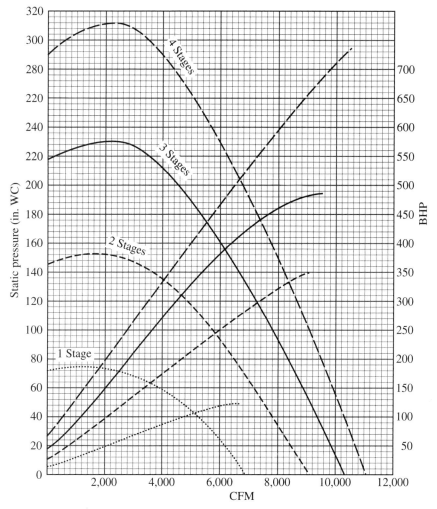

FIGURE 7.42 Performance of 26-in turbo blower at 4800 rpm, diameter ratio of 0.50, 14 radial blades, diffuser ratio of 1.4.

$$\left(\frac{6,900}{16,500}\right)^2 \times 23.3 = 4.1 \text{ inWC}$$

We find that the 2.5 inWC per stage actually produced by the 8-in blower is far below the 4.1 inWC produced by the 18-in blower after conversion for size and speed. There are three reasons for this lower maximum static pressure:

1. The small diffuser ratio of the 8-in blower
2. The multistage configuration of the 8-in blower with the 180° turns, which results in losses, even for large diffuser ratios
3. The smaller size of the 8-in blower, which normally is somewhat less efficient

We now will discuss three multistage units that were designed for three different applications:

1. A 26-in four-stage turbo blower for pneumatic conveying

2. An 8-in three-stage turbo blower for inflating flotation bags

3. A 5-in two-stage turbo exhauster for a vacuum cleaner

Figure 7.41 shows a *26-in four-stage turbo blower* for pneumatic conveying of grain and other materials. The blower wheels have radial blades, as shown in Fig. 7.32. It runs at 4800 rpm, driven by belt drive from a 300-hp, 2400-rpm engine. All parts (engine, blower, cyclone for separating and loading, and various accessories) are mounted on a four-wheel trailer for easy transportation.

Figure 7.42 shows the performance of this 26-in turbo blower at 4800 rpm for one, two, three, and four stages. It shows how the addition of stages affects the performance:

Number of stages	1	2	3	4
Maximum *SP* (inWC)	75	152	231	312
Difference in *SP*		77	79	82

You will notice that the maximum static pressure becomes slightly larger than in proportion with the number of stages because the density of the air becomes somewhat larger after each stage.

FIGURE 7.43 An 8-in three-stage turbo blower showing housing inlet and outlet. *(Courtesy of Pesco Products Division, Borge-Warner Corp., Bedford, Ohio.)*

Figure 7.43 shows an *8-in three-stage turbo blower* that was used to inflate the flotation bags of an army tank to make it amphibious. It runs at 8000 rpm, with direct drive from a 3-hp dc motor. Figure 7.44 shows the performance of this unit. It delivers 150 cfm against 50 inWC. Here the high static pressure is obtained by a combination of multistage design plus increased speed.

Figure 7.45 presents a schematic sketch of a centrifugal fan wheel for a *vacuum cleaner*. These fan wheels usually have six to eight blades, 5 to 6 in o.d. and about ¼ in wide. The blades usually have a strong backward curvature, with blade angles of about 35° at the leading edge and of 35° to 40° at the blade tip. These exhausters for vacuum cleaners have one or two stages and are directly driven by single-phase universal motors of ½, ¾ or 1 hp, running at anywhere between 16,000 and 24,000 rpm. The motor speed varies considerably between free delivery and no delivery, often as much as 4000 rpm. There are various ways to test vacuum cleaners, as will be explained in Chap. 19. A typical vacuum cleaner will deliver about 40 cfm against 40 inWC of vacuum at a maximum efficiency of 40 to 45 percent.

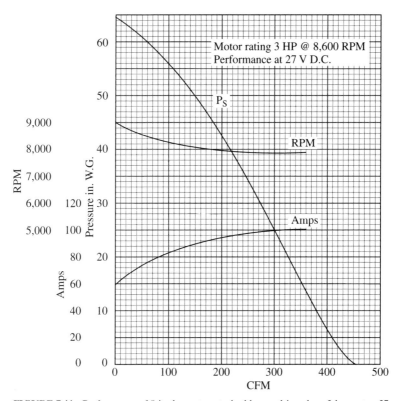

FIGURE 7.44 Performance of 8-in three-stage turbo blower, driven by a 3-hp motor, 27 V dc. Note the considerable variation from free delivery to no delivery for speed and amps.

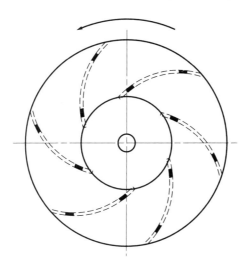

FIGURE 7.45 Schematic sketch of a typical centrifugal fan wheel for a vacuum cleaner. Such wheels usually are 5 to 6 in o.d. and have six to eight BC blades, about ¼ in wide.

High Pressure versus High Vacuum

Most ventilating fans produce maximum pressures of 5 to 15 inWC and perform about the same regardless of whether they are blowing or exhausting. This, however, is no longer true for the high-pressure units we just discussed. Here, the air density becomes larger than 0.075 lb/ft^3 when blowing but smaller than 0.075 lb/ft^3 when exhausting, and this results in larger pressures and smaller vacuums. However, the pressure ratio is the same either way. This can be expressed mathematically (using 1 atm = 407 inWC) as follows:

$$\frac{407 + P}{407} = \frac{407}{407 - V} \tag{7.18}$$

where P and V are the pressure and vacuum in inches of water column. Solving this equation for V, we get

$$V = \frac{407P}{407 + P} \tag{7.19}$$

Figure 7.46 presents Eq. (7.19) graphically, as a curve showing the maximum vacuum produced by an exhauster versus the maximum pressure produced by the same unit

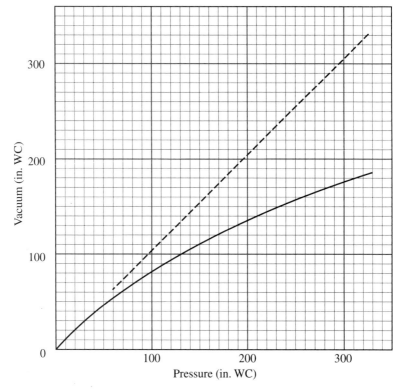

FIGURE 7.46 For a high-pressure unit, the vacuum produced is less than the pressure produced, as indicated by the solid line. The dashes show what the vacuum would be if it were equal to the pressure. The difference between the two curves shows the deficiency.

used as a blower. You will notice that the *deficiency* (the difference between the straight line and the curve) is negligible for pressures up to 20 inWC, but becomes more and more important as we move to higher pressures.

TWO CENTRIFUGAL FANS IN PARALLEL

As mentioned previously, two centrifugal fans in parallel are used occasionally when the air volume required is larger than that available from a certain fan size. Using two fans in parallel rather than one larger fan can have the following advantages:

1. When air is blown into a large space, such as a bin for storing and drying grain, a more even distribution can be obtained by two fans operating in parallel. (This principle applies to axial-flow fans as well.)
2. If one motor should fail, at least the other fan can still be used. In fact, a second fan is sometimes used as a standby, either just to be on the safe side or if later need for additional capacity, due to a change in the system (e.g., a mine), can be anticipated.
3. Two small fans may fit into the space available, whereas one larger fan may not. The conditions here are somewhat similar to those of a DIDW fan replacing a larger SISW fan.
4. Two smaller fans and motors may consume less power, particularly if they operate in the efficient performance range, whereas one larger fan may not.

AF, BC, and BI fans are safe for operation in parallel. FC fans are not recommended, since there is a risk of instability (dip in the pressure curve) or of resonance conditions.

VOLUME CONTROL

In many installations, the fan is selected for maximum output requirements, and means are provided so that the airflow can be reduced at times, either manually or automatically. A discussion of the three most common methods to accomplish this follows.

Variable Running Speed

This method can be applied to any type of fan. The effect of speed reduction on the pressure-volume curve of a BC fan is shown in Fig. 7.47. As mentioned previously, each point on the curve follows the fan laws and moves along a parabolic system characteristic. The fan efficiency remains unchanged, and there is no risk of any shifting into an unstable performance range (in the case of an FC fan). Another advantage is that this method of flow reduction results in the greatest power economy, since the brake horsepower of a fan varies as the third power of the speed. A third advantage is that—because of the reduced speed—the noise level is correspondingly reduced. The disadvantage of this method is that the first cost usually is high, especially if continuous variation is desired.

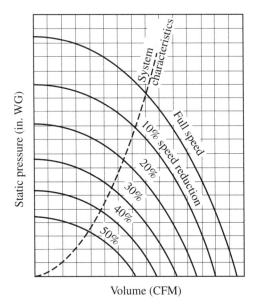

FIGURE 7.47   Volume control by variable running speed. *(From Bleier, F.P., Fans, in Handbook of Energy Systems Engineering. New York: Wiley, 1985. Used with permission.)*

Adjustable Outlet Dampers

A shutterlike mechanism is mounted on the fan outlet. The effect of such a throttle on the pressure-volume curve of an FC fan is shown in Fig. 7.48. Compared with the two other methods, the savings in power consumption are somewhat smaller, but this method is the simplest in construction and the lowest in first cost. It therefore is the one generally used for flow reduction in small sizes, where the effected power sav-

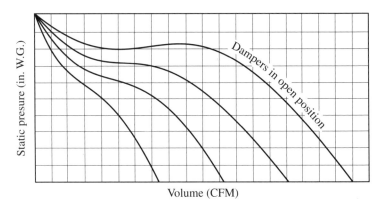

FIGURE 7.48   Volume control by adjustable outlet dampers. *(From Bleier, F.P., Fans, in Handbook of Energy Systems Engineering. New York: Wiley, 1985. Used with permission.)*

TABLE 7.4 Comparison of Some Typical Data Commonly Used in the Various Types of Centrifugal Fans

	Group 1			Group 2							Turbo	
	AF	BC	BI	AH	MH	LS	LSO	RT	FC	RB	Blower	Compressor
Size range (in)	12–132	12–132	12–132	19–104	19–104	19–104	19–104	25–110	2–73	12–122	8–96	8–96
d_1/d_2	0.65	0.6	0.6	0.6	0.6	0.2	0.2	0.5	0.75	0.3	0.4	0.4
	0.8	0.8	0.8	0.6	0.6			0.8	0.9	0.6	0.6	0.6
OA (%)	100	100	100	32	32	32	32	100	100	100	25	15
β_1 (°)	10	10	10	20	30	30	90	20	80	90	30	30
	30	30	30	25				35	120		90	90
β_2 (°)	40	35	35	60	90	90	90	80	120	90	65	65
	50	50	50	70				90	160		90	90
Scroll size (%)	100	100	100	80	80	80	80	90	100	100	90	90
Hag. width (%)	100	100	100	60	60	60	60	75	100	100	50	50
SISW	Yes	Yes	Yes	Yes	Yes	Yes	Yes	Yes	Yes	Yes	Yes	Yes
DIDW	Yes	Yes	Yes	No	No	No	No	Yes	Yes	Yes	Yes	No
Max. temp. (°F)	800	800	800	800	800	1600	1600	800	1500	1200	1100	1100
Contam.	No	Dust	Dust	No	Gran.	Gran.	Gran.	Gran.	No	Gran.	Gran.	Gran.
No. of blades	9	9	9	10	6	6	6	12	24	6	12	12
	12	12	12					24	64	10	24	24
b_{max}/d_2	0.35	0.35	0.35	0.30	0.35	0.35	0.35	0.20	0.55	0.40	0.22	0.22
Shroud, flat or conical	F	F or C	F or C	C	C	None	None	C	F	None	F or C	F or C
SP range (inWC)	5	5	5	5	5	5	5	10	1	20	10	40
	35	60	30	20	20	20	30	40	10	40	50	300
bhp curve	Nonoverload	Nonoverload	Nonoverload	Slight	Overload	Overload	Overload	Overload	Overload	Overload	Overload	Overload
Max. mech. eff. (%)	92	85	80	79	71	67	63	71	65	60	80	55

ings are small at any rate. For medium and large sizes, it shares the field with the other two methods.

Adjustable Inlet Vanes

Not only do adjustable inlet vanes throttle the airflow, they also impart to the entering air stream a spiral motion in the direction of the fan rotation that results in a reduction in air volume, static pressure, and brake horsepower. As to both power savings and first cost, this method is somewhere between the two others. Inlet vanes have the additional advantage of acting as air guides and thereby creating predetermined inlet flow conditions so that disturbances due to inlet turbulence, sometimes caused by inlet boxes, elbows, or other obstructions at the inlet, are minimized.

Inlet vanes also can be used for boosting instead of reducing the output of a centrifugal fan. In this case, they have to produce an inlet spin opposite to the fan rotation. However, this will result in a larger motor horsepower and in a reduced fan efficiency. This method has been used occasionally on large units for mine ventilation if the running speed cannot be increased or for structural or other reasons.

SUMMARY

We have discussed 12 types of centrifugal fans, all having the same operating principle but greatly differing in design, appearance, and performance. Table 7.4 shows some typical data for these 12 types of centrifugal fans.

CHAPTER 8

FAN SELECTION, SPECIFIC SPEED, AND EXAMPLES

SELECTION OF AXIAL-FLOW FANS

Chapter 4, on axial-flow fans, had a section on the selection of axial-flow fans. Since there are only three types of axial-flow fans (propeller, tubeaxial, and vaneaxial), the decision on which type to select for a certain application is easy. If the air is to be moved across a wall, without any duct work, and if the static pressure needed is less than 1 inWC, the obvious solution is a propeller fan, the least expensive of all fans. If duct work is needed, the best fan is either a tubeaxial or a vaneaxial fan. If the duct work is short and not more than 2½ inWC static pressure is needed to overcome the resistance, a tubeaxial fan will be adequate, particularly if the fan exhausts from the system so that the air spin past the fan will be no problem. For static pressures of more than 1 or 2 inWC, a vaneaxial fan might be preferable (see Table 4.2). The static pressure ranges of the three types of axial-flow fans overlap, and the decision in favor of one type of fan often will be made on the basis of first cost versus operating cost.

SELECTION OF CENTRIFUGAL FANS

For centrifugal fans, the selection is more complicated, partly because there are 12 types (see Table 7.4) instead of three types and partly because some types can tolerate hot or contaminated (dust or granular material) air streams and some types can handle only clean air or gas. Some of these 12 types, therefore, may have to be eliminated because of the contamination in some applications. This will narrow down the number of feasible types from 12 to perhaps 5 or 6. As we choose among these remaining types, with their ranges again overlapping, usually several types and sizes will be possible for a certain application.

For a start, recall that the sequence axial-flow, mixed-flow, centrifugal (AF, BC, BI, RT, RB, FC), turbo blower, and multistage indicates the general trend of an increasing pressure-volume ratio (SP/cfm) or of a decreasing volume-pressure ratio (cfm/SP). Furthermore, according to the fan laws, an increase in speed boosts the static pressure more than the air volume, whereas an increase in the wheel diameter D boosts the air volume more than the static pressure. Thus, for pressure fans, we will favor centrifugal fans and high speeds combined with small size. For volume fans, we will favor axial-flow fans and low speeds combined with large size. This does not mean that axial-flow fans can never be used in high-pressure applications. It does mean that they will require a higher tip speed than centrifugal fans in order to produce the same static pressure.

Also, let us keep in mind that axial-flow fans and mixed-flow fans offer a straight-through airflow, whereas centrifugal fans offer a right-angle airflow. One of these configurations may be preferable in a specific application.

The problem now presents itself in the following manner: Known are the required air volume, static pressure, and preferred airflow configuration. Which type and size of fan and which speed should be selected? As mentioned, there is more than one answer to this question, since the operating ranges of the various fan types are overlapping, even at equal size and speed, and still more so if a choice of both size and speed can be made. The fan engineer then might study the rating tables of several fan manufacturers in order to come up with a good selection. A more methodical procedure is based on the concept of specific speed N_s.

SPECIFIC SPEED N_S, SPECIFIC DIAMETER D_S

The specific speed N_s is a characteristic number, not an actual running speed. It can be calculated from a formula that has been derived from the fan laws:

$$N_s = \frac{\text{rpm} \times \sqrt{\text{cfm}}}{(SP)^{3/4}} \tag{8.1}$$

The *specific speed* is defined as the speed at which a certain fan model, with performance converted for size and speed, would have to operate in order to deliver 1 cfm against a static pressure of 1 inWC. In other words, if the required air volume and static pressure were converted in accordance with the fan laws for geometrically similar fans to a certain small size (called the *specific size D_s*) and to a certain high speed (called the *specific speed N_s*), this fan model then would deliver 1 cfm against a static pressure of 1 inWC. Obviously, such a fan would be too small for practical purposes. Its size could be calculated from the following formula:

$$D_s = \frac{D \times (SP)^{1/4}}{\sqrt{\text{cfm}}} \tag{8.2}$$

Note that $SP^{1/4} = \sqrt{\sqrt{SP}}$, $SP^{3/4} = SP^{1/2} \times SP^{1/4} = \sqrt{SP} \times \sqrt{\sqrt{SP}}$. Some calculators provide a means for raising a number to any desired exponent.

The specific speed can be considered a mathematical tool (rather than an actual running speed) for determining the type of fan to be used for a certain application. For our convenience, Eq. (8.1) is more useful in the following form:

$$R = \frac{N_s}{\text{rpm}} = \frac{\sqrt{\text{cfm}}}{(SP)^{3/4}} \tag{8.3}$$

We note that the ratio $R = N_s/\text{rpm}$ classifies fans with respect to their volume-pressure ratio or, to be more accurate, with respect to their $(\text{cfm})^{1/2}/(SP)^{3/4}$ ratio, which is more significant. For volume fans, the ratio R will be large; for pressure fans, it will be small. N_s and N_s/rpm vary over a wide range, as shown by the data below:

		Extreme volume fans		Extreme pressure fans
N_s	varies from	500,000	to	1000
N_s/rpm	varies from	200	to	0.1

Figures 8.1 and 8.2 are graphs showing the ratio $R = N_s/\text{rpm}$ as a function of air volume and static pressure. These graphs permit quick determination of the ratio R if air volume and static pressure (usually at the point of maximum efficiency) are given. Let us now give a number of typical examples, illustrating the selection and application of fans.

EXAMPLES FOR THE SELECTION AND APPLICATION OF FANS

Example 1

Suppose a customer wants a small fan for electronic cooling, direct drive, to deliver 44 cfm against a static pressure of 2 inWC at the point of maximum efficiency. Equation (8.3) or Fig. 8.1 shows that the ratio R will be $R = N_s/\text{rpm} = 3.94$. If 400-Hz electric power is available, we can use a two-pole motor running at 21,000 rpm, so the specific speed will be $N_s = 3.94 \times 21,000 = 82,740$. Table 8.1 shows that a one-stage vaneaxial fan would be a good selection. To determine the size of this vaneaxial fan, refer to Fig. 4.42, showing the performance of two 29-in VAFs running at 1750 rpm. We will use the fan laws plus a trial-and-error method. Let's try a 2-in wheel diameter and convert for size and speed as follows:

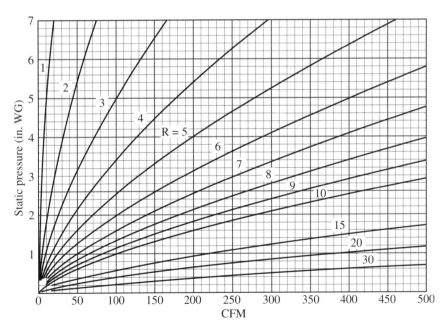

FIGURE 8.1 This graph shows the ratio R = specific speed/fan rpm as a function of air volume and static pressure for small fans.

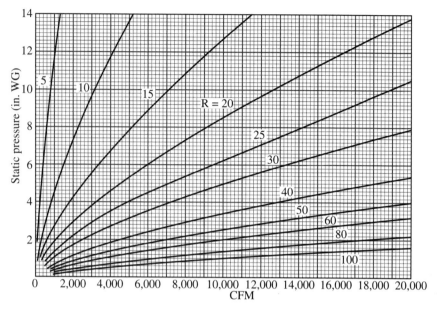

FIGURE 8.2 This graph shows the ratio R = specific speed/fan rpm as a function of air volume and static pressure for large fans.

2 in	29 in	29 in
21,000 rpm	21,000 rpm	1750 rpm
44 cfm	134,140 cfm	11,178 cfm
2 inWC *SP*	421 inWC *SP*	2.92 inWC *SP*

The point 11,178 cfm at 2.92 inWC *SP* is fairly close to the two performances shown in Fig. 4.42 but somewhat beyond the two curves. Let's try a 2¼-in vaneaxial fan:

2.25 in	29 in	29 in
21,000 rpm	21,000 rpm	1750 rpm
44 cfm	94,210 cfm	7851 cfm
2 inWC *SP*	332 inWC *SP*	2.31 inWC *SP*

The point 7851 cfm at 2.31 inWC *SP* is somewhat below the two curves. This is fine. It will take care of the fact that a much smaller fan will deliver slightly less than according to the fan laws. It also will provide a nice safety margin. A hub-tip ratio of about 0.55 will be best so that we will operate near the point of maximum efficiency.

Now it might be interesting to determine what the specific diameter D_s would be (even though we do not need this figure) and to check whether at the specific diameter D_s and at the specific speed N_s the fan would deliver 1 cfm at a static pressure of 1 inWC, as stated in the definition above. From Eq. (8.2) we get

$$D_s = \frac{2.25 \times 2^{1/4}}{\sqrt{44}} = \frac{2.25 \times 1.189}{6.63} = 0.404 \text{ in}$$

TABLE 8.1 Specific Speed Ranges and Other Features of Various Types of Fans

Type of fan	Specific speed range (rpm)	Air volume (cfm)	Static pressure	Fan efficiency	Manufacturing cost	Size and weight
Propeller	500,000–100,000	Large	Low	Low	Low	Small
Tubeaxial	300,000–60,000	Large	Low	Medium	Low	Small
VAF, one stage	130,000–50,000	Large	Medium	High	Medium	Small
VAF, two stages	90,000–35,000	Large	Medium	High	Medium	Medium
Mixed flow	80,000–10,000	Medium	Medium	Low	High	Medium
Wide FC	70,000–25,000	Large	High	Medium	Medium	Large
AF, BC, BI	70,000–20,000	Medium	Medium	High	Medium	Medium
Narrow FC	40,000–10,000	Medium	High	Medium	Medium	Medium
Radial tip	65,000–25,000	Medium	Medium	Medium	Medium	Medium
Radial blades	25,000–10,000	Low	Medium	Low	Medium	Medium
Turbo blower	30,000–5,000	Small	High	Medium	Medium	Medium
Multistage turbo blower	8,000–1,000	Small	High	Low	High	Large

By using the fan laws again, we get

2¼ in	2¼ in	0.404 in
21,000 rpm	82,740 rpm	82,740 rpm
44 cfm	173 cfm	1.001 cfm
2 inWC *SP*	31.0 inWC *SP*	0.999 inWC *SP*

Summarizing, we found in this Example 1 the following:

1. The specific speed N_s is very high (82,740 rpm). It is not an actual running speed but only a mathematical tool for determining the type of fan that will give us the desired air volume and static pressure.
2. The specific diameter D_s is very small (0.404 in). Again, it is not a real fan diameter but only a mathematical tool.
3. If the fan diameter were D_s and the running speed were N_s, this theoretical fan would deliver 1 cfm at a static pressure of 1 inWC, by definition.

Example 1a

We have found that for our requirement of 44 cfm against a static pressure of 2 inWC, the ratio $R = N_s/\text{rpm} = 3.94$. Now let us suppose that 400-Hz electric power is not available, but only 60-Hz electric power is available. The highest possible running speed then will be 3500 rpm, again from a two-pole motor. We then get a specific speed $N_s = 3.94 \times 3500 = 13,790$, a typical figure for a single-stage turbo blower. This, then, will be our selection for these changed conditions. To determine the wheel diameter of this turbo blower, we refer to Fig. 7.33 showing the performance of an 18-in turbo blower at 3500 rpm. Using the fan laws again and trying a 5½-in wheel diameter, we get

5½ in	18 in
3500 rpm	3500 rpm
44 cfm	1542 cfm
2 inWC *SP*	21.4 inWC *SP*

The point 1542 cfm at 21.4 inWC *SP* comes close to the solid line in Fig. 7.33. For a better safety margin, let's use a 5⅝-in wheel diameter.

As we compare the two selections for Examples 1 and 1a, we note the following:

1. The 5⅝-in turbo blower will be larger in size than the 2¼-in vaneaxial fan, but it will produce a lower noise level, due to its much lower tip speed (5154 versus 12,370 fpm).
2. As we lose the pressure-boosting influence of the higher speed, we have to compensate for this loss by selecting a type of fan that inherently has a larger pressure-volume ratio.

Example 2

A fan should be selected for ship ventilation to deliver 10,000 cfm against a static pressure of 4 inWC and to be directly driven by a 10-hp, 1150-rpm motor. Under no condition must the fan overload the motor, and the noise level must be less than 83 dB. From Eq. (8.3) or from Fig. 8.2, we get $R = 35.4$. The specific speed N_s, therefore, is $N_s = 35.4 \times 1150 = 40,710$. Table 8.1 shows that this will be a centrifugal fan with AF, BC, BI, or FC blades. FC blades will have to be eliminated because they probably would overload the motor in the range near free delivery. AF, BC, and BI blades, as mentioned previously, have nonoverloading brake horsepower curves, so any of these will meet the requirement of no motor overload. We will select BI blades because they are more economical than AF or BC blades. We look at the Ammerman BIB (backward-inclined blades) catalog and find that their 30-in BI centrifugal fan will deliver 10,360 cfm against a static pressure of 4 inWC for belt drive at 1194 rpm while consuming 9.31 bhp. For direct drive, the brake horsepower will be 5 to 10 percent less, as mentioned on page 7.2. The Ammerman catalog specifically states that the brake horsepower values shown include the belt-drive losses. Most other fan catalogs state that their brake horsepower values do not include belt drive losses. We also find that our operating point of 10,000 cfm against a static pressure of 4 inWC is close to their point of maximum efficiency. The tip speed will be 8960 fpm, indicating that the noise level requirement also will be met.

Example 2a

If the fan is to be installed in a noisy factory where the noise level requirement can be left out, we would select a 24½-in BI centrifugal fan at 1750 rpm. This fan would have a specific speed $N_s = 35.4 \times 1750 = 61.950$, which is still in the range for a BI centrifugal fan. This fan would be less expensive, due to the smaller size and the higher motor speed, but the Ammerman catalog indicates that this fan would not operate near their point of maximum efficiency. It would deliver 10,410 cfm at a static pressure of 4 inWC, running at 1750 rpm and consuming 11.77 bhp with belt drive. The brake horsepower is high, due to the smaller size (high outlet velocity) and due to the lower fan efficiency at this operating point. The brake horsepower will be too high for a 10-hp motor, even after reducing it for direct drive. A slight reduction in the blade width will have to be made so that the 10-hp, 1750-rpm motor will be adequate. The tip speed for this 24½-in fan at 1750 rpm will be 11,225 fpm, which is 25 percent higher than the 8960-fpm tip speed of the 30-in fan at 1150 rpm, so the noise level will be about 5 dB higher (see Eq. 5.13). The nonoverloading brake horsepower curve will be retained.

Example 3

An inexpensive fan should be selected that will deliver 900 cfm against a static pressure of 1.5 inWC while directly driven by a 1750-rpm motor. This fan will be built into an industrial air heater and should be compact, since the available space is tight. Power consumption and noise level, on the other hand, are of secondary importance. From Eq. (8.3) or from Fig. 8.2, we get $R = 22.1$. For 1750 rpm, the specific speed is $N_s = 22.1 \times 1750 = 38,734$. Table 8.1 shows that this will be a centrifugal fan with AF, BC, BI, or FC blades. Since compactness and low cost are required, we select an FC centrifugal fan.

To determine the wheel diameter of this FC fan, we refer to Fig. 7.31, showing the performance of a 27-in FC centrifugal fan at 570 rpm. Using the fan laws again and trying an 8-in wheel diameter, we get

8 in	27 in	27 in
1750 rpm	1750 rpm	570 rpm
900 cfm	34,599 cfm	11,269 cfm
1.5 inWC *SP*	17.1 inWC *SP*	1.81 inWC *SP*

The point 11,269 cfm at 1.81 inWC *SP* comes close to the solid line in Fig. 7.31 and slightly below it, so this 8-in FC centrifugal fan will be satisfactory. This fan could either blow into the heating coils or exhaust from them. Usually blowing is preferable, because then the fan will handle cold air, so it will handle a larger mass of air. To determine the motor horsepower, we go through a calculation similar to the one shown on page 1.7.

An 8-in centrifugal fan will have an outlet area of 0.366 ft². The outlet velocity, therefore, will be

$$OV = \frac{900}{0.366} = 2459 \text{ fpm}$$

and the corresponding pressures will be

$$VP = \left(\frac{2459}{4005}\right)^2 = 0.38 \quad \text{and} \quad TP = 1.5 + 0.38 = 1.88$$

The air horsepower then will be

$$\text{ahp} = \frac{900 \times 1.88}{6356} = 0.266$$

For a small FC centrifugal fan, we can expect a maximum efficiency of 55 percent, so the brake horsepower will be

$$\text{bhp} = \frac{0.266}{0.55} = 0.48$$

This means, for the point of operation, that a ½-hp, 1750-rpm motor will be adequate, but we had better use a ¾-hp, 1750-rpm motor so that the motor will be safe, even if the fan should operate at a lower static pressure, since FC centrifugal fans do not have a nonoverloading brake horsepower curve.

Example 3a

Suppose the requirements for compactness and low cost were omitted. We then would select a centrifugal fan with BI (instead of FC) blades because BI blades will result in a nonoverloading brake horsepower curve. Then a ½-hp, 1750-rpm motor would be adequate, but the wheel diameter would be larger. To determine this wheel diameter, we refer to Fig. 7.13 showing the performance of a 27-in airfoil centrifugal fan at 1160 rpm. A BI centrifugal fan will have a similar performance, only the fan efficiency will be lower. Using the fan laws and trying a 12¼-in wheel diameter, we get

12¼ in	27 in	27 in
1750 rpm	1750 rpm	1160 rpm
900 cfm	9,637 cfm	6,388 cfm
1.5 inWC *SP*	7.29 inWC *SP*	3.20 inWC *SP*

The point 6388 cfm at 3.20 inWC *SP* is slightly below the curve, so the 12¼-in BI centrifugal fan will be satisfactory.

As we compare the two selections for Examples 3 and 3a, we note the following:

1. The FC centrifugal fan will be more compact, and the fan itself (due to the smaller fan wheel and scroll housing) will be less expensive.

2. The BI centrifugal fan, with its nonoverloading brake horsepower curve, will require only a ½-hp motor instead of a ¾-hp motor, so the total cost of the two selections will be comparable.

3. In view of the preceding, the smaller FC centrifugal fan probably will be the preferable selection.

Example 4

A DIDW fan for induced draft should be selected that will handle air at 610°F at a 700-ft altitude and under these conditions is to deliver 48,000 cfm against a static pressure of 4.5 inWC. To correct the static pressure to standard air density, we need two correction factors, one for temperature and one for altitude. The correction factor for temperature is the ratio of the two absolute temperatures, that is, $(460 + 610)/(460 + 70) = 2.019$. The correction factor for altitude can be obtained from Table 1.1, by interpolation between 0 and 1000 ft of altitude, as $29.92/29.178 = 1.0254$. The product of these two correction factors is $2.019 \times 1.0254 = 2.070$. This is our total correction factor. Converted to standard air density, the specifications, therefore, read 48,000 cfm against a static pressure of $2.070 \times 4.5 = 9.32$ inWC. Each side of the DIDW fan will have to deliver 24,000 cfm against a static pressure of 9.32 inWC.

From Eq. (8.3), we get $R = 29.04$. Estimating a fan speed of 1140 rpm, we get a specific speed $N_s = 29.04 \times 1140 = 33,110$. From Table 8.1, we note that this specific speed can be met by centrifugal fans with AF, BC, BI, FC, and RT blades. RT blades are best for induced draft, as mentioned on page 7.33, so this will be our selection. To determine the wheel diameter, we refer to Fig. 7.30 showing in dotted lines the performance of a 27-in RT centrifugal fan at 1140 rpm. By using the fan laws and trying a 40¼-in wheel diameter, we get

40¼ in	27 in
1140 rpm	1140 rpm
24,000 cfm	7244 cfm
9.32 inWC *SP*	4.19 inWC *SP*

The point 7244 cfm at 4.19 inWC *SP* is slightly below the dotted line in Fig. 7.30 indicating that we have made a satisfactory selection. This is further confirmed by Table 8.2, showing the performance of a 40¼-in RT centrifugal fan from Chicago Blower's catalog. For 24,000 cfm at 9.32 inWC *SP*, we get, by interpolation, 1182 rpm and 47.6 bhp. By dividing by the correction factor 2.070, we find that at the actual operating conditions (610°F and 700-ft altitude), $47.6/2.070 = 23.0$ bhp will be required for each side, or 46.0 bhp for the DIDW fan. Since RT centrifugal fans do not have a nonoverloading brake horsepower curve, we like to have a safety margin in case the actual operating pressure should be somewhat lower than anticipated. A 50-hp motor, therefore, will be selected.

TABLE 8.2 Performance Table for a 40¼-in RT Centrifugal Fan, Belt Drive

Wheel diameter:	40.25 in
Outlet area:	9.32 ft²
Inlet area:	7.72 ft²

Maximum motor size: 200 hp
Maximum allowable speed @70°F: 1988 rpm

CFM	Outlet VEL	8" SP		10" SP		12" SP		14" SP		15" SP		16" SP		17" SP		18" SP		19" SP		20" SP	
		RPM	BHP	RPM	BHP	RPM	BHP	RPM	BHP	RPM	BHP	RPM	BHP	RPM	BHP	RPM	BHP	RPM	BHP	RPM	BHP
15000	1609	1011	25.4	1120	32.4	1224	39.8	1321	49.7												
16000	1716	1019	26.9	1125	34.1	1226	41.7	1327	54.2	1371	58.5	1414	62.9	1457	67.4						
18000	1931	1040	30.0	1141	37.7	1236	45.8	1339	59.1	1381	63.7	1422	68.3	1463	73.0	1503	77.8	1542	82.6	1581	87.6
20000	2145	1065	33.5	1162	41.7	1252	50.3	1356	64.4	1397	69.2	1436	74.1	1475	79.0	1513	84.0	1551	89.0	1588	94.2
22000	2360	1093	37.3	1186	46.0	1273	55.1	1378	70.1	1417	75.1	1455	80.2	1492	85.4	1529	90.6	1565	95.9	1601	101.3
24000	2574	1121	41.6	1214	50.7	1298	60.2	1403	76.1	1441	81.4	1478	86.8	1514	92.2	1549	97.7	1584	103.2	1618	108.8
26000	2789	1150	46.3	1242	55.8	1325	65.8	1430	82.6	1467	88.1	1503	93.7	1538	99.4	1572	105.1	1606	110.9	1639	116.8
28000	3003	1178	51.1	1270	61.4	1354	71.8	1459	89.6	1495	95.4	1530	101.2	1564	107.1	1598	113.1	1631	119.1	1663	125.2
30000	3218	1205	55.7	1299	67.5	1382	78.4	1487	97.2	1523	103.2	1558	109.2	1592	115.3	1625	121.5	1658	127.8	1690	134.1
32000	3432	1232	60.4	1326	73.5	1410	85.6	1516	105.4	1552	111.6	1587	117.8	1621	124.2	1654	130.6	1686	137.1	1717	143.7
34000	3647	1256	66.7	1354	79.1	1438	92.9	1543	114.0	1580	120.6	1615	127.1	1649	133.7	1682	140.3	1714	147.0	1745	153.3
36000	3861	1285	73.9	1380	85.2	1466	99.9	1571	122.4	1608	129.8	1643	136.9	1678	143.8	1711	150.7	1743	157.6	1774	164.6
38000	4076	1326	82.7	1404	93.2	1493	106.4	1599	130.1	1635	138.6	1671	146.6	1705	154.2	1739	161.6	1771	168.9	1802	176.2
40000	4290	1377	93.4	1432	102.0	1518	114.5	1625	138.3	1663	146.7	1699	155.7	1733	164.4	1766	172.6	1799	180.5	1831	188.3
42000	4505	1430	105.4	1472	112.7	1543	124.4														

CFM	Outlet VEL	21" SP		22" SP		23" SP		24" SP		25" SP		26" SP		27" SP		28" SP		29" SP		30" SP	
		RPM	BHP	RPM	BHP	RPM	BHP	RPM	BHP	RPM	BHP	RPM	BHP	RPM	BHP	RPM	BHP	RPM	BHP	RPM	BHP
20000	2145	1619	92.6	1656	97.6	1697	110.0	1732	115.5	1767	121.0	1801	126.5	1839	140.7	1871	146.6	1903	152.5	1935	158.5
22000	2360	1625	99.4	1661	104.7	1705	117.8	1739	123.4	1773	129.1	1806	134.9	1847	149.8	1879	155.9	1910	162.0	1941	168.2
24000	2574	1636	106.7	1671	112.2	1719	126.0	1752	131.9	1784	137.8	1816	143.8	1861	159.4	1891	165.8	1921	172.1	1951	178.5
26000	2789	1652	114.5	1686	120.2	1737	134.7	1768	140.8	1799	147.0	1830	153.2	1878	169.6	1908	176.2	1937	182.8	1966	189.4
28000	3003	1672	122.7	1705	128.7	1758	143.9	1788	150.3	1819	156.7	1849	163.1	1899	180.3	1927	187.1	1956	194.0		
30000	3218	1695	131.4	1727	137.6	1782	153.6	1812	160.2	1841	166.8	1870	173.5	1922	191.5	1950	198.5				
32000	3432	1721	140.6	1751	147.0	1808	163.8	1837	170.6	1866	177.5	1894	184.4	1961	209.3						
34000	3647	1748	150.3	1778	157.0	1849	180.2	1878	187.4	1906	194.6	1933	201.9								
37000	3969	1790	166.1	1820	173.1	1892	198.2	1920	205.8	1948	213.3										
40000	4290	1833	183.4	1863	190.8	1934	218.0	1963	225.8												
43000	4612	1875	202.3	1905	210.2																
46000	4934	1917	221.5	1947	230.4																

Note: Performance shown is for ducted inlet, and outlet with evase discharge.
Source: From Catalog of Blower Corp., Glendale Heights, Illinois.

Example 5

A pressure blower for exhausting the dust particles from a small grinding wheel must be selected. It will be directly driven by a 3500-rpm electric motor, and it should handle 350 cfm against a static pressure of 7 inWC. The outlet velocity should be about 4000 fpm so that the grinding particles can be conveyed through the outlet duct. From Eq. (8.3) or from Fig. 8.1, we get $R = 4.347$. For a running speed of 3500 rpm, the specific speed is $N_s = 4.347 \times 3500 = 15{,}215$. Table 8.1 shows that this will be a centrifugal fan with radial blades. To determine the wheel diameter of this RB centrifugal fan, we refer to Fig. 7.33 dashed lines, showing the performance of an 18-in turbo blower at 3500 rpm. Using the fan laws and trying an 11-in wheel diameter, we get

$$
\begin{array}{ccc}
11' & \longrightarrow & 18'' \\
3500 \text{ RPM} & \longrightarrow & 3500 \text{ RPM} \\
350 \text{ CFM} & \longrightarrow & 1534 \text{ CFM} \\
7'' \text{ SP} & \longrightarrow & 18.7'' \text{ SP} \\
0.673 \text{ BHP} & \longleftarrow & 7.9 \text{ BHP}
\end{array}
$$

The point 1534 cfm at 18.7 inWC *SP* is slightly below the dashed line, so this 11-in RB centrifugal fan will be a satisfactory selection.

The 18-in pressure blower had a back plate and a shroud, but the 11-in pressure blower will only have a back plate. The shroud will be omitted so that there will not be any material buildup from the grind particles. This lack of a shroud will result in smaller air volumes and lower fan efficiencies. The 18-in pressure blower has 12 radial blades, 3 to 2 in wide. Converted for size, the 11-in pressure blower would have 12 radial blades, 1.83 to 1.22 in wide. To compensate for the lack of the shroud, we will use 12 radial blades, 3 to 2 in wide.

We will use a 4-in-diameter duct at the housing inlet and outlet. This will result in an area of 0.0873 ft² and in an air velocity at the housing inlet and outlet of 350/0.0873 = 4009 fpm, as requested. We will extend the radial blades inward to a 4⅞-in i.d., resulting in a diameter ratio of 4.875/11 = 0.443, the same as for the 18-in pressure blower.

The dashed lines in Fig. 7.33 also show that for 1534 cfm, the brake horsepower will be about 7.9. Converted back to an 11-in wheel diameter, this becomes 0.673 bhp. However, this figure will have to be increased for two reasons:

1. The lower efficiency of the 11-in pressure blower without a shroud
2. For a safety margin because of the overloading brake horsepower curve

We will use a 1-hp, 3500-rpm motor.

Example 6

A small belt-driven fan is to be selected that will be built into a small apparatus. It will draw a cooling air stream through some narrow passages and will discharge it into the atmosphere. Thus 60 cfm should be delivered against a static pressure of 0.7 inWC. Moreover, as a special requirement, the fan should be reversible, and the same air volume should be exhausted regardless of the direction of rotation. In view of the small quantities involved, efficiency and power consumption are of minor importance.

Because of the special requirements, this will have to be a centrifugal fan with circumferential discharge, without any scroll housing, similar to the fan shown in Fig.

7.16, except that the blades will have to be radial. This type of fan without a scroll housing is sometimes called a *plug fan*.

From Eq. (8.3) or from Fig. 8.1, we get $R = 10.12$. Let's try belt drive at 2450 rpm. The specific speed then will be $N_s = 10.12 \times 2450 = 24{,}798$. Table 8.1 confirms that an RB centrifugal fan will do the job.

To determine the wheel diameter, we refer to Fig. 7.24, showing the performance of four 26⅛-in centrifugal fans at 1160 rpm, two of them with radial blades, but in the usual configuration with a scroll housing. For circumferential discharge without a scroll housing, the static pressure produced will be somewhat lower. Using the fan laws and trying a 5½-in wheel diameter, we get

5½ in	26⅛ in	26⅛ in
2450 rpm	2450 rpm	1160 rpm
60 cfm	6430 cfm	3044 cfm
0.7 inWC *SP*	15.8 inWC *SP*	3.54 inWC *SP*

The point 3054 cfm at 3.54 inWC *SP* is well below the static pressure curves, so we are on the safe side, even considering the circumferential discharge and the much smaller wheel diameter.

Special features of this selection are quiet operation (no cutoff) and extreme compactness.

CONCLUSION

The six examples just discussed illustrate not only the selection procedure for different types of centrifugal fans but also the various problems and special requirements that are sometimes encountered in the selection process. This will enable readers to make similar decisions when they run into similar problems.

CHAPTER 9
AXIAL-CENTRIFUGAL FANS

FLOW PATTERNS FOR VARIOUS CONFIGURATIONS

Axial-centrifugal fans, also called *tubular centrifugal fans, in-line centrifugal fans,* or *mixed-flow fans,* take a place between the vaneaxial fans and the scroll-type centrifugal fans. Various configurations have become popular for these axial-centrifugal fans. These various configurations differ with respect to the following features:

1. The fan wheel can have a flat back plate. or a conical back plate (as shown in Fig. 9.1).

2. The housing can be cylindrical, square, or barrel-shaped.

3. The drive can be direct drive or belt drive.

There are 12 possible combinations of the 7 features (as can be seen from Table 9.1), but not all of these 12 combinations are practical. Six configurations are shown in Figs. 3.23, 3.24, 3.25a, 3.25b, 3.26, 3.27, 9.2, and 9.3. These are eight (not six) figures because Figs. 3.26, 3.27, and 9.3 all use about the same configuration. Table 9.1 also lists these figure numbers and indicates into which combinations they belong. Let's discuss the eight figures listed in Table 9.1.

Figure 9.2 shows a schematic sketch of a standard centrifugal fan wheel (flat back plate) in a cylindrical housing with direct drive and with guide vanes past the fan wheel. The guide vanes remove the air spin and convert some of the energy loss into additional static pressure, the same as in a vaneaxial fan. The fan wheel probably will have BI blades (for low cost), but it also could have AF or BC blades. It cannot have FC blades because they require a scroll housing for good performance, as mentioned in Chap. 7. The air stream makes a 90° turn at the wheel inlet (as in any centrifugal fan wheel) and leaves the blade tips radially outward (and, of course, with a circumferential spin). It now has to make another 90° turn from radially outward to axial (or rather helical) flow. A conical guide, as shown, will help the air stream make this second 90° turn. The conical guide is sometimes omitted, but this will reduce the efficiency somewhat. In view of the second 90° turn, a diffuser ratio of about 1.5 is recommended for proper flow conditions and for an acceptable fan efficiency. Comparing this in-line fan with a vaneaxial fan, they only have the straight-through flow feature in common. The fan efficiency is lower for the in-line fan, and the compactness of the vaneaxial fan has been lost. As a result of the 1.5 diffuser ratio, a 27-in wheel diameter will require a cylindrical housing of 40½ in diameter.

The next figure listed in Table 9.1 is Fig. 3.23. This is a schematic sketch of a mixed-flow fan wheel with a conical back plate and with a diverging air stream flowing into a cylindrical housing with direct drive. Here the air stream has to make only two 45° turns instead of two 90° turns. As the air stream here diverges conically (rather than radially) outward into the cylindrical housing, a somewhat smaller diffuser ratio of about 1.3 will be adequate. Figure 3.23 shows no guide vanes past the fan wheel, but they could be added and would definitely improve the fan efficiency.

FIGURE 9.1   Cast-aluminum mixed-flow fan wheel showing a conical back plate, a short conical shroud, and 10 backward-curved airfoil blades protruding inward beyond the shroud.

Figure 3.24 shows a BC fan wheel with a flat back plate in a cylindrical housing with belt drive and with outlet guide vanes, a good, inexpensive combination.

Figure 3.25 shows a mixed-flow fan wheel with a conical back plate in a square housing. Two models are shown: model *a* for direct drive and model *b* for belt drive.

Figure 3.26 shows a mixed-flow fan wheel with a conical back plate and with direct drive in a barrel-shaped housing. This is a more expensive spun housing, but it has the advantage of smaller inlet and outlet diameters for connection to smaller-diameters ducts plus good conformity with the shape of the conical fan wheel. Here the guide vanes past the fan wheel are even more important because without them the air spin at the small housing outlet would become excessive, as was explained in Chap. 1.

Figure 3.27 is similar to Fig. 3.26 but has a separate motor chamber to protect the motor from hot or corrosive gases. Figure 9.3 shows the same unit connected to small inlet and outlet ducts, with a clear view of the motor in the separate chamber.

Conclusion

Axial-centrifugal fans have the advantage of easy installation as part of the duct work (like vaneaxial fans). At the same time, they produce more static pressure (and somewhat more air volume) than vaneaxial fans of the same wheel diameter and speed, although not as much as scroll-type centrifugal fans of the same wheel diameter and speed. Unfortunately, their efficiencies are lower than those of vaneaxial fans and scroll-type centrifugal fans, especially in small sizes. FC centrifugal fan wheels cannot be used in this type of unit because, as mentioned previously, FC wheels require a scroll housing for proper performance (see Chap. 7).

TABLE 9.1   Axial-Centrifugal Fans in Various Configurations

	Flat back plate	Conical back plate
Cylindrical housing, direct drive	Fig. 9.2	Fig. 3.23
Cylindrical housing, belt drive	Fig. 3.24	
Square housing, direct drive		Fig. 3.25*a*
Square housing, belt drive		Fig. 3.25*b*
Barrel-shaped housing, direct drive		Figs. 3.26, 3.27, and 9.3
Barrel-shaped housing, belt drive		

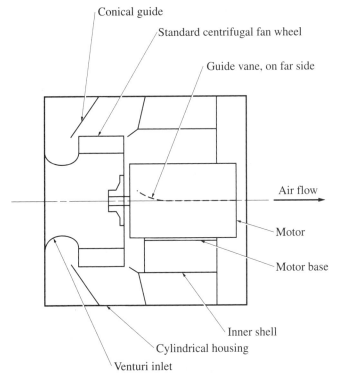

FIGURE 9.2 Axial-centrifugal fan using a standard centrifugal fan wheel plus conical guide and outlet guide vanes in a cylindrical housing.

FIGURE 9.3 Mixed-flow bifurcator fan connected to inlet and outlet ducts of smaller size due to the barrel-shaped housing. *(Courtesy of Bavley Fan, Lau Industries, Lebanon, Ind.)*

PERFORMANCE OF AXIAL-CENTRIFUGAL FANS

Figure 9.4 shows a comparison of the static pressure curves for three types of fans: vaneaxial fans, axial-centrifugal fans, and scroll-type centrifugal fans with airfoil

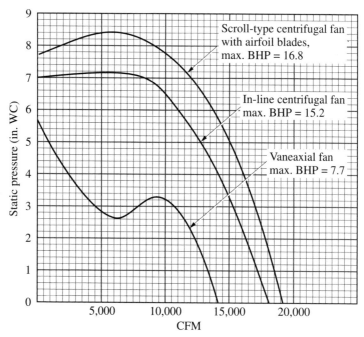

FIGURE 9.4 Comparison of the static pressure curves for three types of fans, all having 27-in wheel diameters and running at 1750 rpm.

blades. For a fair comparison, the three fans have the same 27-in wheel diameter and run at the same 1750 rpm. We note the following:

1. The performance of the in-line centrifugal fan takes a place between the performance of the vaneaxial fan and the scroll-type centrifugal fan, as mentioned earlier.

2. The in-line centrifugal fan performs somewhat, but not too much, below the scroll-type centrifugal fan.

3. The in-line centrifugal fan performs considerably above the vaneaxial fan, but if the vaneaxial fan were to have a 40½-in housing diameter (same as the in-line centrifugal fan has) (with the fan wheel about 40 in o.d.), then the vaneaxial fan would by far outperform the two others. This means that the vaneaxial fan is still the most compact of the three fans. Only the FC centrifugal fan is even more compact, but it has a much lower efficiency.

CHAPTER 10
ROOF VENTILATORS

FOUR WAYS TO SUBDIVIDE ROOF VENTILATORS

Roof ventilators can be seen on many buildings. They have become popular partly because they require little or no duct work and partly because they do not take up valuable space inside the building. They can be subdivided in the following four ways:

1. Exhaust or supply
2. Upblast, radial discharge, or downflow
3. Axial-flow or centrifugal fan wheels
4. Direct or belt drive

Let's discuss these four classifications in more detail.

1. Most roof ventilators are for exhaust from buildings, but some are supply units for makeup air. Exhaust units draw air from the building and emit it either by upblast or by radial discharge. Supply units draw in outside air and blow it down into the building.

2. Radial discharge results in a lower, more pleasing silhouette, with various aluminum spinnings, but upblast is required for the exhaust of grease-laden air from restaurant kitchens because an accumulation of grease on the roof would be a fire hazard. Upblast is also used on some machine shops, foundries, and other buildings where a considerable amount of oil or dust is in suspension in the air. Radial discharge and upblast, of course, can only be used on exhaust units. Supply units obviously require downflow.

3. Axial-flow fan wheels are less expensive and therefore are used in most installations without any duct work. Also, all supply units use axial-flow fan wheels. Centrifugal fan wheels in supply units would be impractical because they would result in a complicated flow pattern with several 90° turns. In exhaust units, centrifugal fan wheels produce more negative static pressure and therefore are often used in connection with duct work. Most of them have BI blades; some have BC blades. Centrifugal fan wheels with airfoil blades are more expensive and in roof ventilators are not of much extra benefit. FC blades, as mentioned previously, will not function without a scroll housing and therefore cannot be used in roof ventilators.

4. Direct drive is simpler and requires less maintenance; it is generally used in small sizes. Belt drive is used in large sizes to avoid expensive low-speed motors and to obtain greater flexibility with regard to air volume and static pressure.

TABLE 10.1 Ten Configurations Used in Roof Ventilators

Configuration no.	Exhaust or supply	Upblast, radial discharge, or downflow	Axial or centrifugal fan wheel	Direct or belt drive	Figure no.
1	Exhaust	Upblast	Axial	Direct	10.1
2	Exhaust	Upblast	Axial	Belt	
3	Exhaust	Upblast	Centrifugal	Direct	10.2
4	Exhaust	Upblast	Centrifugal	Belt	10.3
5	Exhaust	Radial	Axial	Direct	10.4
6	Exhaust	Radial	Axial	Belt	10.5
7	Exhaust	Radial	Centrifugal	Direct	10.6
8	Exhaust	Radial	Centrifugal	Belt	10.7
9	Supply	Downflow	Axial	Direct	10.8
10	Supply	Downflow	Axial	Belt	10.9

TEN CONFIGURATIONS OF ROOF VENTILATORS

By combining the various features discussed, we can obtain 10 practical configurations. (Some combinations, as mentioned earlier, would not be practical.) Table 10.1 is a list of these 10 configurations. They are illustrated in Figs. 10.1 through 10.9.

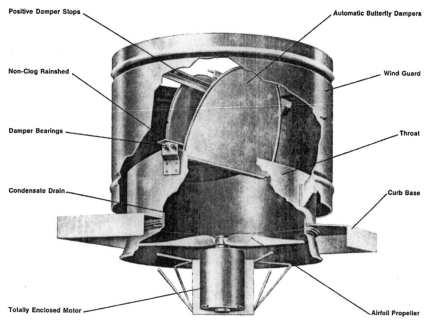

FIGURE 10.1 Exhaust roof ventilator, upblast, with an axial-flow fan wheel and direct drive. Note butterfly dampers to keep out rain, etc. *(Courtesy of Lau Division, Tomkins Industries, Inc., Dayton, Ohio.)*

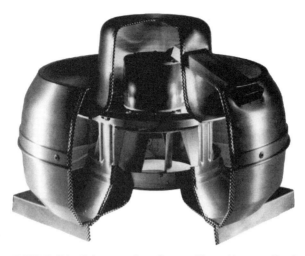

FIGURE 10.2 Exhaust roof ventilator, upblast, with a centrifugal fan wheel and direct drive. *(Courtesy of FloAire, Inc., Bensalem, Pa.)*

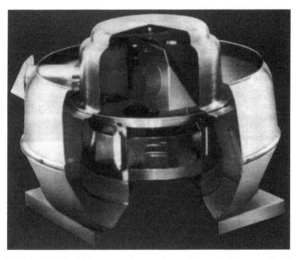

FIGURE 10.3 Exhaust roof ventilator, upblast, with a centrifugal fan wheel and belt drive. *(Courtesy of FloAire, Inc., Bensalem, Pa.)*

The figures indicate that many spinnings are used in the various models of roof ventilators. In large sizes, when these spinnings become too expensive, a fabricated hood can be used instead, as shown in Fig. 10.10. The two flat sides of the hood could be made angular, for better flow conditions.

FIGURE 10.4 Exhaust roof ventilator, radial discharge, with an axial-flow fan wheel and direct drive. *(Courtesy of FloAire, Inc., Bensalem, Pa.)*

FIGURE 10.5 Exhaust roof ventilator, radial discharge, with an axial-flow fan wheel and belt drive. *(Courtesy of FloAire, Inc., Bensalem, Pa.)*

FIGURE 10.6 Exhaust roof ventilator, radial discharge, with a centrifugal fan wheel and direct drive. *(Courtesy of FloAire, Inc., Bensalem, Pa.)*

FIGURE 10.7 Exhaust roof ventilator, radial discharge, with a centrifugal fan wheel and belt drive. *(Courtesy of FloAire, Inc., Bensalem, Pa.)*

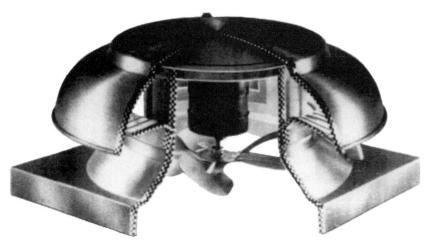

FIGURE 10.8   Supply roof ventilator with an axial-flow fan wheel and direct drive. *(Courtesy of FloAire, Inc., Bensalem, Pa.)*

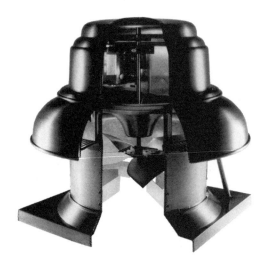

FIGURE 10.9   Supply roof ventilator with an axial-flow fan wheel and belt drive. Note the guide vanes past the fan wheel. *(Courtesy of FloAire, Inc., Bensalem, Pa.)*

FIGURE 10.10 Exhaust roof ventilator, radial discharge, with an axial-flow fan wheel and belt drive, with a fabricated hood in large sizes.

CHAPTER 11
VENTILATION REQUIREMENTS AND DUCT SYSTEMS*

OSHA REGULATIONS

The Occupational Safety and Health Act (OSHA), which became law in 1971, ensures safe and healthful working conditions for all Americans. Much of this act deals with adequate ventilation and noise limitation, important aspects of environmental control. By establishing specific ventilation requirements, such as the elimination of heat, moisture, dust, and harmful vapors, more healthful working conditions are ensured. OSHA pays particular attention to the following:

1. Exhaust of excessive heat and humidity from oven and furnace rooms, laundries, and other drying and washing operations
2. Exhaust of airborne contaminants from polishing, grinding, buffing, and blast cleaning operations
3. Exhaust of paint fumes from spray booths
4. Exhaust systems in areas used for brazing, welding, and cutting operations
5. Introduction of makeup air to relieve negative-pressure buildup caused by air exhaust (This replacement air usually has to be heated or cooled, humidified or dehumidified, and cleaned, by filters or electrostatic precipitators, in order to obtain acceptable working conditions.)
6. Limitation of noise exposure

COMFORT CONDITIONS

Chapter 8 explained how the specific speed can be used in selecting the type of fan (vaneaxial fan, centrifugal fan with airfoil, backward-inclined, forward-curved, or radial-tip blades) for a certain application. However, before we can select a fan type, we have to determine the air volume (cfm) and static pressure required in the application. The air volume will depend on the type of space (office, factory, apartment, school, hospital, auditorium, store, or kitchen) to be ventilated. The static pressure will depend on the duct system required to distribute the air to the various spaces. The comfort of the occupants will depend—in order of importance in most cases—on the following five parameters:

---

* Some of the information presented in this chapter has been taken from Bleier, F. P., *Fan Design and Application Handbook.* (Hopkins, Minn.: Ammerman Company), and from the *Ventilation Guide* (Bensalem, Pa.: FloAire, Inc.)

1. Dry-bulb temperature
2. Air quality
3. Relative humidity
4. Noise level
5. Air movement

Let's discuss these five parameters in more detail.

1. The dry-bulb temperatures desirable for certain spaces are shown below:

Offices	72 to 76°F	Hospital rooms	74 to 76°F
Factories	62 to 68°F	Operating rooms	74 to 78°F
Apartments	74 to 76°F	Auditoriums	68 to 72°F
Classrooms	72 to 75°F	Stores	66 to 68°F
Gymnasiums	60 to 65°F	Kitchens	66 to 68°F

2. For healthful conditions, the air should be free of excessive impurities such as dust, fumes, vapors, and other contaminants.

3. The most desirable relative humidity is 50 percent, but deviations of plus or minus 15 percent are acceptable in most spaces.

4. OSHA specifies the following noise levels as a function of the maximum number of hours of exposure:

Noise, dB	Maximum hours	Noise, dB	Maximum hours
90	8	102	1½
92	6	105	1
95	4	110	½
97	3	115	¼
100	2		

Note: Even these OSHA noise limitations seem on the liberal side. For safety, ear protectors are recommended for noise levels above 94 dB, even for short periods of exposure.

5. Some slight air motion from air outlets such as grills or diffusers is desirable in offices, classrooms, and auditoriums where many sedentary occupants are expected, but high air velocities up to 4000 fpm are used occasionally for cooling the workers in extremely hot factories.

The preceding OSHA requirements pertain to working areas in virtually all commercial and industrial facilities.

FIVE METHODS TO CALCULATE THE AIR VOLUME REQUIRED FOR A SPACE

Coming back to determination of the air volume required to fulfill the OSHA regulations for temperature, heat removal, etc., five methods are in general use for calculating these values. The air volumes may be produced by exhaust fans, by supply fans, or by a combination of both. All five methods are based on the principle of diluting the contaminated air or of heat removal in the space to be ventilated. This space can be described in various ways: volume of the space, floor area, number of occupants, or heat release. Here are the five methods for calculating the air volume required:

1. Air change in the volume of the space
2. Cubic feet per minute of air per square foot of floor area
3. Number of occupants in the space
4. Minimum air velocity through the space
5. Removal of heat released in the space

Which of the five methods should be selected in a specific application will depend on the circumstances, i.e., on the main purpose of the ventilation. Sometimes it will be possible to use two methods and to compare the results. If the two resulting air volumes just about confirm each other, we can be confident that our calculations were correct.

The Air-Change Method

Considered the easiest and most commonly used method of determining the required air volume, the air-change method assumes that all the air within a specified space must be changed within a given time span. The recommended rate of change is based on fan industry experience and on health regulations. This method uses the equation

$$V \text{ (ft}^3) = Q \text{ (ft}^3\text{/min)} \times M \text{ (minutes per change)} \qquad (11.1)$$

where V = volume of the space to be ventilated
Q = air volume (cfm) the fan must deliver against the static pressure of the system
M = rate of air change, expressed by the number of minutes it will take to replace all the air in the space

Solving this equation for Q, the required fan cubic feet per minute (cfm), we get

$$\text{Fan cfm} = V/M \qquad (11.2)$$

This means the required fan cfm will have to be larger if M is smaller, i.e., if the desired air change should be faster. This, of course, is the way one would expect it to be. Table 11.1 gives the rates of air change recommended for various types of buildings. Here are some examples, using Eq. (11.2) and the data given in Table 11.1.

Example 1. What fan cfm is required for an office, $25 \times 40 \times 10$ ft high, that needs an air change every 5 minutes? From Eq. (11.2), we get

$$\text{cfm} = \frac{25 \times 40 \times 10}{5} = 2000 \text{ cfm}$$

TABLE 11.1   Minutes per Air Change
Recommended for Various Types of Buildings

Type of building	Minutes per air change for adequate ventilation	
	No smoking	Smoking
Assembly halls	10–15	4–10
Bakeries	2–3	
Banks	15–30	5–15
Boiler rooms	2–4	
Classrooms	3–10	
Club rooms	6–7	3–4
Churches	10–15	
Dance halls	7–10	3–5
Engine rooms	1–3	
Foundries	1–3	
Garages	7–10	
Hospital rooms	10–15	
Kitchens	2–5	
Laundries	2–3	
Machine shops	5–10	
Offices	10–15	4–8
Paint shops	1–2	
Photo dark rooms	4–6	
Pig houses	6–10	
Poultry houses	6–10	
Residences	30–60	15–30
Restaurants	5–10	3–5
Ships storage	2–3	
Swimming pools	2–3	
Theaters	4–7	2–3
Toilets		1–4
Transformer rooms	2–5	
Warehouses	2–10	

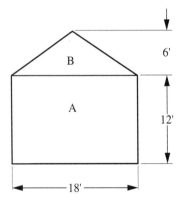

Example 2.   A warehouse building with dimensions as shown is to be ventilated. The building is 40 ft long. Calculate the required fan cfm.

Volume $A = 40 \times 18 \times 12 = 8640$ ft^3

Volume $B = 40 \times 18 \times 3 = 2160$ ft^3

Building volume $= 8640 + 2160 = 10,800$ ft^3

From Table 11.1, we find that for warehouses the recommended minutes per air change will range from 2 to 10. We select a value of 10 and get

$$\text{cfm} = \frac{10,800}{10} = 1080 \text{ cfm}$$

The cfm per Square Foot of Floor Area Method

This method of determining air volume (cfm) is sometimes used for large gathering areas such as auditoriums, restaurants, classrooms, or other areas where more ventilation is required. Here, the calculation goes as follows:

$$\text{Required cfm} = \text{floor area (ft}^2) \times \text{cfm per square foot} \qquad (11.3)$$

Table 11.2 gives the cubic feet per minute per square foot recommended for various types of spaces.

Example 3. Calculate the required ventilation air for a large classroom that is 45 ft wide by 60 ft long. The floor area to be ventilated will be $45 \times 60 = 2700$ ft^2. In accordance with Table 11.2, 2.0 cfm/ft^2 is recommended for classrooms. From Eq. (11.3), therefore, we get

$$\text{Required cfm} = 2700 \text{ ft}^2 \times 2.0 \text{ cfm/ft}^2 = 5400 \text{ cfm}$$

Note that this is the air volume required regardless of the ceiling height.

Example 4. The dining room and kitchen area of a restaurant are to be ventilated. The dining room has a floor area of 950 ft^2, while the kitchen has an area of 525 ft^2. What is the required cfm? According to Table 11.2, for the dining room, 2.0 cfm/ft^2 is recommended, but for the kitchen, 3.0 cfm/ft^2 is recommended. The total required cfm, therefore, will be

$$950 \times 2.0 + 525 \times 3.0 = 1900 + 1575 = 3475 \text{ cfm}$$

Note that the kitchen requires less air volume, despite its higher square footage, because it has a smaller floor area.

TABLE 11.2 Cubic Feet per Minute per Square Foot Recommended for Various Types of Spaces

Type of space	Recommended cfm per square foot
Auditoriums	2.0
Auto repair rooms	1.5
Classrooms	2.0
Elevators	1.0
Gymnasiums	1.5
Locker rooms	0.5
Photo dark rooms	0.5
Public toilets	2.0
Restaurant dining rooms	2.0
Restaurant kitchens	3.0
Retail stores	0.3
Swimming pools	0.5
Warehouses	1.0

The cfm per Occupant Method

This method again (same as method 2) is often used for determining the air volume (cfm) required for spaces such as auditoriums and conference rooms where large numbers of occupants can be expected. However, it is also used for smaller spaces with few occupants, such as hospital rooms, operating rooms, and prison cells. Here, we calculate

$$\text{Required cfm} = \text{number of persons} \times \text{cfm per occupant} \qquad (11.4)$$

Table 11.3 gives the minimum cubic feet per minute per occupant recommended for various types of spaces.

Example 5. An auditorium has a seating capacity of 500 persons. On most occasions, all seats are taken, plus an additional 10 percent of people are standing. How much ventilation is required? Using Eq. (11.4), with 15 cfm per occupant from Table 11.3, we get

$$\text{Required cfm} = 550 \times 15 = 8250 \text{ cfm}$$

Example 6. How much ventilation air is required for a conference room for 30 persons, assuming no other loads except lighting, which may be negligible? Using Eq. (11.4), with 20 cfm per occupant from Table 11.3, we get

$$\text{Required cfm} = 30 \times 20 = 600 \text{ cfm}$$

The Minimum Air Velocity Method

For more severe conditions, as encountered in spaces contaminated by fumes or particulate matter, considerably more ventilation is required, and the minimum air

TABLE 11.3 Minimum Cubic Feet per Minute per Occupant Recommended for Various Types of Spaces

Type of space	Minimum cfm per occupant
Auditoriums, churches	15
Beauty shops	25
Bowling alleys	25
Classrooms	15
Conference rooms	20
Dance halls	25
Dry cleaning plants	30
Hospital rooms	25
Laboratories	20
Libraries	15
Operating rooms	30
Physical therapy rooms	15
Prison cells	20
Public toilets	50
Railroad cars	15
Restaurant dining rooms	20
Restaurant kitchens	20
Retail stores	15
Smoking lounges	60
Supermarkets	15

velocity method is generally used. In this method, the building is considered to be a large rectangular duct with a cross section equal to its width W times its height H. The ventilating air is drawn through this duct at a velocity V of 150 to 250 fpm. Even though these are small air velocities, the resulting air volumes are large, due to the large cross sections. Here we can calculate

$$\text{Required cfm} = W \times H \times V \tag{11.5}$$

For buildings up to 100 ft long, a minimum air velocity of $V = 150$ fpm is used. For buildings 100 to 200 ft long, $V = 200$ fpm is used. For longer buildings, $V = 250$ fpm is used.

Example 7.   Let's consider a laboratory $25 \times 40 \times 10$ ft high containing toxic fumes. The cross section of the building is 25×10 ft $= 250$ ft^2. Since the building is only 40 ft long, we will use an air velocity of 150 fpm. Using Eq. (11.5),

$$\text{Required cfm} = 25 \times 10 \times 150 = 37{,}500 \text{ cfm}$$

Note: If we were calculating the required ventilation air by method 1 or 2, we would get much smaller air volumes. For example, the air-change method, even with 1 minute per air change, would give us

$$\text{cfm} = \frac{25 \times 40 \times 10}{1} = 10{,}000 \text{ cfm}$$

The cfm per square foot method, even using a high 5.0 cfm/ft^2, would give us even less:

$$25 \times 40 \times 5.0 = 5000 \text{ cfm}$$

For contaminated spaces, therefore, the minimum air velocity method is definitely needed.

The Heat-Removal Method

This method is used whenever the main purpose of the ventilation is removal of the heat being released into the space. This might be heat produced by machinery or simply by occupants, lights, and sun radiation. Here we have to estimate the British thermal units per hour produced by the various heat sources. We also have to estimate what the average outside temperature t_2 and the maximum tolerable inside temperature t_1 will be. After all these data have been established, we can calculate

$$\text{Required cfm} = \frac{\text{Btu/h}}{1.085(t_1 - t_2)} \tag{11.6}$$

Example 8.   Let's consider a boiler room, where 200,000 Btu is produced per hour. The average outside temperature t_2 is 77°F. The maximum tolerable inside temperature t_1 is 85°F. From Eq. (11.6), we can calculate

$$\text{Required cfm} = \frac{200{,}000}{1.085(85 - 77)} = 23{,}041 \text{ cfm}$$

Example 9. Calculate the required ventilation for a classroom 45 ft wide by 60 ft long, the same as in Example 3, but using the heat-removal method instead of the floor-area method. This is a large classroom. It can accommodate 180 students, each being a source of metabolic heat of about 430 Btu/h. The heat gain from maximum occupancy, therefore, will be $180 \times 430 = 77,400$ Btu/h. The lights will add another 1000 Btu/h. Assuming a temperature difference of 14°F between the inside and the outside air, we can calculate, from Eq. (11.6),

$$\text{Required cfm} = \frac{77,400 + 1000}{1.085 \times 14} = 5161 \text{ cfm}$$

This compares with 5400 cfm obtained by the floor-area method, a difference of only 4 percent.

FOUR METHODS TO DESIGN DUCT SYSTEMS AND TO DETERMINE THE REQUIRED STATIC PRESSURE

After we have calculated the required air volume for a certain space, by using one of the five methods we have described, we now have to design the duct system for distributing the air to or from several spaces, as desired. We also have to determine the required static pressure to be produced by the fan. This will be the static pressure, either positive (for supply) or negative (for exhaust), required to overcome the resistance of the duct system, consisting of a main duct and usually several branch ducts. Each duct will present some resistance due to friction. Usually there also will be some additional resistances, such as elbows, hoods, filters, heating and cooling coils, air inlets, diffusers, dampers, and other possible components. Dampers are a waste of energy but are needed for balancing the system, i.e., for obtaining the air volume desired for each space. The fan, therefore, must produce some extra static pressure so that the balancing can be done by means of the dampers, which will reduce the air volume to the desired amount in each space.

Four methods are in general use for designing the duct systems and for calculating the required static pressure. These four methods are

1. Equal friction
2. Velocity reduction
3. Constant velocity
4. Static regain

Which of these four methods should be selected for a specific application will depend on the circumstances, such as the required number of branch ducts and the permissible duct velocities. Here we often run into conflicting requirements, since high air velocities (3000 to 6000 fpm) result in smaller duct diameters (less space and lower first cost) but at the same time result in higher noise levels and higher static pressures (higher operating costs). Table 11.4 shows the ranges that are used for the duct velocities, regardless of which of the four methods is used.

Here are some general guidelines plus examples for the application of the four methods of designing a duct system and of calculating the required static pressure. In order to keep the duct friction low, round ducts are best and square ducts are sec-

TABLE 11.4 Recommended and Maximum Duct Velocities (fpm)

	Residences, schools, theaters, public buildings		Commercial and industrial buildings		High-velocity systems	
	Recommended	Maximum	Recommended	Maximum	Recommended	Maximum
Main duct	700–1300	1600	1200–1800	2200	2500	6000
Branch ducts	600–1200	1300	800–1000	1800	2000	4500
Filters	250–300	350	300–350	350	350	350
Heating and cooling coils	450–500	600	600	700	600	700
Air washers	500	500	500	500	500	500

ond. If rectangular ducts are used, they preferably should not be too far from square. We will make our sample calculations for round ducts.

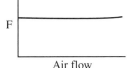

The Equal-Friction Method

This method is applicable primarily to systems using moderate duct velocities, from 1000 to 2000 fpm, and preferably to layouts where the branch ducts are approximately symmetrical with respect to air volume and air velocity in corresponding branch ducts. The method is commonly used for both exhaust and supply systems. It is called the *equal-friction method* because the duct diameters and air velocities are selected for equal friction per 100 ft of duct length. This friction loss per 100 ft usually is 0.1 to 0.2 inWC. The larger figure, as mentioned, results in smaller duct diameters and in larger air velocities. The duct diameter D can be calculated from the following formula:

$$D^5 = \left(\frac{\text{cfm}}{K}\right)^2 \tag{11.7}$$

where D is the duct diameter in inches, cfm is the desired airflow in the duct (either a main or a branch duct), and K is a constant depending on the selected friction loss F per 100 ft of duct length. The value of K can be taken from Table 11.5 or from Fig. 11.1.

Equation (11.7) indicates the following:

1. The duct diameter D can be calculated. No cut-and-try method is needed for determining the value of D.

TABLE 11.5 Constant K of Eq. (11.7) for Various Friction Losses F (inWC) per 100 ft of Duct Length

F	0	0.1	0.2	0.3	0.4	0.5	1.0
K	0	1.427	2.019	2.474	2.857	3.194	4.517
F		2.0	3.0	4.0	5.0	6.0	7.0
K		6.388	7.823	9.032	10.10	11.06	11.90

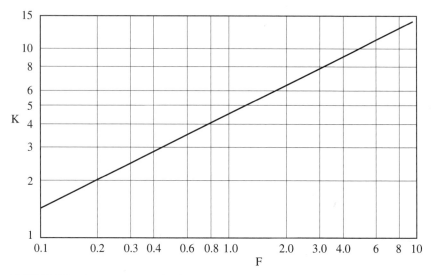

FIGURE 11.1 Constant K of Eq. (11.7) versus friction loss F (inWC) per 100 ft of duct length

2. The duct diameter D depends only on two variables: the desired air volume and the selected friction loss F per 100 ft of duct length.

3. A larger air volume obviously will result in a larger duct diameter D. A larger friction loss F per 100 ft of duct length will result in a larger constant K and therefore in a smaller duct diameter D.

Figure 11.2 shows a line diagram of an exhaust system with eight branch ducts. It shows the required air volume (cfm) at each air inlet and the lengths of the main duct and of the branch ducts. Selecting a friction loss of 0.2 inWC per 100 ft of duct length, we get from Table 11.5 a K value of 2.019. Equation (11.7) now becomes

$$D^5 = \left(\frac{\text{cfm}}{2.019}\right)^2 \tag{11.8}$$

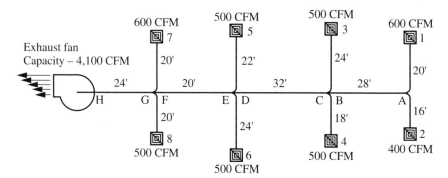

FIGURE 11.2 Line diagram for an exhaust system.

TABLE 11.6 Calculation of Duct Diameters D and Friction Losses F for the Exhaust System in Fig. 11.2 Using a Friction Loss of 0.2 inWC per 100 ft of Ducting

			Duct diameter D (in)		Duct area A (ft^2)	Duct Velocity V (fpm)	Duct velocity pressure VP (inWC)	Duct length L (ft)	Ratio $\dfrac{12L}{D}$	Duct friction loss F (inWC)
Section	cfm	D^5	Calculated	Rounded off						
A-1	600	88,314	9.8	10	0.545	1100	0.076	20	24.0	0.035
A-2	400	39,251	8.3	8	0.349	1146	0.082	16	24.0	0.038
A-B	1000	245,317	12.0	12	0.785	1273	0.101	28	28.0	0.055
C-3	500	61,329	9.1	9	0.442	1132	0.080	24	32.0	0.050
C-4	500	61,329	9.1	9	0.442	1132	0.080	18	24.0	0.037
C-D	2000	981,267	15.8	16	1.396	1432	0.128	32	24.0	0.060
E-5	500	61,329	9.1	9	0.442	1132	0.080	22	29.3	0.046
E-6	500	61,329	9.1	9	0.442	1132	0.080	24	32.0	0.050
E-F	3000	2,207,852	18.6	19	1.969	1524	0.145	20	12.6	0.036
G-7	600	88,314	9.8	10	0.545	1100	0.076	20	24.0	0.035
G-8	500	61,329	9.1	9	0.442	1132	0.080	20	26.7	0.042
G-H	4100	4,123,776	21.0	21	2.405	1705	1.181	24	13.7	0.048

From Eq. (11.8), we can calculate the various duct diameters D, as shown in Table 11.6.

Note: Formulas used: D^5 from Eq. (11.8). D (in) is the fifth root of D^5. If your calculator does not provide a key for this, use logarithms or the data shown in Figs. 11.3 and 11.4

Once the air volume and the duct diameter D have been determined, the duct area A and the duct velocity V can be calculated as follows:

$$\text{Duct area } A = \left(\frac{D}{24}\right)^2 \pi$$

$$\text{Duct velocity } V = \frac{\text{cfm}}{A}$$

$$\text{Duct velocity pressure } VP = \left(\frac{V}{4005}\right)^2$$

The duct velocities should be checked against the velocities shown in Table 11.4 to make sure these calculated velocities conform with the ranges shown. If these calculated duct velocities should be outside the recommended range, the selected friction loss F per 100 ft of duct length may have to be revised.

The duct length is taken from Fig. 11.2:

$$\frac{\text{Duct length}}{\text{Duct diameter}} = \frac{12L}{D}$$

$$\text{Duct friction } F = 0.0195 \frac{12L}{D} VP$$

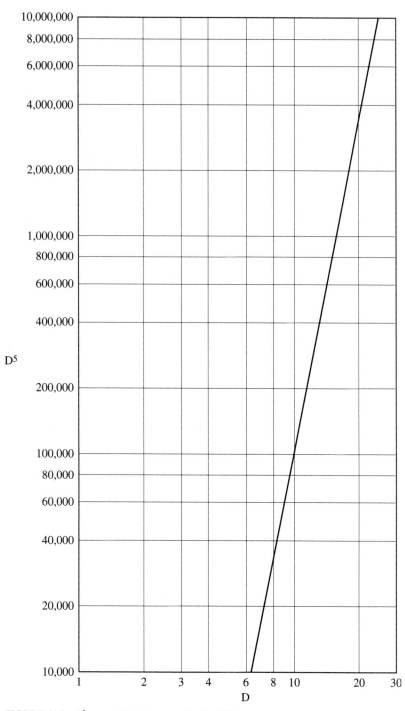

FIGURE 11.3 D^5 versus D for the range $D = 7$ to 25 in.

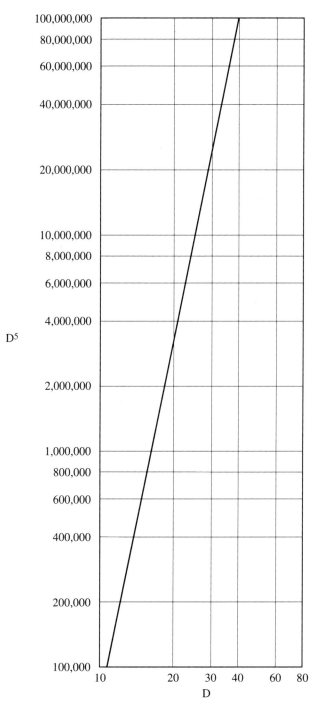

FIGURE 11.4 D^5 versus D for the range $D = 10$ to 40 in.

From the last column in Table 11.6, we now can calculate the static pressure needed to overcome the friction of the ducts. To do this, we add the duct frictions F for the various sections of the main duct (A-B, C-D, E-F, and G-H) plus the duct friction of the most distant branch duct with the largest friction loss (in our case A-2). This will give us a total friction loss of 0.237 inWC. The other branch ducts join the main duct at various points and are more or less in parallel with this most distant branch duct. They are not in series and therefore will not add to the total friction loss. However, this 0.237 inWC is just the total duct friction. We will have to add the resistances of the air inlets (about 0.10 inWC) and the elbows (25 to 60 percent of the VP, depending on the angle and the sharpness of the bend) and provide some reserve to allow for the losses due to the dampers for balancing (about 0.4 inWC). Including these additional resistances, the fan will have to deliver 4100 cfm against a static pressure of approximately 1 inWC. This, then, will be the requirements for which the fan will be selected.

The Velocity-Reduction Method

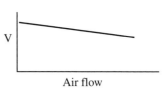

This method is best suited for sizing the duct work for supply and exhaust systems with numerous branches of different lengths. In the case of supply systems, we first select from Table 11.4 a velocity for the beginning of the main duct, after the fan outlet. This starting velocity should be fairly high (as long as it is in the range shown in Table 11.4), because it will become progressively smaller at the various branch takeoffs. This is the reason why this is called the *velocity-reduction method*.

In the case of exhaust systems, the same name is used, even though it really should be called the *velocity-increase method*. Here we start out with the selection (from Table 11.4) of a small velocity for the beginning of the branch ducts and let the velocities become progressively larger as we move through the duct work toward the fan inlet.

Again, the friction losses are proportional to the velocity pressures in the duct, and dampers are used for balancing. Compared with the equal-friction method, the dampers are even more important in the velocity-reduction method because here we have less symmetry of the branch ducts, so the dampers will have to do more adjusting.

Figure 11.5 shows a line diagram for a supply system with eight branch ducts. It shows the lengths of the main duct sections and of the branch ducts and the required air volume at each air outlet. Table 11.7 shows the various duct velocities we select and the duct diameters and friction losses we obtain by calculation.

The data in Table 11.7 were obtained as follows: The duct velocities V were selected for a gradual reduction, but in accordance with the ranges shown in Table 11.4.

$$\text{Duct velocity pressure } VP = \left(\frac{V}{4005}\right)^2$$

$$\text{Duct area } A = \frac{\text{cfm}}{V}$$

$$\text{Duct diameter } D = 13.54\sqrt{A}$$

$$\frac{\text{Duct length}}{\text{Duct diameter}} = \frac{12L}{D}$$

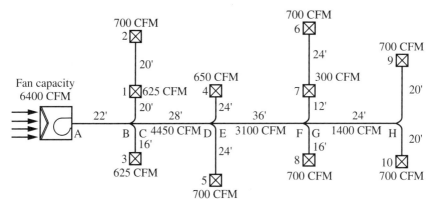

FIGURE 11.5   Line diagram for a supply system.

$$\text{Duct friction } F = 0.0195\frac{12L}{D}VP$$

Duct lengths L are taken from Fig. 11.5.

From the last column in Table 11.7, we now can calculate the static pressure needed to overcome the friction of the ducts. To do this, we add the duct frictions F for the various sections of the main duct (A-B, C-D, E-F, and G-H) plus the duct friction of the most distant branch duct with the largest friction loss (in our case $H - 9$). This will give us a total friction loss of 0.240 inWC. Again, we have to add the resistances of the air outlets, the elbows, and the dampers and will arrive at a fan requirement of 6400 cfm against a static pressure of about 1 inWC.

TABLE 11.7   Calculation of Duct Diameters D and Friction Losses F for the Supply System in Fig. 11.5 Using the Velocity-Reduction Method

Section	cfm	Duct velocity V fpm	Duct velocity pressure VP (inWC)	Duct area A (ft²)	Duct diameter D (in) Calculated	Rounded off	Duct length L (ft)	Ratio $\dfrac{12L}{D}$	Duct friction loss F (inWC)
A-B	6400	2000	0.249	3.20	24.2	24	24	12.0	0.058
C-D	4450	1800	0.202	2.47	21.3	21	28	16.0	0.063
E-F	3100	1600	0.160	1.94	18.8	19	36	22.7	0.071
G-H	1400	1100	0.075	1.27	15.3	15	24	19.2	0.028
B-1	1325	1100	0.075	1.20	14.9	15	20	16.0	0.023
F-7	1000	1100	0.075	0.91	12.9	13	12	11.1	0.016
1-2	700	900	0.050	0.78	11.9	12	20	20.0	0.020
D-5	700	900	0.050	0.78	11.9	12	24	24.0	0.023
6-7	700	900	0.050	0.78	11.9	12	24	24.0	0.023
F-8	700	900	0.050	0.78	11.9	12	16	16.0	0.016
H-9	700	900	0.050	0.78	11.9	12	20	20.0	0.020
H-10	700	900	0.050	0.78	11.9	12	20	20.0	0.020
D-4	650	900	0.050	0.72	11.5	12	24	24.0	0.023
B-3	625	900	0.050	0.69	11.3	11	16	17.5	0.017

The Constant-Velocity Method

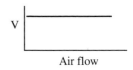

V

Air flow

This method is used for pneumatic conveying systems (supply or exhaust), where higher duct velocities must be maintained to keep the material (such as sawdust, corn, etc.) in suspension. In these systems, there often is only one duct, but occasionally there are a main duct and one or two branch ducts. The duct velocities here are much higher, between 3000 and 7000 fpm, depending on the weight of the material particles to be transported. The duct velocities must be high enough to prevent the material from settling instead of moving along with the air stream. A cyclone is often used to separate the material from the airflow before it passes through the fan and gets damaged. The resistance of the cyclone is considerable. It ranges from one to five velocity pressures in the inlet duct to the cyclone. Table 11.8 shows the duct velocities required for various materials.

Figure 11.6 shows a line diagram for an exhaust system with one main duct and one branch duct. It shows the lengths of the ducts and the required air volume (cfm) at each duct section.

Table 11.9 shows the various velocities we select and the duct diameters and friction losses we obtain by calculation. As a result of the large conveying velocities, the friction losses are much higher, too.

The data in Table 11.9 were obtained as follows: The duct velocities V were selected in accordance with Table 11.8.

$$\text{Duct velocity pressure } VP = \left(\frac{V}{4005}\right)^2$$

TABLE 11.8　Material-Conveying Data

Material	Approximate weight (lb/ft^3)	Average conveying velocity (fpm)	Suction to pickup (inWC)
Ashes from coal	30	5500	3.0
Beans	28	6000	4.0
Buffing dust		3700	2.5
Cement	100	7000	5.0
Cork	14	3000	1.5
Corn (shelled)	45	5500	3.5
Cotton (dry)	5	3500	2.0
Grinding dust	30	4500	2.0
Lime (hydrated)	30	5500	3.0
Malt	35	4800	3.0
Metal turnings		4500	2.5
Mineral wool	12	3500	2.0
Paper cuttings	20	5000	3.0
Rags (dry)	30	4000	2.5
Saw dust (dry)	12	3500	2.5
Shavings (light)	9	3500	2.5
Shavings (heavy)	24	4000	3.0
Wheat	46	6000	4.0

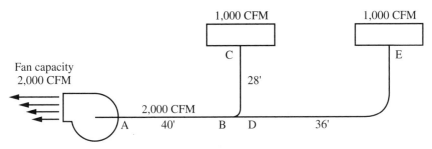

FIGURE 11.6   Line diagram for an exhaust system for pneumatic conveying.

$$\text{Duct area } A = \frac{\text{cfm}}{V}$$

$$\text{Duct diameter } D = 13.54\sqrt{A}$$

Duct length L is taken from Fig. 11.6.

$$\frac{\text{Duct length}}{\text{Duct diameter}} = \frac{12L}{D}$$

$$\text{Duct friction loss } F = 0.0195\frac{12L}{D}VP$$

From the last column in Table 11.9, we now can again calculate the static pressure needed to overcome the friction of the duct system by adding the duct friction losses F for the sections $A\text{-}B$ and $D\text{-}E$. This will give us a total friction loss of 2.14 inWC. To this we have to add the resistances of the material pickup (about one velocity pressure), of the elbow (about one-half velocity pressure), and of the cyclone (about three velocity pressures) and will arrive at a fan requirement of 2000 cfm against a static pressure of about 6.6 inWC.

The Static-Regain Method

This method has certain advantages when an equal amount of fresh air is to be delivered to numerous spaces, as shown in Fig. 11.7. It is based on the principle

TABLE 11.9   Selection of Duct Velocities and Calculation of Duct Diameters and Friction Losses for an Exhaust System for Pneumatic Conveying as in Fig. 11.6

Section	Air volume (cfm)	Duct velocity V (fpm)	Duct velocity pressure VP (inWC)	Duct area A (ft²)	Duct diameter D (in) Calculated	Rounded off	Duct length L (ft)	Ratio $\frac{12L}{D}$	Duct friction loss F (inWC)
$A\text{-}B$	2000	4000	1.0	0.50	9.6	10	40	48.0	0.94
$B\text{-}C$	1000	4000	1.0	0.25	6.8	7	28	48.0	0.94
$D\text{-}E$	1000	4000	1.0	0.25	6.8	7	36	61.7	1.20

of static regain, which was discussed in Chap. 1. Static regain will occur whenever an airflow slows down. When this happens, about half the difference in velocity pressures is regained as additional static pressure (the other half is lost to friction and turbulence). This lower air velocity may be the result of an expanding cone, as discussed in Chap. 1. Or it may be the result of a branch duct bleeding off some of the airflow while the diameter of the main duct remains the same, as will be discussed now.

In the first three methods we described, the diameter of the main duct was reduced at each branch duct takeoff in view of the air volume reduction at the takeoff point. In the static-regain method, the diameter of the main duct is kept constant throughout its length. This results in a reduction in the main-duct velocity at each branch takeoff and in a static regain at this point, which in turn is used to overcome the duct friction in the subsequent main-duct section until the next branch takeoff. The static pressure at each takeoff point, therefore, will remain about the same.

In the velocity-reduction method, a relatively high air velocity was selected for the first section of the main duct, and the corresponding duct diameter was calculated. For the subsequent sections of the main duct, both the duct velocities and the duct diameters were reduced. In the static-regain method, again, a high duct velocity is selected for the first section of the main duct, and the corresponding duct diameter is calculated. However, for the subsequent sections, the diameter of the main duct is maintained, and the duct velocities, therefore, are reduced even more than in the velocity-reduction method. This increased difference in velocity pressures gives us more room for static recovery.

Figure 11.7 shows a line diagram for a supply system with five branch ducts. It shows the lengths of the various sections and the required air volume at each air outlet. Table 11.10 shows the various duct velocities, the duct diameters, the friction losses, and the static regains.

The data in Table 11.10 were obtained as follows: The required air volumes were taken from Fig. 11.7. The duct velocity V for section A-B was selected as 2200 fpm. The corresponding duct velocity pressure was calculated as

$$VP = \left(\frac{V}{4005}\right)^2$$

$$\text{Duct area } A = \frac{\text{cfm}}{V}$$

$$\text{Duct diameter } D = 13.54\sqrt{A}$$

Duct length L for section A-B was taken from Fig. 11.7.

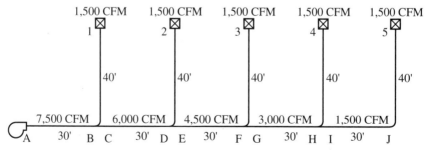

FIGURE 11.7   Line diagram for a supply system with five branch ducts.

TABLE 11.10 Selection and Calculation of Velocities, Diameters, Friction Losses, and Static Regains

Section	Air volume (cfm)	Duct velocity V (fpm)	Duct pressure VP (inWC)	Duct area A (ft²)	Duct diameter D (in) Calculated	Rounded off	Duct length L (ft)	Ratio $\frac{12L}{D}$	Duct friction loss F (inWC)	Static regain $\frac{1}{2}\Delta VP$ (in WC)
A-B	7500	2200	0.302	3.409	25.0	25	30	14.4	0.085	
										0.055
C-D	6000	1760	0.193						0.054	
										0.042
E-F	4500	1320	0.109						0.031	
										0.031
G-H	3000	880	0.048						0.013	
										0.018
I-J	1500	440	0.012	3.409	25.0	25	30	14.4	0.003	
B-1	1500	800	0.040	1.875	18.5	19	40	25.3	0.020	
D-2										
F-3										
H-4										
J-5	1500	800	0.040	1.875	18.5	19	40	25.3	0.020	

$$\frac{\text{Duct length}}{\text{Duct diameter}} = \frac{12L}{D}$$

$$\text{Duct friction loss } F = 0.0195 \frac{12L}{D} VP$$

For the second line (section C-D), we did not select a duct velocity V, as we did for the first line (section A-B), but we kept the same 25-in duct diameter and the same 3.409-ft² duct area. From this we calculated the duct velocity $V = $ cfm/A and the $VP = (V/4005)^2$ for the second line. Next, we entered the duct length L and calculated the duct friction loss F for the second line. We now could calculate the static recovery (last column) as one-half the difference between the velocity pressures of line 1 minus line 2. We continued the same way for the remaining three lines.

Now that we have completed the first five lines of Table 11.10, we can calculate the static pressures available at the various takeoff points along the main duct, tentatively assuming that the fan will deliver 7500 cfm against a static pressure of 1 inWC.

Point A (past the fan): $SP = 1.000$
Point B (ahead of branch duct B-1): $SP = 1.000 - 0.085 = 0.915$ inWC
Point C (past branch takeoff B-1): $SP = 0.915 + 0.055 = 0.970$ inWC
Point D (ahead of branch duct D-2): $SP = 0.970 - 0.054 = 0.916$ inWC
Point E (past branch takeoff D-2): $SP = 0.916 + 0.042 = 0.958$ inWC
Point F (ahead of branch duct F-3): $SP = 0.958 - 0.031 = 0.927$ inWC
Point G (past branch takeoff F-3): $SP = 0.927 + 0.031 = 0.958$ inWC
Point H (ahead of branch duct H-4): $SP = 0.958 - 0.013 = 0.945$ inWC
Point I (past branch takeoff H-4): $SP = 0.945 + 0.018 = 0.963$ inWC
Point J (ahead of branch duct J-5): $SP = 0.963 - 0.003 = 0.960$ inWC

We note that the various static recoveries tend to make up for the friction losses in the different main duct sections. As a result, the static pressure available for the airflow into the various branch ducts is almost the same, varying only from 0.916 to 0.960 inWC, a variation of less than 5 percent.

Coming back to Table 11.10, we now select a velocity for the five branch ducts, say, 800 fpm, and calculate the remaining data in Table 11.10 using the same formulas. We note the following:

1. The friction loss F in each branch duct is only 0.02 inWC, much smaller than the available static pressure.

2. The resistance of the air outlets (0.10 inWC) and of the elbows (0.02 inWC) still will be much less than the available static pressure.

3. No dampers will be needed, due to the even distribution of the air stream and due to the almost constant static pressure available at each branch duct takeoff.

4. The tentative estimate of a static pressure of 1 inWC was on the safe side by far. A fan delivering 7500 cfm against 1 inWC *SP* or slightly less will be acceptable.

CHAPTER 12
AGRICULTURAL VENTILATION REQUIREMENTS*

TYPES OF VENTILATION FOR AGRICULTURAL REQUIREMENTS

The following three types of ventilation are used for agricultural requirements:

1. Ventilation of animal shelters
2. Recirculation of heated air in incubator houses
3. Drying and aeration systems for grain

Ventilation of Animal Shelters

Animal shelters need ventilation, just as buildings occupied by humans, but the ventilation of animal shelters is simpler. It can be accomplished without any duct work. Direct-drive propeller fans are mounted in one building wall, as shown in Fig. 3.1. Evenly spaced fans exhaust air from the building. Openings for the air inlet are provided in the opposite wall. The required air volume can be calculated by the air-change method. Table 11.1 recommends 6 to 10 minutes per air change for pig houses and poultry houses.

Poultry Houses. Poultry houses are usually ventilated by a series of 10- or 12-in propeller fans with direct drive from a 1/25-hp two-speed motor equipped with thermostats that shut the fan off when the inside temperature drops below 40°F. One hundred chickens give off about 4 gal of moisture per day plus some ammonia fumes caused by droppings. The fans remove the moisture and the fumes. As an alternate method for calculating the required air volume, 2 cfm per bird is recommended, so 500 cfm will take care of 250 birds. The number of birds thus determines the number and size of propeller fans to be installed. Figure 12.1 shows the outside of a poultry house; Figure 12.2 shows the inside.

Dairy Barns for Cows. Dairy barns are usually ventilated by one or two propeller fans with direct drive from two-speed motors, as shown in Table 12.1. Fans will be selected to exhaust from 40 to 160 cfm per cow, depending on the climate. In regions with severe winters, 40 to 80 cfm per cow is adequate. Regions with mild winters will allow 120 to 160 cfm per cow. A thermostat again will provide control of the tem-

---

* Some of the information presented in this chapter was taken from booklets published by Farm Fans, Inc., Indianapolis, Indiana, and Chick Master Incubator Company, Medina, Ohio.

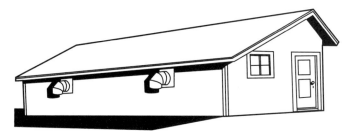

FIGURE 12.1   Outside of a poultry house, showing two exhaust elbows for the discharging airflow.

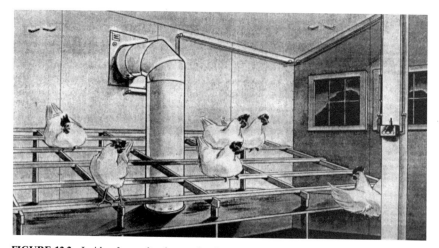

FIGURE 12.2   Inside of a poultry house, showing several hens and an optional vertical duct for exhausting air from the bottom of the poultry house.

perature in the barn. The thermostat is usually set at 50°F. No heat other than that generated by the animals is needed to warm the barn. Figure 12.3 shows the inside of a dairy barn.

Incubator Houses for Hatching Eggs

Incubator houses handle 100,000 eggs and keep them for 18 days at 100°F for artificial incubation. Several axial-flow fans are located near the ceiling of the incubator

TABLE 12.1   Required Ventilation for Dairy Barns

Fan diameter (in)	Motor hp	Air volume at high speed (1140 rpm)	Air volume at low speed (860 rpm)	Number of cows
12	1/25	980	585	6 to 16
16	1/8	2000	1290	13 to 33
16	1/4	2750	1780	20 to 45

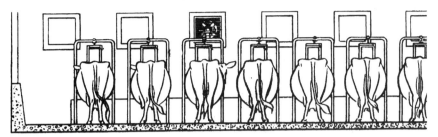

FIGURE 12.3 Inside of a dairy barn, showing seven cows in their stalls and one exhaust fan in the wall.

house and blow the warm air down into the center aisle. The air stream then turns 90° to both sides, where the egg carriers are located. It passes through them and returns to the top of the fans. This flow pattern results in the desired recirculation and ensures an even temperature distribution for all eggs. Six vaneaxial fans are recommended because they will produce enough pressure to force the air through the narrow passages between the egg trays.

Two small openings in the ceiling are provided to allow the carbon dioxide gas released by the hatching eggs to escape. Figure 12.4 shows the inside of an incubator house. Figure 12.5 shows an egg carrier containing 15 egg trays. There are 18 egg carriers in the incubator house, 9 on each side of the center aisle. The egg carriers are on wheels so that they can be rolled into the incubator house and through the center aisle to their places on each side of the center aisle.

FIGURE 12.4 Inside view of an incubator showing the egg carriers on each side of the center aisle and some fans at the top. (*Courtesy of Chick Master Incubator Company, Medina, Ohio.*)

Grain Drying and Aeration Systems

When corn is harvested, it has a moisture content of 26 to 30 percent. This makes it perishable, because if the corn were stored at this high moisture content, it would soon deteriorate due to mold growth and insect activity. The shelled corn (kernels removed from the cob), therefore, is dried by heated air, forced either through a large grain drying bin or through a mobile batch dryer. Figure 12.6 shows both these structures. Fan-heater combinations blow the heated air through the grain until the moisture content is reduced to about 13 percent. This may take 1 or 2 months in a bin containing large quantities of corn but only a few days in a mobile batch dryer that handles much less grain. Similar procedures are used for other types of grain, such as wheat, rice, and soybeans.

After the drying process has been completed, the corn is stored and aerated; i.e., unheated air is forced through the storage bin, again to prevent mold growth and insect activity. This aeration process is continued for as long as the grain is stored in the bin.

FIGURE 12.5 Egg carrier on wheels, as used in an incubator. *(Courtesy of Chick Master Incubator Company, Medina, Ohio.)*

The airflow is produced by various types of fans, all using direct drive at 3450 or 1750 rpm. These fans may be tubeaxial fans (for pressures up to 3 inWC), vaneaxial fans (for pressures up to 5 inWC), or BI centrifugal fans (for pressures up to 14 inWC). Figure 12.7 shows some vaneaxial fans for grain drying, with heaters on the outlet side and with wheel diameters of 24 in (3450 rpm), 28 in (3450 rpm), and 36 in (1750 rpm). Figure 12.8 shows a BI centrifugal fan for grain drying with wheel diameters of 20, 24, and 27 in, all running at 1750 rpm. Here the heaters can be located either on the outlet side or on the inlet side, since the motor is not in the air stream. The heaters sometimes use electric power but more often burn propane or natural gas. In order to select the right fan for each application, we have to know the air volume and the static pressure required. These data can be determined from the size of the bin, the depth of the grain in the bin, the type and amount of grain, and its moisture content. Let's see how this is done.

FIGURE 12.6 Sketch showing a centrifugal fan blowing heated air into the bottom section of a grain bin and three vaneaxial fans blowing heated air through a batch dryer. *(Courtesy of Farm Fans, Inc., Indianapolis, Ind.)*

Determination of the Air Volume Required for Drying Shelled Corn

Three simple steps have to be taken to determine the required air volume.

1. As a first step, we have to determine the amount of corn (in bushels) in

Model U524-B	Model U1028-B	Model U1036-A

Model U524-B
For bin drying systems.
- 5-7 HP, 24" Diameter, 3450 RPM
- 2,000,000 BTU per hour capacity
- Humidistat or thermostat included
- Approximate weight: 320 lbs.
- Magnetic motor control included

Model U1028-B
High capacity dryer for deeper grain depths.
- 10-13 HP, 28" Diameter, 3450 RPM
- 3,000,000 BTU per hour capacity
- Humidistat or thermostat included
- Approximate weight: 400 lbs.
- Magnetic motor control included

Model U1036-A
For bin drying systems.
- 10-16 HP, 36" Diameter, 1750 RPM
- 2,500,000 BTU per hour capacity
- Humidistat or thermostat included
- Approximate weight: 700 lbs.
- Magnetic motor control included

Model U724-B
For bin drying systems.
- 7-9 HP, 24" Diameter, 3450 RPM
- 2,000,000 BTU per hour capacity
- Humidistat or thermostat included
- Approximate weight: 350 lbs.
- Magnetic motor control included

Model U1036-H
For higher heat capacity in special drying systems.
- 10-16 HP, 36" Diameter, 1750 RPM
- 4,000,000 BTU per hour capacity
- Thermostat included
- Approximate weight: 700 lbs.
- Magnetic motor control included

FIGURE 12.7   Various sizes of vaneaxial fans for grain drying. *(Courtesy of Farm Fans, Inc., Indianapolis, Ind.)*

the bin. Grain bins are cylindrical in shape. The bin diameters usually range from 18 to 60 ft, but sometimes they are even larger, up to 90 ft. The depth of the grain usually is between 8 and 60 ft but sometimes is as deep as 100 ft. From the bin diameter and the grain depth, we can calculate the amount of grain in the bin from the formula

$$\text{Number of bushels} = 0.6313 \times (\text{bin diameter})^2 \times \text{grain depth} \qquad (12.1)$$

or we can find it from the graphs shown in Figs. 12.9 and 12.10.

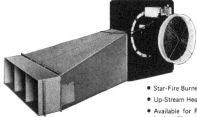

FFC-H HEATERS

- Star-Fire Burner for Efficient Combustion.
- Up-Stream Heater for Heat Mixing.
- Available for Propane Vapor, Propane Liquid (with Vaporizer), or Natural Gas.
- Direct Fired Heater for Mounting on Centrifugal Fan Inlet.
- Full Safety Controls, with Burner Shut-down Flame Failure Control.
- Full Electric Ignition.
- Heater Control by Humidistat or Thermostat.
- Includes Humidistat or Thermostat.
- Transition and Collar Available.

HEATER MODEL NUMBER	USE WITH FAN MODEL NO.	HEATER DIAMETER	BURNER CAPACITY	TRANSITION NUMBER
FFC-H20B	FFC-520B	20"	750,000 BTU/Hr.	T20-CB
FFC-H24B	FFC-1024B	24"	1,500,000 BTU/Hr.	T24-CB
FFC-H24B	FFC-1524B	24"	1,500,000 BTU/Hr.	T24-CB
FFC-H27B	FFC-2027B	28"	2,000,000 BTU/Hr.	T27-CB

FIGURE 12.8   BI centrifugal fan for grain drying, with the heater on the inlet side and the transformation piece on the outlet side. *(Courtesy of Farm Fans, Inc., Indianapolis, Ind.)*

2. The second step is to determine the required air volume per bushel, which depends on the moisture content of the corn. The graph in Fig. 12.11 shows that the minimum required air volume per bushel for shelled corn varies from 5 cfm (for 30 percent moisture) down to 1 cfm (for 22 percent moisture). Please note that these are minimum values.

3. As a third step, we multiply the number of bushels (from item 1) by the air volume per bushel (from item 2). This gives us the total air volume required.

The bins then have a storage capacity of 1600 to 500,000 bushels (1 bushel equals 1.244 ft^3). The air (heated or unheated) is blown into a low space between the floor of the bin and a perforated partition about 12 to 15 in above the floor. From this space, the airflow passes through the perforated partition and up through the grain, finally leaving through some openings at the top of the bin.

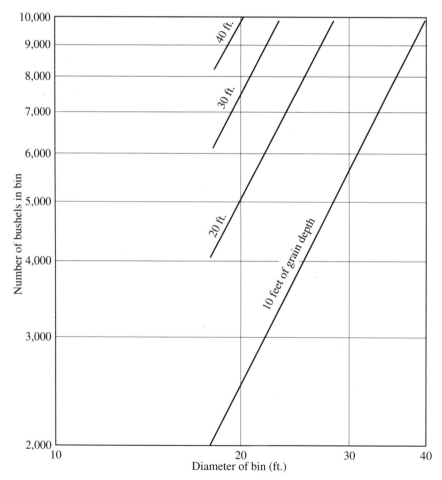

FIGURE 12.9 Number of bushels in bins versus bin diameters (up to 40 ft) for various grain depths.

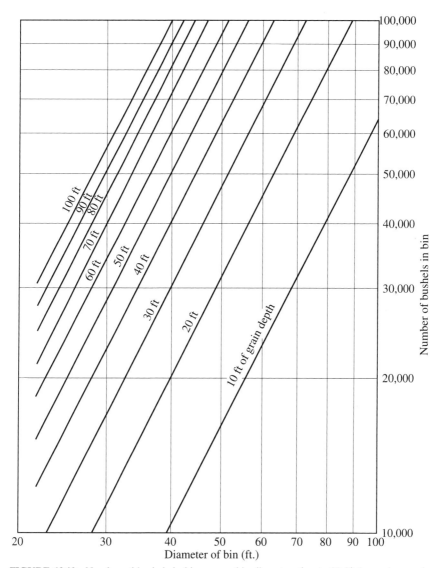

FIGURE 12.10   Number of bushels in bins versus bin diameters (up to 90 ft) for various grain depths.

The amount of heat depends on the size of the bin (or batch dryer), on the type of grain, on its moisture content, and on the relative humidity of the air. When air is heated, it becomes dryer; i.e., its relative humidity is lowered. The amount of heat required varies from 0.75 to 5 million Btu per hour. The temperature rise at the heater outlet is between 70 and 150°F, but as the air stream passes through the grain, the temperature drops rapidly. In other words, the drying is done primarily by the air

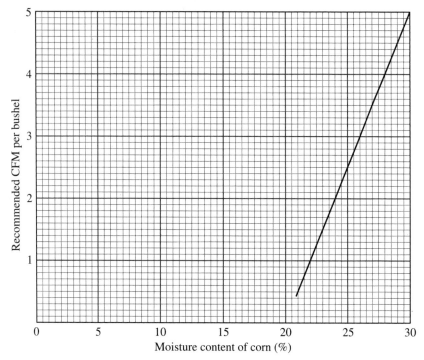

FIGURE 12.11   Recommended minimum air volume (cfm) per bushel for shelled corn of 20 to 30 percent initial moisture content.

stream having a lower humidity than the grain. The amount of air is the important parameter.

Determination of the Static Pressure Required

The static pressure needed to force the air stream through the grain depends on the depth of the grain in the bin, the air volume per bushel, and the type of grain. Figure 12.12 shows the relationship for shelled corn. You will notice that for 2 cfm per bushel, the static pressure rises from 1.3 inWC for 8 ft of depth to 6.3 inWC for 16 ft of depth. In other words, as you double the depth, the static pressure needed becomes 4.8 times as high, as shown by the parabolic curves in Fig. 12.12. These parabolic curves may be surprising at first glance. As the grain depth doubles from 8 to 16 ft, you would not expect the static pressure to become 4.8 times as high but only 2 times as high. This would be true if the total air volume through the grain were to remain the same, regardless of the depth. However, the total air volume does not remain the same, the air volume *per bushel* remains the same. If the grain depth is doubled from 8 to 16 ft, more bushels are added; in fact, the number of bushels is doubled. Thus, if the air volume per bushel remains the same, the total air volume and the air velocity through the grain are doubled, too. The static pressure required to force this doubled air volume through the grain, therefore, will more than double, as shown by the parabolic curves in Fig. 12.12.

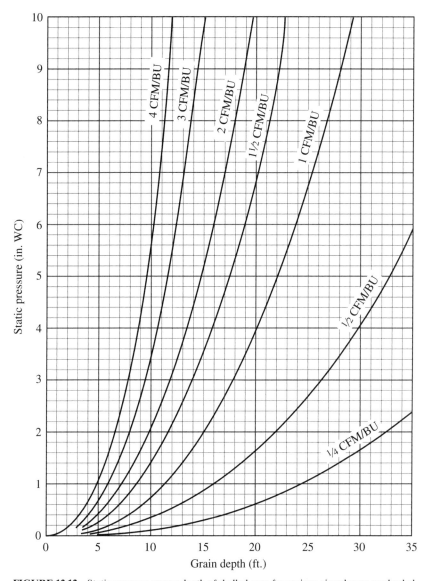

FIGURE 12.12 Static pressure versus depth of shelled corn for various air volumes per bushel.

Selection of the Proper Fan

After the required air volume and static pressure have been determined, the proper fan now can be selected. Table 12.2 shows the performance for the four vaneaxial fans, and Table 12.3 shows the performance for the four BI centrifugal fans, as mentioned previously. The corresponding performance curves are shown in Figs. 12.13 and 12.14.

TABLE 12.2 Performance of Vaneaxial Fans for Grain Drying

Model number	hp	Diameter (in)	rpm	Air volume (cfm) at indicated static pressure (inWC)								
				1	1½	2	2½	3	3½	4	5
U524-B	5–7	24	3450	11,050	10,400	9,600	8,650	7,500	6,300	5,200	3,100
U724-B	7–9	24	3450	13,200	12,400	11,650	10,800	9,800	8,750	7,200	4,500
U1028-B	10–13	28	3450	19,400	18,400	17,300	16,300	15,100	14,000	12,800	9,500
U1036-A	10–16	36	1750	26,000	24,400	22,600	20,600	18,100	12,600	10,700	6,700

Source: Courtesy of Farm Fans, Inc., Indianapolis, Ind.

TABLE 12.3 Performance of BI Centrifugal Fans for Grain Drying

Model number	Wheel diameter (in)	Air volume (cfm) at indicated static pressure (inWC)					
		1	2	3	4	5	6
FFC-520B	20	8,300	7,500	6,600	5,100	2,000	—
FFC-1024B	24	13,600	13,000	12,000	11,000	10,000	8,500
FFC-1524B	24	16,000	15,100	14,200	13,100	11,800	10,000
FFC-2027B	27	18,800	17,900	17,000	16,000	15,000	13,700

Example 1. Suppose we have a 30-ft-diameter bin with a 15-ft depth of shelled corn with a moisture content of 22 percent. From Fig. 12.11, we know that a minimum of 1 cfm per bushel is required. From Fig. 12.9 or Eq. (12.1), we find that this bin contains a total of 8520 bushels. At 1 cfm per bushel, we need 8520 cfm. From Fig. 12.12, we find that it requires a static pressure of 2.0 inWC to move 1 cfm per bushel through the 15 ft of depth. We therefore need a fan to deliver 8520 cfm against a static pressure of 2.0 inWC. From Table 12.2 or Fig. 12.13, we find that Farm Fans makes a 24-in vaneaxial fan with direct drive from a 5- to 7-hp, 3450-rpm motor that will deliver 9600 cfm against a static pressure of 2.0 inWC, about 13 percent more than required. Since the bin contains a total of 8520 bushels, we will provide 1.13 cfm per bushel. We will exceed the required minimum and are on the safe side.

Example 2. Suppose the 15 ft of corn in our 30-ft-diameter bin has a 23 percent (instead of a 22 percent) moisture content, requiring a minimum of 1.5 cfm per

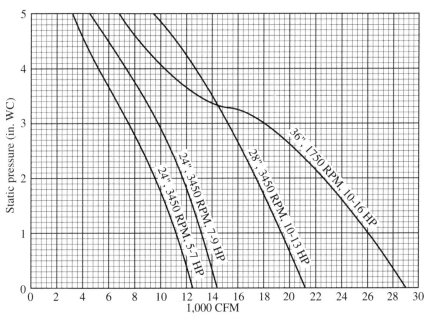

FIGURE 12.13 Performance of four sizes of vaneaxial fans. *(Courtesy of Farm Fans, Inc., Indianapolis, Ind.)*

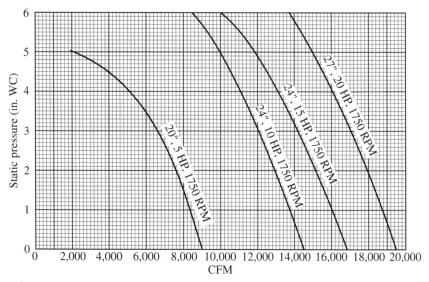

FIGURE 12.14 Performance of four sizes of BI centrifugal fans. *(Courtesy of Farm Fans, Inc., Indianapolis, Ind.)*

bushel or a total of 12,780 cfm. From Fig. 12.12, we find that this will require a higher static pressure of 3.6 inWC. From Table 12.2 or Fig. 12.13, we find that the 28-in vaneaxial fan at 3450 rpm with the 10- to 13-hp motor will deliver 13,600 cfm against a static pressure of 3.6 inWC, slightly (6 percent) more than the required minimum. From Table 12.3 or Fig. 12.14, we find that we also could select the 24-in BI centrifugal fan at 1750 rpm with a 15-hp motor. This fan would deliver 13,550 cfm against a static pressure of 3.6 inWC, almost the same as the 28-in vaneaxial fan. We probably will prefer the 24-in BI centrifugal fan because it will be quieter owing to the 1750 rpm instead of 3450 rpm. Earlier we explained that centrifugal fans are used for the high-pressure applications, and this is generally true. Here, however, we find that a pressure requirement that is too high for a vaneaxial fan is not the only condition where a centrifugal fan is preferable.

Conclusion

Bins of larger diameters require more air volume. Higher moisture content will require more air volume and also higher static pressure. Greater grain depth will require a higher static pressure as well as more air volume.

CHAPTER 13
CROSS-FLOW BLOWERS

REVIEW

In reviewing some of the past chapters, we note the following: Chaps. 4, 7, 9, and 10 discussed axial-flow fans, centrifugal fans, axial-centrifugal fans, and roof ventilators. These are the types of fans that are used in large ventilating systems, as described in Chap. 11.

If you will look at Chap. 3, which contained a short description of the different types of fans, you will notice that there are two other types of fans left: cross-flow blowers and vortex blowers. These two types cannot handle the large air volumes needed in large ventilating systems, but they are good solutions for certain special applications. We will discuss the cross-flow blowers in this chapter and the vortex blowers in Chap. 15.

FLOW PATTERN AND APPEARANCE

In the cross-flow blower, as mentioned in Chap. 3, the airflow passes twice through FC blading, first inward and then outward, as shown in Fig. 3.29. One might question how this is possible, and if it is possible, why the same thing does not happen in regular FC fan wheels having about the same blading. In order to explain this seeming inconsistency of inward and outward flow, let us analyze the pattern of airflow through a cross-flow blower, as shown in Fig. 13.1. We start with the outward airflow.

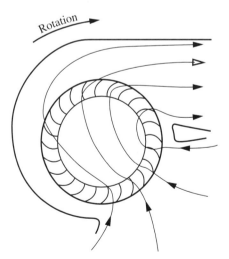

FIGURE 13.1 Schematic sketch of a cross-flow blower showing the airflow lines through the unit.

13.1

The air stream leaves the blade tips and flows through the scroll housing outlet, the same as in a regular FC fan unit. This produces a negative static pressure inside the fan wheel. This negative static pressure requires replacement of the air leaving the housing outlet. In a regular FC unit, with single or double inlet, this air is replaced by the airflow entering axially through the inlets on one or both sides. In the cross-flow blower, however, these inlets are closed. The negative static pressure inside the fan wheel, therefore, tries to pull in some air from another inlet. This other inlet is provided by an opening in the housing scroll, as shown in Fig. 13.1. This opening is less than 90° apart from the housing outlet, and it extends all the way across the unit, the same as the housing outlet. Since the FC blades are forwardly inclined, they help push the air stream inward, even though centrifugal force does not support them in this endeavor (in fact, it tries to prevent it), but the negative pressure inside does.

This simple explanation can be further illustrated by a detailed description of the flow lines across the fan wheel from the inward flow to the outward flow, as shown in Fig. 13.1. These flow lines are not straight, but they are curved as shown. The airflow enters the housing inlet and the fan wheel radially inward and then is deflected by the blades in the direction of the fan rotation. Next, the curved flow lines pass across the inside of the fan wheel, toward some blades near the housing outlet, where they are again deflected in the direction of the fan rotation and leave the unit through the housing outlet.

The diameter ratio d_1/d_2 is between 0.7 and 0.8, compared with 0.75 to 0.90 for regular FC wheels. In other words, the cross-flow blower blades tend to extend farther inward.

Figure 13.2 shows a typical cross-flow blower. You will notice the large inlet area located close to the outlet area, giving the airflow ample space (more than 180°) to pass through the fan wheel.

Figure 13.3 presents a schematic sketch, giving the principal dimensions for a typical cross-flow blower with a 3-in wheel outside diameter and 28 FC blades. The dimensions for other sizes will be in geometric proportion.

FIGURE 13.2   Cross-flow blower showing the closed housing ends and the large housing inlet at the blade tips.

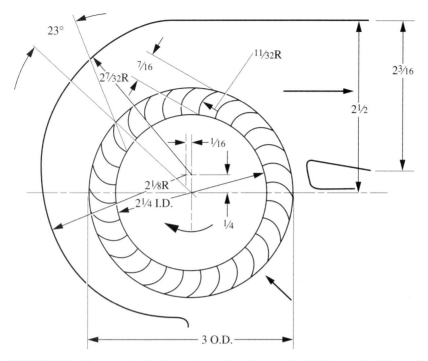

FIGURE 13.3 Schematic sketch of typical cross-flow blower with 28 blades and a 3-in outside diameter.

PERFORMANCE OF CROSS-FLOW BLOWERS

Figure 13.4 shows in solid lines the performance for this 3-in cross-flow blower with an 11-in blade width. For comparison, the performance for a regular 3-in FC centrifugal fan is plotted in dashed lines. This comparison indicates the following:

1. The maximum static pressure produced is the same for both fans. (Some manufacturers claim a higher static pressure for the cross-flow blower.)
2. The cross-flow blower delivers more air volume because of its larger blade width, but the air volume per inch of blade width is considerably smaller for the cross-flow blower.
3. The maximum efficiency is much lower for the cross-flow blower. This is due to the airflow passing through the blading twice, with each pass producing abrupt changes in the direction of the airflow, resulting in turbulence.

ADVANTAGES AND DISADVANTAGES

Summarizing, we can say that the cross-flow blower has the following two advantages:

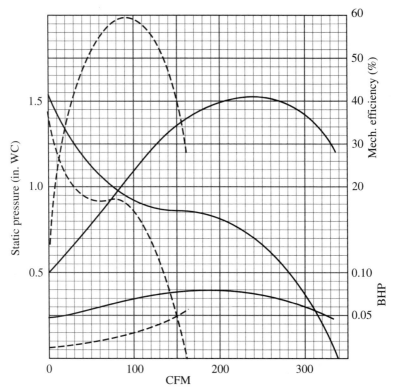

FIGURE 13.4 Comparison of performances for two units with 3-in wheel diameters running at 3500 rpm (*solid lines:* cross-flow blower per Fig. 13.3, 11-in blade width; *dashes:* typical FC centrifugal fan, 1.8-in blade width).

1. A long and narrow outlet area, which can be a desirable feature for such applications as air curtains, dry blowers in a car wash, hand dryers, electric heaters, electronic cabinets, room air conditioners, automotive air conditioners, and copy machines

2. A relatively large air volume, since the axial width b is limited only by structural considerations (not by aerodynamic considerations), compared with FC wheels, where the maximum blade width is limited (by aerodynamic considerations) to $b_{max} = 0.60D$, and with BI wheels, where it is limited to $b_{max} = 0.46D$

On the other hand, the cross-flow blower has the following four disadvantages:

1. For most ventilating applications, the geometric configuration is not suitable.

2. The efficiency is low. The maximum efficiency usually is between 35 and 45 percent.

3. The structural strength is low; thus the maximum permissible tip speed is limited.

4. The manufacturing cost is relatively high due to difficulties in assembly and balancing.

CHAPTER 14
FLOW COEFFICIENT AND PRESSURE COEFFICIENT

REVIEW

The fans we have discussed thus far can be divided into three basic groups: axial-flow fans, centrifugal fans (including axial-centrifugal fans), and cross-flow blowers. Roof ventilators are not basically different; they also can be classified as either axial-flow or centrifugal fans.

TWO METHODS FOR COMPARING PERFORMANCE

If we want to compare the performance of different types of fans with regard to air volume, static pressure, and so on, we can convert their performances to the same wheel diameter and speed and plot the resulting performance curves in a combined graph. This is a simple and perfectly acceptable method of comparison, and we have used this method in several graphs so far in this book.

Another method that is sometimes used for comparing performances of different fan types is to derive two dimensionless coefficients from the air volume and static pressure at the point of maximum efficiency and for standard air density, one for flow and one for pressure. Greek letters are used as symbols for these two coefficients: ϕ for the flow coefficient and ψ for the pressure coefficient. These two coefficients preferably should be functions of only four variables: the wheel diameter D in inches, the fan speed in revolutions per minute, the flow rate in cubic feet per minute, and the static pressure in inches of water column.

Both these methods are accurate, except for minor deviations due to the size effect, mentioned on page 5.5.

Pressure Coefficient

The pressure coefficient ψ can be calculated from the following simple formula:

$$\psi = \frac{2.39 \times 10^8 \times SP}{D^2 \times \text{rpm}^2} \qquad (14.1)$$

This formula actually is a function of only D, speed (rpm), and static pressure, as desired, and it can be used for any type of fan.

Flow Coefficient

Determination of the flow coefficient ϕ is more complicated. For centrifugal fans, ϕ depends only on D, speed (rpm), and air volume (cfm), but for axial-flow fans, ϕ depends also on the hub-tip ratio d/D, and for cross-flow blowers, ϕ depends also on the blade width b. This means that three different formulas are needed for the three different fan types. Obviously, this makes the comparison more difficult than in the simple method of conversion to equal size and speed.

For axial-flow fans, the formula for the flow-coefficient ϕ is

$$\phi = K \frac{\text{cfm}}{D^3 \times \text{rpm}} \tag{14.2}$$

where the constant K depends on the hub-tip ratio d/D. This is shown in Fig. 14.1. As the hub-tip ratio increases, so does the constant K and the slope of the curve.

For centrifugal fans, the formula for the flow coefficient ϕ is simply

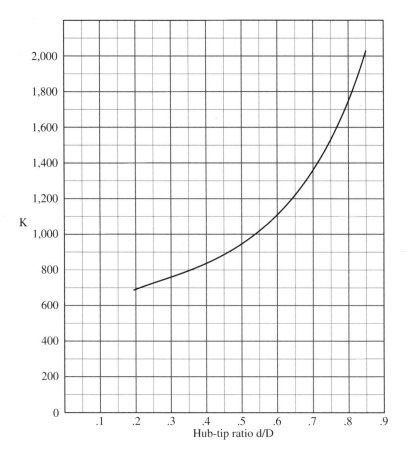

FIGURE 14.1 Graph showing the constant K versus the hub-tip ratio d/D for axial-flow fans.

$$\phi = \frac{694 \times \text{cfm}}{D^3 \times \text{rpm}} \tag{14.3}$$

For cross-flow blowers, the formula for the flow coefficient ϕ is

$$\phi = \frac{550 \times \text{cfm}}{D^2 \times b \times \text{rpm}} \tag{14.4}$$

where b is the long axial width of the blades.

The coefficients for flow and pressure are a representation of the fan design, regardless of size and speed. If the performance is converted by the fan laws to a different size and speed, the coefficients ϕ and ψ will remain the same.

Examples

Let us calculate the flow coefficient ϕ and the pressure coefficient ψ for some of the fans we have discussed in previous chapters.

Example 1. We start with Fig. 6.6, showing the performances for three fans *a, b,* and *c.*

(*a*) Let us first analyze the performance of the lower-pressure 29-in vaneaxial fan at 1750 rpm with a hub-tip ratio d/D of 0.52. At the point of maximum efficiency, this fan delivers 10,880 cfm at a static pressure of 2.1 inWC. From Fig. 14.1, for d/D = 0.52, we get a constant K of 955. The flow coefficient then will be

$$\phi = 955 \, \frac{10,880}{29^3 \times 1750} = 0.243$$

and the pressure coefficient will be

$$\psi = \frac{2.39 \times 10^8 \times 2.1}{29^2 \times 1750^2} = 0.195$$

(*b*) The other 29-in vaneaxial fan (for the higher *SP*) also runs at 1750 rpm, but the hub-tip ratio is 0.68. At the point of maximum efficiency, it delivers 8600 cfm at a static pressure of 3.4 inWC. For $d/D = 0.68$, we get a constant K of 1270 and a flow coefficient of

$$\phi = 1270 \, \frac{8600}{29^3 \times 1750} = 0.256$$

The pressure coefficient will be

$$\psi = \frac{2.39 \times 10^8 \times 3.4}{29^2 \times 1750^2} = 0.316$$

Note that the pressure coefficient for the second fan is larger than that for the first fan due to the larger static pressure of the second fan. Note also that the flow coefficient for the second fan is slightly larger than that for the first fan, even though the air volume for the second fan is smaller. The reason for this is that the larger hub-tip ratio of the second fan results in a larger constant K. In other words, the flow coefficient ϕ is not simply a measure of the air volume but is a measure of the axial air velocity through the annular area between hub and tip. Thus, for a larger hub-tip ratio, the annular area becomes smaller, the air velocity becomes larger, and this tends to make ϕ larger, even if the air volume is somewhat smaller.

(c) Next, we look at the third curve in Fig. 6.6, showing the performance of a 15-in mixed-flow (axial-centrifugal) fan at 3500 rpm. At the point of maximum efficiency, this fan delivers 3400 cfm at a static pressure of 6 inWC. The flow coefficient will be

$$\phi = \frac{694 \times 3400}{15^3 \times 3500} = 0.200$$

and the pressure coefficient will be

$$\psi = \frac{2.39 \times 10^8 \times 6}{15^2 \times 3500^2} = 0.520$$

Note that the pressure coefficient ψ of the axial-centrifugal fan is much larger than that for either of the two vaneaxial fans, even though all three fans have approximately the same tip speed. This could be expected, since centrifugal fans (even axial-centrifugal fans) will produce higher static pressures, even at the same tip speed.

Note also that the flow coefficient of the axial-centrifugal fan is somewhat smaller than that for either of the two vaneaxial fans, but it is not as much smaller as could be expected in view of the much smaller air volume (only 3400 cfm instead of 8600 or 10,880 cfm). This small, but larger than expected, ϕ for the axial-centrifugal fan is caused by the much smaller size (only 15 in instead of 29 in). The purpose of Fig. 6.6 was not to compare fans of equal size and speed but to compare vaneaxial fans with an axial-centrifugal fan of about equal tip speed.

Example 2.   Now we look at Fig. 7.30 showing four static pressure curves for centrifugal fans, all converted to 27-in wheel diameters and running at 1140 or 570 rpm. Let's select a point on each curve that is close to the maximum efficiency and calculate the flow and pressure coefficients for each of these four points.

(a) The FC fan at 1140 rpm delivers 24,000 cfm at a static pressure of 7.9 inWC. The flow coefficient will be

$$\phi = \frac{694 \times 24,000}{27^3 \times 1140} = 0.742$$

and the pressure coefficient will be

$$\psi = \frac{2.39 \times 10^8 \times 7.9}{27^2 \times 1140^2} = 1.99$$

(b) The FC fan at 570 rpm delivers 12,000 cfm at a static pressure of 1.98 inWC. The flow coefficient will be

$$\phi = \frac{694 \times 12,000}{27^3 \times 570} = 0.742$$

and the pressure coefficient will be

$$\psi = \frac{2.39 \times 10^8 \times 1.98}{27^2 \times 570^2} = 2.00$$

As mentioned previously, the two coefficients for a fan will remain the same, regardless of the speed.

(c) The BI fan at 1140 rpm delivers 6000 cfm at a static pressure of 3.9 inWC. The flow coefficient will be

$$\phi = \frac{694 \times 6000}{27^3 \times 1140} = 0.186$$

and the pressure coefficient will be

$$\psi = \frac{2.39 \times 10^8 \times 3.9}{27^2 \times 1140^2} = 0.984$$

Note that for the BI fan, both coefficients are much smaller than for the FC fan (but, as mentioned previously, the efficiency is higher for the BI fan).

(d) The RT fan at 1140 rpm delivers 7000 cfm at a static pressure of 4.8 inWC. The flow coefficient will be

$$\phi = \frac{694 \times 7000}{27^3 \times 1140} = 0.217$$

and the pressure coefficient will be

$$\psi = \frac{2.39 \times 10^8 \times 4.8}{27^2 \times 1140^2} = 1.211$$

Note that for the RT fan both coefficients are larger than for the BI fan but still far below those of the FC fan.

Example 3. We now look at Fig. 9.4, showing three static pressure curves, one for a vaneaxial fan and two for centrifugal fans, all converted to 27 in and 1750 rpm. Again, we will select a point close to the maximum efficiency on each curve and calculate the flow and pressure coefficients for each of these three points.

(a) The 27-in vaneaxial fan at 1750 rpm delivers 11,500 cfm at a static pressure of 2.56 inWC. With a hub-tip ratio of 0.618, we get a constant K of 1120 and a flow coefficient

$$\phi = \frac{1120 \times 11,500}{27^3 \times 1750} = 0.374$$

The pressure coefficient will be

$$\psi = \frac{2.39 \times 10^8 \times 2.56}{27^2 \times 1750^2} = 0.274$$

(b) The 27-in axial-centrifugal fan at 1750 rpm delivers 11,000 cfm at a static pressure of 6.02 inWC. The flow coefficient will be

$$\phi = \frac{694 \times 11,000}{27^3 \times 1750} = 0.222$$

and the pressure coefficient will be

$$\psi = \frac{2.39 \times 10^8 \times 6.02}{27^2 \times 1750^2} = 0.644$$

(c) The 27-in airfoil centrifugal fan at 1750 rpm delivers 11,000 cfm at a static pressure of 7.4 inWC. The flow coefficient will be

$$\phi = \frac{694 \times 11,000}{27^3 \times 1750} = 0.222$$

and the pressure coefficient will be

$$\psi = \frac{2.39 \times 10^8 \times 7.4}{27^2 \times 1750^2} = 0.792$$

Example 4. We now look at Fig. 13.4, showing two sets of performance curves, one for a 3-in FC centrifugal fan and one for a 3-in cross-flow blower, both running at 3500 rpm. The FC fan has a blade width of 1.8 in, and the cross-flow blower has a considerably larger blade width of 11 in

(a) At the point of maximum efficiency, the FC fan delivers 94 cfm at a static pressure of 0.90 inWC. The flow coefficient will be

$$\phi = \frac{694 \times 94}{3^3 \times 3500} = 0.690$$

and the pressure coefficient will be

$$\psi = \frac{2.39 \times 10^8 \times 0.90}{3^2 \times 3500^2} = 1.951$$

(b) At the point of maximum efficiency, the cross-flow blower delivers 240 cfm at a static pressure of 0.69 inWC. The flow coefficient will be

$$\phi = \frac{550 \times 240}{3^2 \times 11 \times 3500} = 0.381$$

and the pressure coefficient will be

$$\psi = \frac{2.39 \times 10^8 \times 0.69}{3^2 \times 3500^2} = 1.50$$

Note that the two coefficients for the FC fan are close to the ones we obtained for the FC fans in Example 2. Note also that the pressure coefficient for the cross-flow blower is slightly lower than that for the FC fan. The flow coefficient for the cross-flow blower is smaller than that for the FC fan, despite the much larger blade width because, as mentioned previously, the air volume per inch of blade width is smaller for the cross-flow blower.

COMPARISON OF COEFFICIENTS FOR VARIOUS FAN TYPES

Table 14.1 shows a comparison of the various coefficients we obtained in the preceding examples. Note that the largest coefficients are obtained with FC fans (at the expense of lower efficiency). The lowest pressure coefficients are for vaneaxial fans

TABLE 14.1 Comparison of Flow and Pressure Coefficients
for Various Types of Fans

Example no.	Type of fan	Hub-tip ratio	Blade width (in)	Flow coefficient ϕ	Pressure coefficient ψ
1a	VAF	0.52		0.243	0.195
1b	VAF	0.68		0.256	0.316
1c	Axial-centrifugal			0.200	0.520
2a	FC centrifugal			0.742	1.99
2b	FC centrifugal			0.742	2.00
2c	BI centrifugal			0.186	0.984
2d	RT centrifugal			0.217	1.211
3a	VAF	0.62		0.374	0.274
3b	Axial-centrifugal			0.222	0.644
3c	Airfoil centrifugal			0.222	0.792
4a	FC centrifugal			0.690	1.95
4b	Cross-flow		11	0.381	1.50

(which have high efficiencies). The cross-flow blower has the second highest pressure coefficient and the RT centrifugal is third.

PLOTTING ψ VERSUS ϕ INSTEAD OF STATIC PRESSURE VERSUS AIR VOLUME

Review

Table 14.1 listed the coefficients ϕ and ψ that we obtained at the point of maximum efficiency for various types of fans, such as vaneaxial fans, centrifugal fans (BI, RT, FC), axial-centrifugal fans, and cross-flow blowers. By comparing the values of these coefficients, we can judge the relative potentials of these fan types regarding air volume and static pressure. Going one step further, we can get an even better insight if we calculate ϕ and ψ not only for the point of maximum efficiency but also for a number of points on the performance curves and then plot ψ versus ϕ instead of static pressure versus air volume. This has been done in Fig. 14.2, which was derived from Fig. 6.6.

Comparison of the Two Graphs

In comparing the two graphs of Figs. 6.6 and 14.2, we find some similarities and some dissimilarities between them. Let's analyze them in detail. As in Table 14.1, we will call the 52 percent vaneaxial fan *fan 1a*, the 68 percent vaneaxial fan *fan 1b*, and the axial-centrifugal (mixed-flow) fan *fan 1c*. Here are two similarities:

1. The three fans have maximum pressure coefficients that are almost in proportion to the maximum pressures.

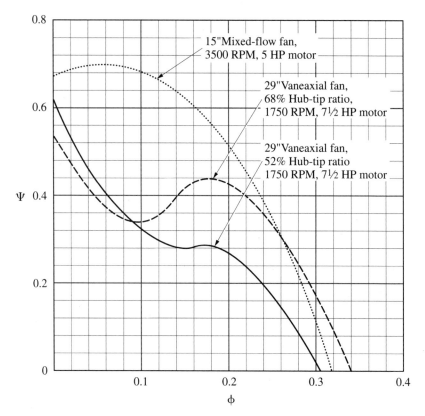

FIGURE 14.2   Pressure coefficient ψ versus flow coefficient ϕ for the three fans of Fig. 6.6.

2. The shapes of the pressure-coefficient curves are quite similar to the shapes of the pressure curves.

Here are four dissimilarities:

1. While the air volumes at free delivery are spread out over a wide range, the flow coefficients at free delivery are close to each other.

2. The air volumes at free delivery follow the sequence *c, b,* and *a* (mixed-flow fan smallest, 68 percent VAF in the middle, 52 percent VAF largest). For the flow coefficients, however, the sequence is *a, c,* and *b* (52 percent VAF smallest, mixed-flow fan in the middle, 68 percent VAF largest).

3. For the two vaneaxial fans, the two pressure curves intersect in their operating ranges, but the two pressure-coefficient curves do not intersect in the operating range. The 68 percent vaneaxial fan has larger ϕ values than the 52 percent vaneaxial fan in the entire operating range, and even quite a bit beyond, due to the fact that the ϕ values are a measure of the axial air velocity through the annular area, as mentioned previously.

4. The pressure curve of the mixed-flow fan intersects both vaneaxial fans in the stalling range, but the pressure-coefficient curve of the mixed-flow fan intersects the 68 percent VAF in the operating range; it intersects the 52 percent VAF nowhere, since it is above it over the entire range.

Summary

Plotting static pressure versus air volume provides the performance for a specific fan design at a certain size and speed. Plotting ψ versus ϕ is more general. It provides the performance for a line of fans of a specific design. In other words, as long as the different sizes of a line are in geometric proportion, the graph of ψ versus ϕ will apply to all sizes of the line (with some minor inaccuracies due to the size effect).

If a graph of ψ versus ϕ, such as shown in Fig. 14.2, is given for the line, we can derive the performance for any size and speed simply by solving Eqs. (14.1) through (14.4) for air volume or static pressure and inserting the desired values for D and speed for a number of points. These equations then will become

$$SP = \frac{D^2 \times \text{rpm}^2 \times \psi}{2.39 \times 10^8} \tag{14.1a}$$

$$\text{cfm} = \frac{D^3 \times \text{rpm} \times \phi}{K} \tag{14.2a}$$

$$\text{cfm} = \frac{D^3 \times \text{rpm} \times \phi}{694} \tag{14.3a}$$

$$\text{cfm} = \frac{D^2 \times b \times \text{rpm} \times \phi}{550} \tag{14.4a}$$

CHAPTER 15
VORTEX BLOWERS

REVIEW

Chapter 3 presented some short descriptions of the different types of fans. Chapter 13 pointed out that among the different types of fans, there are two special types that cannot handle the large air volumes needed in large ventilating systems but which are good solutions for certain special applications. These two special types are the cross-flow blower and the vortex blower. The cross-flow blower was discussed in detail in Chap. 13. The vortex blower will be discussed in this chapter.

FLOW PATTERN AND APPEARANCE

In the vortex blower, as illustrated in Fig. 3.30, the air enters through an inlet pipe and leaves about 330° later through an outlet pipe, with a separator baffle between inlet and outlet to prevent a short circuit. During these 330°, the air circles around in a torus-shaped space. On one side of the torus are the rotating fan blades, throwing the air outward, as is normally done in a centrifugal fan. The other side of the torus guides the airflow back inward so that it must reenter at the leading edges of the rotating blades. In this way, the air passes through the rotating fan blade section 8 to 16 times for each revolution. As a result, the pressure produced is very high, as if it were a multistage centrifugal fan, but the unit is considerably more compact than a multistage centrifugal fan. This, of course, is a great advantage in many applications.

The disadvantage of the vortex blower is its low efficiency (about 30 percent maximum), caused by much turbulence as the airflow rotates in the torus. Because of the low efficiency, a large motor horsepower is required. The vortex blower, therefore, is used in applications where compactness is more important than fan efficiency.

PRINCIPAL DIMENSIONS
OF THE VORTEX BLOWER

The ratio of the torus diameter to the wheel diameter usually is between 0.22 and 0.30. The inlet and outlet pipes have an inside diameter about one-half the torus diameter. The number of blades can be from 18 to 42, depending on the shape of the blades. Some designs use axial-radial blades; this requires many blades. The blades, however, can be tilted (as in an axial-flow fan) to deflect the air toward the other side of the torus; this will be more expensive in production, but the number of blades then can be reduced.

PERFORMANCE OF THE VORTEX BLOWER

Figure 15.1 shows the performance of a typical vortex blower. Note the following four features: low air volume, high static pressure, nonoverloading BHP curve, and low mechanical efficiency.

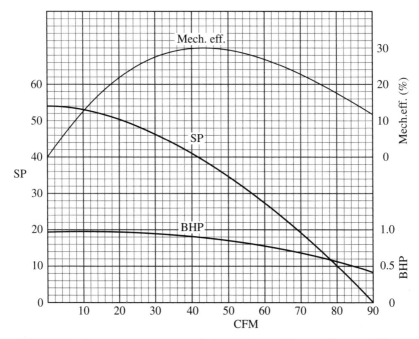

FIGURE 15.1   Performance curves for a typical vortex blower, 11-in wheel diameter, 3500 rpm.

In view of the high static pressure, it would be interesting to compare this performance with that of another high-pressure fan. Thus we select the four-stage turbo blower of Fig. 7.42 and convert its performance curves to 11-in wheel diameter and 3500 rpm, the same as for the vortex blower. This comparison is shown in Fig. 15.2. In order to get it on a graph, we had to compress the air volume scale. This comparison indicates the following:

1. The vortex blower delivers less air volume but produces a higher static pressure than a four-stage turbo blower of equal wheel diameter and speed.

2. The vortex blower will only need a 1-hp, 3500-rpm motor, while the four-stage turbo blower will require at least a 2-hp, 3500-rpm motor due to the much larger air volume and an overloading BHP curve.

3. The vortex blower has a maximum efficiency of 30 percent, compared with 52 percent for the four-stage turbo blower.

4. The vortex blower obviously will be more compact than the four-stage turbo blower.

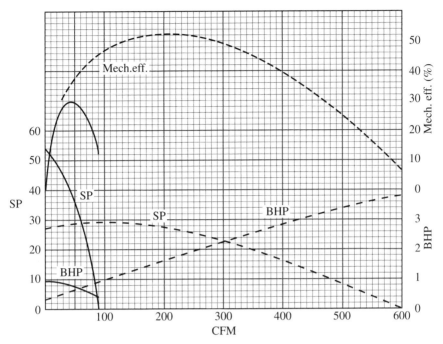

FIGURE 15.2 Comparison of performance curves for two fans, 11-in wheel diameter, 3500 rpm (*solid lines:* vortex blower of Fig. 15.2; *dashes:* four-stage turbo blower of Fig. 7.42 converted to 11-in wheel diameter and 3500 rpm).

These are our observations in comparing the vortex blower with a turbo blower. We should mention, however, that the vortex blower is sometimes compared with positive-displacement compressors, such as reciprocating pistons, rotary vanes, lobed rotors, and screw types. In such comparisons, the vortex blower delivers more air volume, produces less static pressure, and consumes less BHP. You might say that the performance of the vortex blower is between that of the multistage turbo blower and that of the positive-displacement compressor.

Please note that both Figs. 15.1 and 15.2 show the performance for blowing, i.e., with all the resistance on the outlet side. For exhausting, the negative suction pressure will be somewhat lower, in accordance with Eq. (7.19) and Fig. 7.46. For example, the maximum static pressure of 54 inWC for the vortex blower will become 47.7 inWC for suction.

NOISE LEVEL OF THE VORTEX BLOWER

Due to the excessive turbulence, the vortex blower produces a high noise level, which, however, is somewhat reduced by sound-attenuating sleeves at the inlet and outlet. In addition to these internal mufflers that are part of the unit, some external mufflers are sometimes used when extraquiet operation is desired.

APPLICATIONS OF THE VORTEX BLOWER

Here are some typical applications for the vortex fan, some for blowing and some for exhausting:

Aeration and agitation of fluids in tanks
Waste-water treatment
Pneumatic conveying
Sand blasting
Dust collection
Chemical processing
Vacuum hold-down
High-velocity air knives
Combustion air

CHAPTER 16

VARIOUS METHODS TO DRIVE A FAN

PRIME MOVERS

There are four types of prime movers for driving a fan:

1. Electric motors (the most commonly used)
2. Engines (see Fig. 7.41)
3. Turbines (compressed air or steam)
4. Compressed-air jets (see Fig. 4.61)

TYPES OF MOTOR DRIVES

There are three ways that can be used for an electric motor to drive a fan:

1. Belt drive
2. Direct drive
3. Gear drive

Here are the pros and cons of these three methods:

1. Belt drive is prevalent in large sizes, for two reasons:
   a. Flexibility: Any speed can be obtained (while direct drive is confined to a few speeds, such as 3450, 1740, 1150, and 860 rpm).
   b. Expensive low-speed motors are avoided.
2. Direct drive is prevalent in small sizes, for four reasons:
   a. Small sizes normally run at higher speeds, so there is no need for expensive low-speed motors.
   b. Direct drive results in lower cost, since no extra shafts, bearings, bearing supports, and sheaves are needed.
   c. Direct drive results in somewhat better fan efficiencies, since belt losses are avoided.
   d. Direct drive requires minimum maintenance.
3. Gear drive is used for very high fan speeds that would result in excessive belt slip and belt losses.

TYPES OF ELECTRIC MOTORS USED TO DRIVE FANS

There are seven types of electric motors used to drive fans. The first two types are for three-phase power; the other five types are for single-phase power.

1. Three-phase squirrel-cage motors are used most of the time whenever the fan requires 1 hp or more. These are the most frequently used fan motors. They have the advantages of high efficiency and low cost. Sometimes they are built as two-speed motors. The second speed requires a separate winding. It can be 50 or 67 percent of the top speed.

2. Three-phase wound-rotor motors are sometimes used for adjustable-speed arrangements, but they are more expensive.

3. Single-phase, split-phase induction motors, with an auxiliary winding for starting only, are used for requirements up to ½ hp. They are somewhat less efficient but the best we can get for single-phase power. They have the disadvantage of a high starting current.

4. Single-phase, permanent-split-capacitor motors are used for requirements up to ⅓ hp. They use a capacitor instead of the auxiliary winding to start the motor. They are less efficient but have the advantage of a low starting current. They are often used for direct drive.

5. Single-phase, capacitor-start and capacitor-run motors are used for requirements up to ¾ hp. They have an auxiliary winding with a capacitor in series, resulting in a high starting torque for heavy wheels.

6. Single-phase, shaded-pole motors are used for small fans up to ⅙ hp. They have the advantage of low cost but the disadvantages of low efficiency (about 30 percent) and of high slip, resulting in a high starting current. At full load, they run at 1550 rpm (instead of 1740 rpm) and at 1040 rpm (instead of 1150 rpm).

7. Single-phase universal motors are used for requirements up to 1 hp, for vacuum cleaners and other high-speed applications. They have the advantage of high speed with direct drive but the disadvantages of slightly lower efficiencies and of commutator brushes that have to be adjusted and replaced at times.

8. Single-phase, inside-out induction motors are different in that the rotating part is on the outside instead of on the inside. These motors occasionally are used in propeller fans and in axial-centrifugal fans.

SUMMARY

Summarizing, we can say

1. Most large fans have belt drive from three-phase electric motors.

2. Most small fans have direct drive from single-phase electric motors.

3. Multistage turbo blowers sometimes have belt drive from electric motors or from engines.

4. Single-stage turbo blowers, running at very high speeds, have gear drive from electric motors.

CHAPTER 17
FANLESS AIR MOVERS

PRINCIPLE OF OPERATION

In Chap. 1 we discussed Bernoulli's theorem, which in its simplest form says: When the velocity of a fluid (such as water or air) increases (as in a converging cone), the static pressure will decrease. A well-known application of this phenomenon is in the water-stream air pump, as shown in Fig. 17.1. Here, a stream of water passes through the converging end of a pipe and leaves it at an accelerated velocity, thereby creating a negative static pressure nearby, i.e., a pressure smaller than atmospheric pressure. This in turn draws in some air through an inlet pipe.

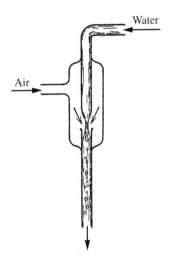

Water

Air

FIGURE 17.1 Schematic sketch of a water-stream air pump drawing in some air through the inlet pipe.

The same principle is used in the fanless air mover, as shown in Figs. 17.2 and 17.3. Here, compressed air takes the place of the water. It first enters an annular chamber on the outside of a nozzle that is venturi-shaped on the inside. The compressed air then penetrates into the inside of the nozzle through 12 small holes, producing 12 high-velocity jets that create a negative static pressure inside the nozzle. This in turn induces the main air stream from the inlet side of the nozzle. A long outlet diffuser improves the flow by gradually slowing down the discharge velocity of the combined airflow consisting of a mixture of compressed air and induced air.

The compressed-air jets penetrate the inside of the nozzle at an angle of about 25° to the axis. The diameters of the jet holes vary from about $\frac{1}{16}$ in (for a 3-in i.d. nozzle) to about $\frac{1}{8}$ in (for a 9-in i.d. nozzle). The volume of the induced air is 30 to 40 times the volume of the compressed air. This is called the *induction ratio*. This ratio can be compared to a fan efficiency, but it is not an efficiency in percentage; it is a volume ratio.

PERFORMANCE

The fanless air mover can be tested like a fan, and it performs like a fan. The air volume delivered gradually decreases as the static pressure increases. Figure 17.4 shows the performance of a 6-in air mover at 80 lb/in² of compressed-air pressure. It is interesting to note the considerable static pressure developed by the unit.

FIGURE 17.2 Fanless air mover with 6-in i.d. inlet nozzle and long outlet diffuser. *(Courtesy of Texas Pneumatic Tools, Inc., Reagan, Texas.)*

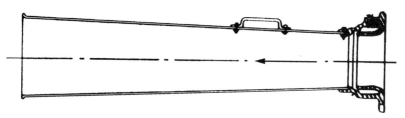

FIGURE 17.3 Schematic sketch of a typical fanless air mover showing the compressed-air chamber and the angular jet holes. *(Courtesy of Texas Pneumatic Tools, Inc., Reagan, Texas.)*

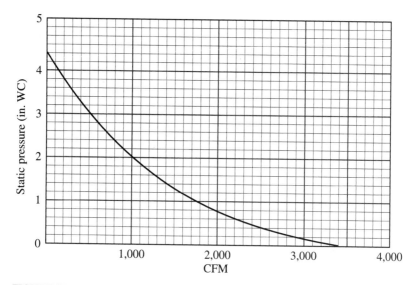

FIGURE 17.4 Performance of a typical 6-in fanless air mover.

CHAPTER 18

PERFORMANCE TESTING OF FANS

The performance of a fan can be presented in a curve sheet in which all or some of the following seven variables are plotted against air volume (cfm): static pressure *SP*, total pressure *TP*, brake horsepower bhp, motor input (kW), mechanical efficiency *ME*, static efficiency *SE,* and sound level (dB). The size and speed of the fan usually are noted on top. Many such curve sheets were already shown.

These performance curves may be the result of tests (performed in a laboratory) or they may be predicted performance curves, derived from test performance curves on some more or less similar fan units. The laboratory tests should be run in accordance with a test code. Such test codes have been published by two American societies (as well as by some foreign organizations). One test code has been published jointly by the American Society of Mechanical Engineers (ASME) and by the American National Standards Institute (ANSI). This code covers primarily the testing of high-pressure units, sometimes referred to as *blowers* and *compressors*. The other test code has been published jointly by Air Movement and Control Association (AMCA) and American Society of Heating, Refrigeration and Air Conditioning Engineers (ASHRAE). It is known as the *AMCA Test Code* and is accepted and used by most American fan manufacturers.

DESCRIPTION OF THE AMCA TEST CODE

The *AMCA Test Code,* publication AMCA 210-85, is a 60-page booklet, entitled *Laboratory Methods of Testing Fans for Rating.* The booklet starts with definitions, symbols, units of measurement, description of instruments, methods, and setups, calculations and formulas, tables and charts for air density and other quantities, sketches of pitot tubes and pitot tube traverses, and dimensions of straighteners, transformation pieces, and nozzles for air volume measurement.

Next, the booklet shows schematic drawings of 10 different setups for testing fans: on outlet ducts, on inlet ducts, on outlet chambers, on inlet chambers, and on some combinations of ducts and chambers. This probably is the most important part of the *AMCA Test Code,* since it describes the various setups for testing a fan.

Also given are a sample graph showing typical fan performance curves and some appendices about metric units, derivations, fan laws, uncertainties, and references.

VARIOUS LABORATORY TEST SETUPS

This section will reproduce and discuss four of the ten test setups shown in the *AMCA Test Code.* Figure 18.1 shows the setup with an outlet duct, i.e., of a fan blowing into a round duct; Figure 18.2a shows the setup with an inlet duct; Figure 18.2b shows the setup with an outlet chamber; and Figure 18.3 shows the setup with an

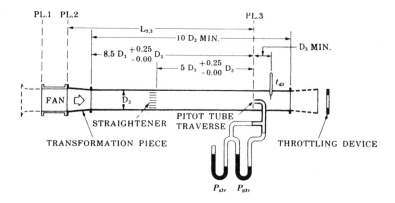

FLOW AND PRESSURE FORMULAE

$$P_{v3} = \left(\frac{\Sigma \sqrt{P_{v3r}}}{n}\right)^2 \qquad P_v = P_{v3}\left(\frac{A_3}{A_2}\right)^2\left(\frac{\rho_2}{\rho_3}\right)$$

$$V_3 = 1096 \sqrt{\frac{P_{v3}}{\rho_3}} \qquad P_{t1} = 0$$

$$Q_3 = V_3 A_3 \qquad P_{t2} = P_{s3} + P_{v3} + f\left(\frac{L_{2,3}}{D_{h3}} + \frac{L_e}{D_{h3}}\right)P_{v3}$$

$$Q = Q_3\left(\frac{\rho_3}{\rho}\right) \qquad P_t = P_{t2} - P_{t1}$$

$$P_{s3} = \frac{\Sigma P_{s3r}}{n} \qquad P_s = P_t - P_v$$

NOTES

1. Dotted lines on fan inlet indicate an inlet bell and one equivalent duct diameter which may be used for inlet duct simulation. The duct friction shall not be considered.

2. Dotted lines on the outlet indicate a diffuser cone which may be used to approach more nearly free delivery.

FIGURE 18.1 Outlet duct setup per AMCA showing test fan blowing into transformation piece and into test duct with flow straightener, pitot tube traverse, manometers reading static pressure and velocity pressure, thermometer, and throttling device. Note the optional inlet bell (or equivalent short inlet duct) and optional outlet diffuser.

inlet chamber. Outlet ducts and outlet chambers are preferable for testing fans used primarily for blowing, such as forced-draft centrifugal fans or supply roof ventilators. Inlet ducts and inlet chambers will be used for testing exhaust fans, such as exhaust roof ventilators, induced-draft centrifugal fans, and most tubeaxial fans. Test ducts can be used when the test fan will produce sufficient pressure to overcome the friction of the test duct, as will be discussed in detail later.

Test chambers are needed for testing low-pressure fans, such as propeller fans. Figure 18.4 is a photograph of an inlet duct with a centrifugal roof ventilator exhausting from it. If this were an axial roof ventilator, it would not produce sufficient static pressure to overcome the friction of the test duct. It would have to be tested on an inlet chamber where a booster fan is provided to overcome the the friction of the screens and nozzles.

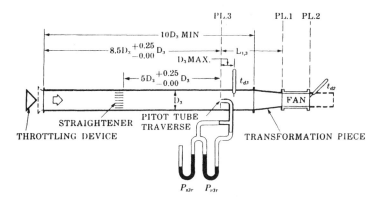

FLOW AND PRESSURE FORMULAE

$$P_{v3} = \left(\frac{\Sigma\sqrt{P_{v3r}}}{n}\right)^2$$

$$V_3 = 1096\sqrt{\frac{P_{v3}}{\rho_3}}$$

$$Q_3 = V_3 A_3$$

$$Q = Q_3\left(\frac{\rho_3}{\rho}\right)$$

$$P_{s3} = \frac{\Sigma P_{s3r}}{n}$$

$$P_v = P_{v3}\left(\frac{A_3}{A_2}\right)^2\left(\frac{\rho_3}{\rho_2}\right)$$

$$P_{t1} = P_{s3} + P_{v3} - f\left(\frac{L_{1,3}}{D_{h3}}\right)P_{v3}$$

$$P_{t2} = P_v$$

$$P_t = P_{t2} - P_{t1}$$

$$P_s = P_t - P_v$$

NOTES
1. Dotted lines on inlet indicate an inlet bell which may be used to approach more nearly free delivery.
2. Dotted lines on fan outlet indicate a uniform duct 2 to 3 equivalent diameters long and of an area within ± 0.5% of the fan outlet area and a shape to fit the fan outlet. This may be used to simulate an outlet duct. The outlet duct friction shall not be considered.

FIGURE 18.2a   Inlet duct setup per AMCA showing test fan exhausting from test duct with throttling device, flow straightener, pitot tube traverse, manometers reading velocity pressure and negative static pressure, thermometer, and transformation piece. Note the short outlet duct for static recovery.

ACCESSORIES USED WITH OUTLET OR INLET TEST DUCTS

Figure 1.4 showed a photograph of an outlet test duct with two supports for pitot tube traverses and an orifice ring at the outlet end of the test duct. If this were an inlet test duct, the orifice ring would be fastened to the inlet end of the duct. A set of about 12 such orifice rings (with different inside diameters) can be used as a throttling device to vary the air volume and static pressure of the test fan and to produce different test points on the performance curves.

Another throttling device sometimes used is a conical structure that can be moved axially to and from the end of the test duct, thereby varying the size of the duct outlet opening for the air stream to pass through. The advantage of the conical

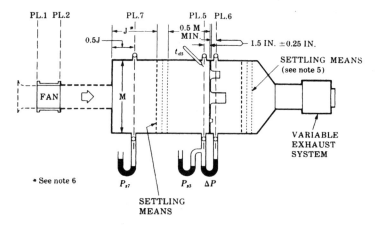

FLOW AND PRESSURE FORMULAE

$$Q_s = 1096\ Y\sqrt{\Delta P/\rho_5}\ \Sigma(CA_6)$$

$$Q = Q_s\left(\frac{\rho_s}{\rho}\right)$$

$$V_2 = \left(\frac{Q}{A_2}\right)\left(\frac{\rho}{\rho_2}\right)$$

$$P_{v2} = \left(\frac{V_2}{1096}\right)^2\rho_2$$

$$P_v = P_{v2}$$

$$P_{t1} = 0$$

$$P_{t2} = P_{s7} + P_v$$

$$P_t = P_{t2} - P_{t1}$$

$$P_s = P_t - P_v$$

NOTES

1. Dotted lines on fan inlet indicate an inlet bell and one equivalent duct diameter which may be used for inlet duct simulation. The duct friction shall not be considered.

2. Dotted lines on fan outlet indicate a uniform duct 2 to 3 equivalent diameters long and of an area within ± 0.5% of the fan outlet area and a shape to fit the fan outlet. This may be used to simulate an outlet duct. The outlet duct friction shall not be considered.

3. The fan may be tested without outlet duct in which case it shall be mounted on the end of the chamber.

4. Variable exhaust system may be an auxiliary fan or a throttling device.

5. The distance from the exit face of the largest nozzle to the downstream settling means shall be a minimum of 2.5 throat diameters of the largest nozzle.

6. Dimension J shall be at least 1.0 times the fan equivalent discharge diameter for fans with axis of rotation perpendicular to the discharge flow and at least 2.0 times the fan equivalent discharge diameter for fans with axis of rotation parallel to the discharge flow.

7. Temperature t_{d7} may be considered equal to t_{ds}.

FIGURE 18.2b Outlet chamber setup per AMCA showing test fan blowing into short duct (for static recovery) and into nozzle chamber with manometer (reading static pressure), settling means, thermometer, partition carrying multiple nozzles, manometers reading static pressure and pressure drop across the nozzles, transition cone, and booster fan exhausting from the chamber. Note the optional inlet bell (or equivalent short inlet duct).

structure is that it provides an infinite number of throttling positions. The advantage of the orifice rings is that each of the 12 throttling positions is easily reproducible for checking the test data for a test point.

Figure 18.1 shows an axial fan blowing into a test duct, but it also could be a centrifugal fan. The air stream leaving the fan outlet first passes through a transition cone (in the case of an axial fan) or through a transformation piece from rectangular

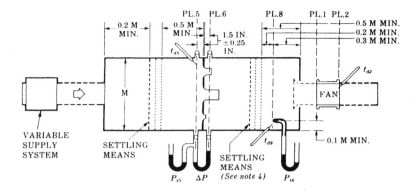

FLOW AND PRESSURE FORMULAE

$$Q_s = 1096 \, Y\sqrt{\Delta P/\rho_s} \, \Sigma(CA_6) \qquad P_v = P_{v2}$$

$$Q = Q_s \left(\frac{\rho_s}{\rho}\right) \qquad\qquad P_{t1} = P_{t8}$$

$$V_2 = \left(\frac{Q}{A_2}\right)\left(\frac{\rho}{\rho_2}\right) \qquad\qquad P_{t2} = P_v$$

$$\qquad\qquad\qquad\qquad\qquad P_t = P_{t2} - P_{t1}$$

$$P_{v2} = \left(\frac{V_2}{1096}\right)^2 \rho_2 \qquad\qquad P_s = P_t - P_v$$

NOTES

1. Dotted lines on fan inlet indicate an inlet bell and one equivalent duct diameter which may be used for inlet duct simulation. The duct friction shall not be considered.

2. Dotted lines on fan outlet indicate a uniform duct 2 to 3 equivalent diameters long and of an area within ± 0.5% of the fan outlet area and a shape to fit the fan outlet. This may be used to simulate an outlet duct. The outlet duct friction shall not be considered.

3. Variable supply system may be an auxiliary fan or throttling device.

4. The distance from the exit face of the largest nozzle to the downstream settling means shall be a minimum of 2.5 throat diameters of the largest nozzle.

FIGURE 18.3 Inlet chamber setup per AMCA showing test fan exhausting from nozzle chamber with booster fan, short connecting duct, thermometer, settling means, partition carrying multiple nozzles, manometers reading pressure drop across the nozzles and static pressure (positive near free delivery, negative near no delivery), thermometer, and pitot tube with manometer reading negative static pressure. Note the optional inlet bell (or equivalent short inlet duct) and optional short outlet duct for static recovery.

to round (in the case of a centrifugal fan) leading to a round duct. The duct area must be within plus or minus 5 percent of the fan outlet area. The test duct must be at least 10 diameters long. The transition cone or the transformation piece must have small angles with the duct axis: 7.5° will be the maximum for converging elements and 3.5° for diverging elements, as indicated in Fig. 18.5.

A number of pitot tube traverses are made to determine the air volume at each test point. These traverses are made at a location in the test duct where the turbulence of the fan outlet or of the duct inlet has been smoothed out somewhat, after the air stream has passed through 8½ diameters of ducting and through a flow

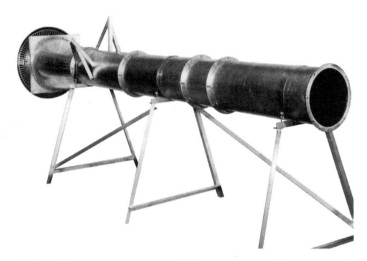

FIGURE 18.4   Inlet duct setup per AMCA showing a centrifugal roof ventilator exhausting from the duct plus a transition cone and also showing two supports for pitot tube traverses.

straightener, sometimes called an *egg-crate straightener.* Figure 18.6 shows the dimensions of the flow straightener.

Figure 18.7 shows the radial locations of the traverse points. You will notice that 3 traverses are made for each test point, with 8 points for each traverse, for a total of 24 traverse points. It might be interesting to look also at Fig. 18.8, which shows the location of the traverse points in accordance with a 1952 Test Code, which was published by the National Association of Fan Manufacturers (NAFM), the predecessor of AMCA. You will notice that the older test code specified only 2 traverses of 10 points each for a total of 20 traverse points. These traverse points represented five areas (one circle and four annular areas) of equal area.

For each traverse point, the static pressure and the velocity pressure are measured. The static pressures are almost constant across the duct, but the velocity pressures and the velocities vary in a parabolic shape, as was shown in Fig. 1.3. Since the velocities are proportional to the square roots of the velocity pressures (see Eq. 1.3), these square roots of the velocity pressures are averaged and then squared back to obtain the velocity pressure corresponding to the average velocity.

INSTRUMENTS

The following instruments are used in running a fan test:

1. A barometer is required, since the barometric pressure is needed to calculate the air density. Mercury barometers will give the most accurate readings but have to be corrected for temperature. Aneroid barometers are less accurate and less expensive. They have an adjustment for calibration for altitude.

2. A psychrometer, consisting of two thermometers, is used to measure the ambient dry-bulb and wet-bulb temperatures. The wet-bulb thermometer has a wet wick

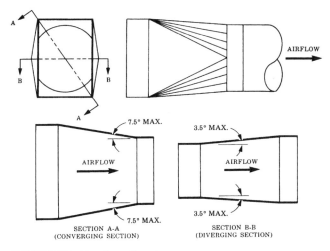

FIGURE 18.5 Transformation piece.

NOTE: All Dimensions shall be within ± 0.005D
except y which shall not exceed 0.005D.

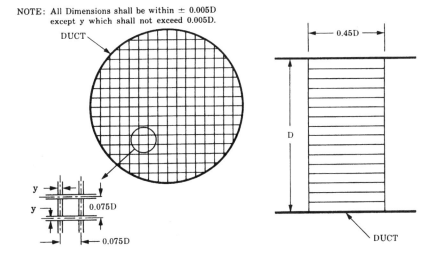

FIGURE 18.6 Flow straightener.

over the mercury end. An air stream passing over the wet wick results in some water evaporation, which lowers the temperature. This air stream can be provided by a small separate fan or by spinning the thermometers around as in a sling psychrometer.

3. Some dry-bulb thermometers are used to measure the air temperature at various locations.

4. A tachometer or a strobotac is used to measure the fan speed. On very small motor horsepowers, a tachometer will slow down the motor; thus a strobotac is preferable. A strobotac, however, has to be calibrated carefully each time in order to give accurate readings.

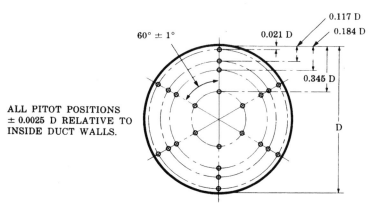

ALL PITOT POSITIONS
± 0.0025 D RELATIVE TO
INSIDE DUCT WALLS.

D IS THE AVERAGE OF FOUR MEASUREMENTS AT TRAVERSE PLANE AT 45°
ANGLES MEASURED TO ACCURACY OF 0.2% D. TRAVERSE DUCT SHALL BE ROUND
WITHIN 0.5% D AT TRAVERSE PLANE AND FOR A DISTANCE OF 0.5 D ON EITHER
SIDE OF TRAVERSE PLANE.

FIGURE 18.7   Traverse points in a round duct per AMCA, 1985.

5. A pitot tube, as shown in Fig. 18.9, is pointed into the airflow. It has two outlets to measure the total pressure and the static pressure at any point in the air stream. The velocity pressure, being the difference between total pressure and static pressure, can be read directly on a manometer, with one side connected to the total-pressure outlet and the other side connected to the static-pressure outlet of the pitot tube. In other words, the manometer does the subtracting.

6. Two manometers are needed for a duct test. One manometer is connected to measure the velocity pressure, as mentioned in item 5 and as also pictured in Figs. 18.1 and 18.2. The other manometer is connected to measure the static pressure, which is positive in an outlet duct but negative in an inlet duct. In a chamber test, one manometer measures the static pressure (positive in an outlet chamber, negative in an inlet chamber), and the other manometer measures the pressure drop across the nozzles. For measuring small pressures up to 3 or 4 inWC, the manometers usually are inclined for greater accuracy. These manometers are sometimes called *draft gauges,* a term originating from the draft of a chimney stack. For larger pressures, up to 6 or 8 inWC, a vertical tube is adequate. For even higher pressures, vertical manometers are used, often filled with so-called heavy liquids or even with mercury, having a density 13.6 times that of water. Figure 18.10*a* shows a setup for testing a small, direct-connected cen-

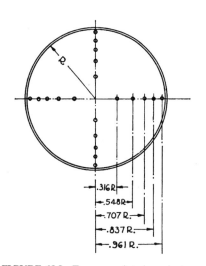

FIGURE 18.8   Traverse points in a duct per NAFM, 1952.

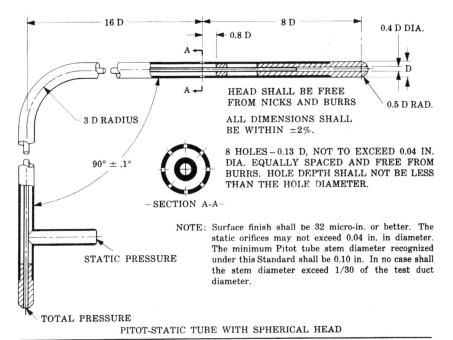

8 HOLES – 0.13 D, NOT TO EXCEED 0.04 IN.
DIA. EQUALLY SPACED AND FREE FROM
BURRS. HOLE DEPTH SHALL NOT BE LESS
THAN THE HOLE DIAMETER.

NOTE: Surface finish shall be 32 micro-in. or better. The
static orifices may not exceed 0.04 in. in diameter.
The minimum Pitot tube stem diameter recognized
under this Standard shall be 0.10 in. In no case shall
the stem diameter exceed 1/30 of the test duct
diameter.

PITOT-STATIC TUBE WITH SPHERICAL HEAD

ALL OTHER DIMENSIONS ARE THE SAME
AS FOR SPHERICAL HEAD PITOT-STATIC
TUBES.

$\dfrac{X}{D}$	$\dfrac{V}{D}$	$\dfrac{X}{D}$	$\dfrac{V}{D}$
0.000	0.500	1.602	0.314
0.237	0.496	1.657	0.295
0.336	0.494	1.698	0.279
0.474	0.487	1.730	0.266
0.622	0.477	1.762	0.250
0.741	0.468	1.796	0.231
0.936	0.449	1.830	0.211
1.025	0.436	1.858	0.192
1.134	0.420	1.875	0.176
1.228	0.404	1.888	0.163
1.313	0.388	1.900	0.147
1.390	0.371	1.910	0.131
1.442	0.357	1.918	0.118
1.506	0.343	1.920	0.109
1.538	0.333	1.921	0.100
1.570	0.323		

ALTERNATE PITOT-STATIC TUBE WITH ELLIPSOIDAL HEAD

FIGURE 18.9   Pitot-static tubes.

trifugal fan blowing into an outlet duct 10 diameters long. The figure also shows two
pitot tube supports (in the background), two draft gauges with rubber tubes for connection with the pitot tube, an aneroid barometer, a strobotac, an industrial analyzer,
and a variable rheostat. Figure 18.10*b* shows a setup for testing a small FC centrifugal fan with belt drive for cooling a projector lamp. It is tested with an inlet duct 10
diameters long. The figure also shows the pitot tube, two draft gauges (one of them
an inclined-vertical combination), and a strobotac. Note that the suction end of the

Pilot tube support   Draft gauges   Aneroid barometer

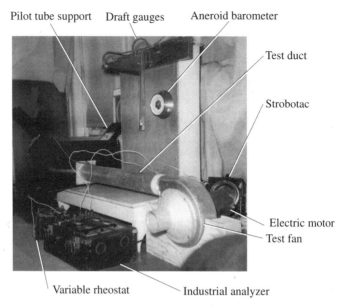

FIGURE 18.10a   Small centrifugal fan blowing into an outlet test duct.

FIGURE 18.10b   Small FC centrifugal fan exhausting from an inlet test duct.

static pressure gauge is used because the static pressure in the inlet duct will be negative. A strobotac is needed for measuring the fan speed because a tachometer would consume too much power and would slow down the small motor.

7. A micromanometer or hook gauge is used for calibrating the manometers and draft gauges. A typical hook gauge is shown in Fig. 18.11, taken from the old NAFM test code. A micromanometer is similar, only larger, for higher pressures. Either one is connected in parallel with the manometer to be calibrated so that the same pressure can be applied to both instruments. The hook gauge and the micromanometer are very accurate, due to the wide cylinders and the threaded-screw arrangement with the vernier scales, as shown in Fig. 18.11. Figure 18.12 shows several draft gauges connected in parallel with a hook gauge for calibration.

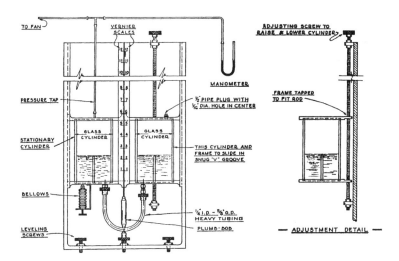

FIGURE 18.11 Typical hook gauge per NAFM, 1952.

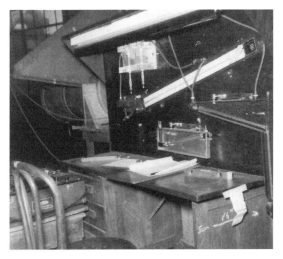

FIGURE 18.12 Draft gauges and hook gauge connected in parallel for calibration.

8. Some electrical instruments are needed for measuring volts, amps, and watts. These can be separate instruments, as shown in Fig. 18.13, or they can be combined in an industrial analyzer, as shown in Fig. 18.10b, containing voltmeter, ampmeter, kilowatt meter, and power factor meter. The analyzer can be used for single-phase or three-phase motors. To calculate the brake horsepower for each test point, a calibrated motor is needed, with a calibration curve showing motor efficiency versus kilowatt input. The brake horsepower then can be calculated according to Eq. (1.9a). Most fans over 1 hp are driven by three-phase motors.

FIGURE 18.13 Mobile table containing electrical instruments and water barrel for variable resistance due to steel plates partly immersed into the water.

9. Another method for determining the brake horsepower is by means of an electric dynamometer, as is often used for testing centrifugal fans. An electric dynamometer is somewhat similar to an electric motor, having stationary field windings on the outside and rotating armature windings on the inside. The electric dynamometer also has an armature assembly on the inside, rotating with the shaft and carrying the fan wheel. In other words, the dynamometer not only measures the brake horsepower, but it is also the prime mover, driving the fan wheel. The field on the outside of the dynamometer, however, has separate bearings and tries to turn, too, under the influence of the rotating armature assembly. However, a radial rod attached to the outside keeps it from turning, since the end of the rod pushes against a scale that measures the restraining force F in pounds. With the radius R (from the shaft center to the touch point on the scale), the brake horsepower of the fan can be calculated as

$$\text{bhp} = \frac{F \times R \times \text{rpm}}{63,025} \tag{18.1}$$

where F is in pounds and R is in inches.

10. Still another method for determining the brake horsepower is by means of a torquemeter, as shown in Figs. 18.14 and 18.15. This is a complicated and expensive instrument containing some strain gauges, electric and electronic components, and various digital displays. The torquemeter is placed between the electric motor and the fan wheel and therefore is subject to torsion. For good alignment, flexible couplings or universal joints are placed on each side of the torquemeter, plus self-aligning bearings, to prevent overheating and damage of the torquemeter. Calibration before each use takes one-half hour or more. The torquemeter has digital displays of torque, speed, and brake horsepower on pushing various buttons. No calculations are needed, but extreme caution in the assembly setup is required.

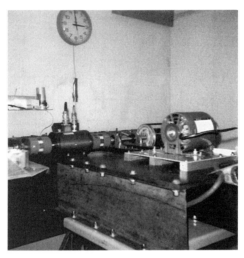

FIGURE 18.14 Torquemeter placed between the electric motor and the test fan, with bearings and flexible couplings on each side. *(Courtesy of FloAire, Inc., Bensalem, Pa.)*

11. Sound level. Let's look again at Fig. 4.39 showing the performance of a vaneaxial fan with four curves (static pressure, brake horsepower, mechanical efficiency, and sound level) plotted against air volume (cfm). We know by now that the air volume is important for proper ventilation, that the static pressure is important to overcome the resistance of the duct system, that the brake horsepower is important for the proper motor load, and that the mechanical efficiency is important to keep the brake horsepower and the operating cost down. Just as important for the comfort of the occupants, however, is the sound level produced by the fan. Sound is measured in decibels. It can vary from 40 dB for an almost noiseless condition to 70 dB for a still comfortable condition to 100 dB for an uncomfortable condition. Any noise above 110 dB is dangerous for the hearing system and requires ear protection.

The preceding decibel figures may be misleading because they give the impression that there is not too much difference in noise between, say, 70 and 100 dB. Actually, there is a considerable difference in the sound power. The decibel system tends

FIGURE 18.15 Torquemeter connected to inlet of a roof ventilator. Note the long shaft extension to ensure an unobstructed airflow into the fan housing. *(Courtesy of FloAire, Inc., Bensalem, Pa.)*

to reduce the apparent difference because decibels are the logarithm of the power ratio and thereby reduce the spacing between sound powers.

The decibels at a certain place are measured with a microphone located at that place connected to a sound level meter, as shown in Figs. 18.16 and 18.17. This is an electronic instrument reading directly in decibels. The AMCA has published several booklets explaining the methods for measuring sound, among them AMCA-

FIGURE 18.16 Microphone and sound level meter measuring the decibels at a certain distance from a vaneaxial fan with an oversize venturi inlet.

FIGURE 18.17 Acceptance test for customers checking sound level of a vaneaxial fan. Note tarpaulins used for reducing sound reverberation.

300-94, entitled *Reverberant-Room Method for Sound Testing of Fans,* and AMCA-301-90, entitled *Methods for Calculating Fan Sound Ratings from Laboratory Test Data.*

Two properties of sound should be pointed out here:

1. Sound levels normally diminish with increasing distance from the sound source. This, of course, is obvious for wide-open spaces. In a small, enclosed space, however, this decrease in sound may be small or even nonexistent, especially if the surfaces of the walls, ceiling, and floor are reverberant.

2. The distribution of the total sound power over various frequencies often is uneven. For example, a whistle will have high frequencies, while thunder will have predominantly low frequencies. Or a vaneaxial fan running at a high speed will produce a musical note of higher frequency in the operating range than in the stalling range, where a low-frequency rumbling noise will predominate.

Figure 18.18 is a photograph showing the inlet side of a test fan blowing into an outlet chamber with a short rectangular duct between the outlet of the test fan and the inlet of the chamber for static recovery. Figure 18.19 is a photograph showing the same test fan from the drive side with a view of the torquemeter for measuring the brake horsepower. Figures 18.20 and 18.21 are inside views of the chamber, one showing the nozzle inlets and one showing the nozzle outlets. Figure 18.22 shows a small, portable nozzle chamber for measuring from 1 to 600 cfm. It can be used as an outlet chamber or as an inlet chamber. Figure 18.23 shows an inlet chamber with a roof ventilator exhausting from it. Note the booster fans on the inlet side of the chamber. Figure 18.24 shows a view of a fan test laboratory with several test ducts and a test chamber in the background.

From all these figures it is obvious that a test chamber is a more elaborate setup that is more expensive than a test duct. Nevertheless, the decision on which of the 10 AMCA test setups (ducts and chambers) should be built will depend on various considerations that now will be discussed.

FIGURE 18.18   Airfoil centrifugal fan, set up for a test, blowing into an outlet chamber per AMCA. View showing the fan inlet side and a short rectangular outlet duct for static recovery. *(Courtesy of Ammerman Division, General Resource Corporation, Hopkins, Minn.)*

In general, the setup for testing a fan should simulate the way the fan presumably will be used in an actual installation. For example, if the fan is expected to blow into a system, it should be tested as a blower. On the other hand, if the fan is expected to exhaust from a system, it should be tested as an exhauster. For low to moderate static pressures, the fan will perform the same as a blower and as an exhauster, but for higher pressures, there will be a difference in performance.

The next decision is between a test duct and a test chamber. Here are a few points to consider in this decision:

1. Since a test duct will be considerably less expensive to build, it might be preferable if only one fan is to be tested or possibly two fans of only slightly different sizes so that the same test duct can be used for both fans, with transitions between fan and duct. If many fans of different sizes are to be tested, several test ducts might be needed, and this might be more expensive than a chamber that would accommodate all the different sizes.

FIGURE 18.19   Airfoil centrifugal fan, set up for a test, blowing into an outlet chamber per AMCA. View from the drive side showing a torquemeter for measuring the brake horsepower. *(Courtesy of Ammerman Division, General Resource Corporation, Hopkins, Minn.)*

2. In a simple duct test without a booster fan (some do have one), the performance curves will not extend all the way to free delivery because of the duct friction.

FIGURE 18.20 Inside view of test chamber showing nozzle inlets of various sizes. *(Courtesy of Ammerman Division, General Resource Corporation, Hopkins, Minn.)*

For a chamber test, due to the booster fan, the performance curves will extend all the way to free delivery. For fans producing considerable static pressure (such as centrifugal fans or vaneaxial fans), the performance curves from a duct test can be extrapolated to free delivery, but for propeller fans and axial-flow roof ventilators, the first duct test point might be in the strongly curved portion of the static pressure curve, and an extrapolation therefore might be inaccurate; thus a setup with a booster fan will be needed.

3. For propeller fans and tubeaxial fans, the outlet air has a spin that might distort the performance if an outlet duct were used. A chamber, therefore, would be preferable for testing propeller fans. A chamber or an inlet duct could be used for testing tubeaxial fans.

4. All but one of the duct tests require a pitot tube traverse for each test point, while most chamber tests require only one pressure-drop reading. A duct test with a pitot tube traverse, therefore, will take about 3 hours; a chamber test will take less than 1 hour. Over 3 hours, test conditions may change. A chamber test, therefore, usually will be more accurate.

FIGURE 18.21 Inside view of test chamber showing nozzle outlets of various sizes. *(Courtesy of Ammerman Division, General Resource Corporation, Hopkins, Minn.)*

TEST PROCEDURE

Prior to running a test, test forms should be prepared in which the test data will be entered. Unless a computer program is available, the test form also should provide the formulas for calculating the air volume, pressures, brake horsepowers, and efficiencies that will be plotted as the performance curves of the fan.

After the fan to be tested has been set up and turned on, let the motor warm up for at least 10 minutes. During this time, the manometers should be

FIGURE 18.22 Two sections of a small nozzle chamber showing five small nozzles for measuring from 1 to 600 cfm. Can be used as an outlet chamber or as an inlet chamber.

checked against a micromanometer, and readings should be taken (and entered in the test form) of the barometric pressure and the ambient temperatures near the fan inlet for dry-bulb and wet-bulb by using a sling psychrometer or a similar device. From these data, the air ratio can be determined, i.e., the ratio of standard air density 0.075 lb/ft^3 divided by the ambient air density, as taken from tables shown in the

FIGURE 18.23 Inlet chamber with a roof ventilator exhausting from it for measuring up to 27,000 cfm. View shows three access doors and one of two booster fans, 10 hp each, blowing air into the chamber. *(Courtesy of FloAire, Inc., Bensalem, Pa.)*

FIGURE 18.24   View of a fan test laboratory with several test ducts and with a test chamber in the right background.

AMCA Test Code. Figure 18.25 shows a graph from which the air ratio can be read directly. The air ratio will be used in the test calculations for correcting the various pressures and brake horsepowers to standard air density of 0.075 lb/ft³.

Next, we read and enter for each test point the motor volts, amps, and watts (if the motor is calibrated) or the brake horsepower (if a torquemeter is used) or the weight and arm length (if a reaction dynamometer is used). We also read and enter the fan speed, the various air temperatures, the static pressure, and the pressure drop (in case of a chamber test) or the velocity pressures (in case of a duct test). As mentioned, 10 test points are adequate. They should be fairly evenly spaced from free delivery to no delivery.

TEST FORMS

Tables 18.1 through 18.5 show test forms for outlet ducts, inlet ducts, outlet chambers, and inlet chambers. These forms can be used when no computer programs are at hand. They use some approximations, but for all practical purposes, they are accurate enough. The booster fan will be on for the first two or three points, with various damper positions, resulting in different values for the air volume and the static pressure. The remaining test points will be obtained again by various damper positions, but with the booster fan not powered (it will coast, however, due to the airflow passing through it). The first 12 to 15 lines (down to the heavy line indicated) show the various test data obtained, except the motor efficiency, which is obtained from the motor calibration, i.e., from a curve of motor efficiency versus watts input. (Some different data, as mentioned earlier, will be entered if a torquemeter or a dynamometer is used.) Below the heavy line we show various formulas used to calculate the data for plotting the performance curves. The air densities at various planes of the test chamber are calculated and used to correct the test data obtained.

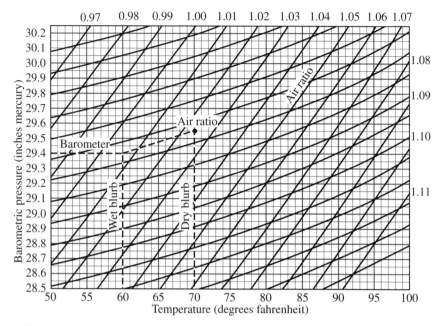

FIGURE 18.25 Graph for determining the air ratio from barometric pressure, dry-bulb temperature, and wet-bulb temperature.

In Tables 18.4 and 18.5, two formulas are shown for calculating cfm$_5$, i.e., the air volume ahead of the partition carrying the nozzles. One formula is for a 5-in nozzle, using the factor 147. The other formula is for a 12-in nozzle, using the factor 844. For other nozzle sizes, the factor can be calculated easily, since it is in proportion with the square of the nozzle inside diameter. For example, for an 18-in nozzle, the factor will be $(18/12)^2 \times 844 = 1899$. The total air volume will be the sum of the various air volumes of the nozzles that are open during the test. Occasionally, it may be necessary, for good accuracy, to close some of the nozzles when the air volume becomes smaller as the test progresses.

SUMMARY

The *AMCA Test Code* specifies 10 setups for testing fans. Which setup should be used depends on the type of fan to be tested, the type of application, and the number of fans to be tested. Test forms for various setups should provide the formulas needed, or else a computer program can be used. The resulting data then are plotted, showing pressures, brake horsepower, and efficiencies versus air volume.

TABLE 18.1 Pitot Tube Traverse Test Form

Descr. of test fan:
Fan outlet area A_{FO} =
Tested by: Date:
Outlet or inlet duct
Duct i.d. D_F =
Duct area A_P =

Orif. i.d. No.	p_v	$\sqrt{p_v}$	p_s	p_v	$\sqrt{p_v}$	p_s	p_v	$\sqrt{p_v}$	p_s	p_v	$\sqrt{p_v}$	p_s	p_v	$\sqrt{p_v}$	p_s	p_v	$\sqrt{p_v}$	p_s
1			p_s			p_s			p_s			p_s			p_s			p_s
2																		
3			rpm			rpm			rpm			rpm			rpm			rpm
4																		
5			Volts			Volts			Volts			Volts			Volts			Volts
6																		
7																		
8																		
9			Amps			Amps			Amps			Amps			Amps			Amps
10																		
11																		
12																		
13			Ave. amps			Ave. amps			Ave. amps			Ave. amps			Ave. amps			Ave. amps
14																		
15			Watts			Watts			Watts			Watts			Watts			Watts
16																		
17			Bar			Bar			Bar			Bar			Bar			Bar
18																		
19			dB			dB			dB			dB			dB			dB
20																		
21			WB			WB			WB			WB			WB			WB
22																		
23			AR			AR			AR			AR			AR			AR
24																		
Cent.																		
Sum	↓	→		↓	→		↓	→		↓	→		↓	→		↓	→	
Ave.																		

Remarks:

TABLE 18.2 Calculations for Outlet Duct Test

Size and type of fan:

Tested by:	Date:	L = length of duct from
Calcul. by:	Date:	Pitot to fan incl. transf.
		Drive:

Orifice i.d.

Air ratio AR

p_s

$\sqrt{p_v}$

p_v

$p_t = p_s + p_v$

$p_{f+c} = 0.02 \left(\dfrac{L}{D} + 4 \right) p_v$

$p_t + p_{f+c}$

Volts

Amps

Watts

rpm

$\overline{P_{vP} = AR \times p_v}$

$V = 4005 \sqrt{AR} \sqrt{p_v}$

$\text{cfm} = A_p \times V$

$P_t = AR \left(p_t + p_{f+c} \right)$

$P_{vF} = (A_p/A_{FO})^2 P_{vP}$

$P_s = P_t - P_{vF}$

Watts = $AR \times$ watts

Motor eff EE

$\text{bhp} = EE \times \text{watts}/746$

$\text{ahp}_t = \text{cfm} \times P_t/6356$

Total eff. = ahp_t/bhp

Corrected to RPM

cfm

P_s

bhp

Remarks:

TABLE 18.3 Calculations for Inlet Duct Test

Size and type of fan:			
Tested by:	Date:	L = length of duct from	
Calcul. by:	Date:	Pitot to fan incl. transf.	Drive:
Orifice i.d.			
Air ratio AR			
p_s			
$\sqrt{p_v}$			
p_v			
$p_t = p_s + p_v$			
$p_f = 0.02\left(\frac{L}{D}\right)p_v$			
$p_t + p_f$			
Volts			
Amps			
Watts			
rpm			
$P_{vP} = AR \times p_v$			
$V = 4005\sqrt{AR}\sqrt{p_v}$			
cfm = $A_p \times V$			
$P_{vF} = (A_p/A_{FO})^2 P_{vP}$			
$P_t = AR\,(p_t + p_f) + P_{vF}$			
$P_s = P_t - P_{vF}$			
Watts = $AR \times$ watts			
Motor eff. EE			
bhp = $EE \times$ watts/746			
ahp_t = cfm $\times P_t/6356$			
Total eff. = ahp_t/bhp			
Corrected to RPM cfm			
Corrected to RPM P_s			
Corrected to RPM bhp			

Remarks:

TABLE 18.4 Outlet Chamber Test Form

Tested by:	Date:	Fan $OA =$	Drive:
Type and size of fan tested:		Open nozzles:	

Booster fans	On	Off
Outlet damper setting		
Barometric pressure BP		
DB_1 (fan inlet)		
WB_1 (fan inlet)		
DB_5 (ahead of nozzles)		
Volts (average)		
Amps (average)		
1/watts		
watts		
Motor efficiency EE		
bhp $= (EE \times \text{watts})/746$		
Fan wheel rpm		
sp_7 (in WC) (in chamber past fan)		
pd (in WC) (across nozz.)		
$AR = 0.075/\rho_1$ (fr.graph)		
$\rho_1 = 0.075/AR$		
$K_T = (460 + DB_1)/(460 + DB_5)$		
$K_P = (BP + SP/13.62)/BP$		
$\rho_5/\rho_1 = K_T \times K_P$		
$\rho_5 = \rho_1 \times \rho_5/\rho_1$		
$PD = AR \times pd$		
5-in no.: $\text{cfm}_5 = 147 \sqrt{pd/\rho_5}$		
12-in no.: $\text{cfm}_5 = 844 \sqrt{pd/\rho_5}$		
Total cfm_5		
$\text{cfm}_1 = \text{cfm}_5 \, (\rho_5/\rho_1)$		
$OV = \text{cfm}_5/OA$		
$VP_{FO} = (OV/1096)^2 \, \rho_5$		
$SP = sp_7 \times 0.075/\rho_5$		
$TP = SP + VP_{FO}$		
BHP $= AR \times$ bhp		
AMP $= \text{cfm}_1 \times TP/6356$		
Mech. eff. $= AHP/BHP$		

TABLE 18.5 Inlet Chamber Test Form

Tested by:	Date:	Fan OA =	Drive:	
Type and size of fan tested:		Open nozzles:		

Booster fans	On		Off
Inlet damper setting			
Barometric pressure BP_a (ambient)			
$DB_{ambient}$			
$WB_{ambient}$			
DB_2 (fan outlet)			
DB_5 (ahead of nozzles)			
Volts (average)			
Amps (average)			
1/watts			
watts			
Motor efficiency EE			
bhp = $(EE \times$ watts$)/746$			
Fan wheel rpm			

sp_8 (in chamber, ahead of fan) (neg.)
pd (across nozzles)
$AR = 0.075/\rho_{ambient}$ (from graph)
$\rho_{ambient} = 0.075/AR$
$K_P = (BP_a - |SP_8| + pd)/BP_a$
$\rho_5 = \rho_a \times K_P$
5-in nozzle: cfm$_5 = 147 \times \sqrt{pd/\rho_5}$
12-in nozzle: cfm$_5 = 844 \times \sqrt{pd/\rho_5}$
Total cfm$_5$
$(BP_a - |sp_8|)/BP_a$
$\rho_1 = \rho_a \times (BP_a - |sp_8|)/BP_a$
cfm = cfm$_5$ (ρ_5/ρ_1)
$\rho_a/\rho_2 = (460 + DB_2)/(460 + DB_a)$
$\rho_2 = \rho_a \times (\rho_2/\rho_a)$
Fan outlet $V_2 = ($cfm$/OA)(\rho_1/\rho_2)$
$vp_2 = \rho_2 (v_2/1096)^2$
$tp = vp_2 + |sp_8|$
ahp = cfm $\times tp/6356$
Mech. eff. = ahp/bhp
Inlet cfm = $AR \times$ cfm
$SP = AR \times sp_8$
BHP = $AR \times$ bhp

CHAPTER 19
VACUUM CLEANERS

REVIEW

Figure 7.45 showed the design of a small centrifugal fan wheel as used in vacuum cleaners. These fan wheels usually have six to eight backward-curved blades, 5 to 6 in o.d., and about ¼ in wide. Most vacuum cleaners have two stages, but some have only one stage. They are directly driven by single-phase, high-speed universal motors of ½, ¾, or 1 hp.

CONFIGURATIONS

Various configurations of the parts can be used, but regardless of the configuration, the small turbo exhauster is the heart of a vacuum cleaner. Figure 19.1 shows the most common configuration used in tank-type vacuum cleaners. Here the air stream first enters the floor nozzle and then passes through a pickup pipe, a flexible hose, a filter bag, the exhaust fan and the motor, and finally leaves the vacuum cleaner housing. Looking at the two-stage exhauster shown in Fig. 19.1, note that the diffuser ratio is only 1.14. In other words, the space available past the blade tips (for the airflow to diffuse and to make a 180° turn) is rather limited. This is done because in a vacuum cleaner compactness is more important than fan efficiency. This configuration has the advantage that the air passing through the fan and motor is clean, so the fan and motor will never get plugged up by the dust. The configuration has the disadvantage that the airflow cooling the motor is already heated from adiabatic compression in the fan so that motor cooling is less effective.

In another configuration used occasionally, the locations of motor and fan are reversed. Here the air stream passes through the flexible hose, the filter bag, and the motor and then through the exhaust fan. This sequence has the advantage that the air passing through the motor is still cold and the motor cooling is thus more effective. It has the disadvantage that the air entering the fan is warm (from the motor) and the vacuum produced, therefore, is smaller. The fan efficiency is lower, too, due to inlet turbulence.

Still another configuration places the fan ahead of the motor (as in Fig. 19.1), but the main air stream does not pass through the motor, it bypasses the motor and leaves the vacuum cleaner housing ahead of the motor. A separate centrifugal fan wheel, mounted on the same motor shaft and located between the motor and the main fan, draws some cooling air through the motor and blows it out together with the main air stream leaving the vacuum cleaner housing. This results in good motor cooling and in a good vacuum, but it also results in an increased manufacturing cost and power consumption.

Other configurations avoid the tank altogether and use a hand-held upright structure containing the exhaust fan and the motor. The filter bag may be located ahead of fan and motor (for clean air) or past the fan, on the outside of the vacuum cleaner, as

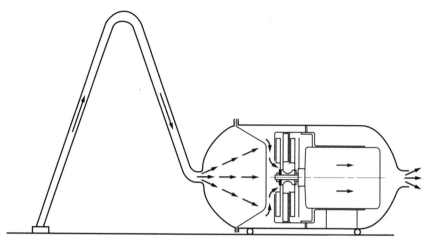

FIGURE 19.1   Schematic sketch of vacuum cleaner with small two-stage turbo exhauster and filter bag ahead of exhauster inside tank.

shown in Fig. 19.2, for easy accessibility. The advantage of these configurations is light weight, lower cost, and elimination of the flexible hose, which consumes a portion of the vacuum produced by the exhaust fan. Another advantage of this configuration is that the outside filter bag can be made larger, for a reduced pressure loss. In this configuration, a one-stage exhaust fan may be sufficient, since the larger filter bag plus elimination of the flexible hose will reduce the total required vacuum. On the other hand, when the filter bag is located past the fan, the fan blades will require steeper tip angles to prevent the dust from sticking to the blades. This might result in a somewhat lower fan efficiency, which, however, is a minor disadvantage in view of the small motor horsepowers and the already low efficiencies.

TESTING

Vacuum cleaners could be tested with a small inlet duct, but such a duct would be too small for a pitot tube traverse. One could just measure the velocity pressure in the center of the duct, calculate the corresponding maximum duct velocity, and estimate that the average duct velocity would be about 91 percent of the maximum velocity in the center of the duct. As in any duct test, one would use a throttling device (such as a set of orifice rings), obtain a number of test points, calculate the air volume, static pressure, etc., and plot the performance curves. This method would not be 100 percent accurate, but the deviations might be acceptable.

Vacuum cleaners also could be tested on a small inlet chamber, if it is strong enough and will not collapse at a suction pressure of 50 inWC. A small nozzle of 1½ in i.d. would register a pressure drop of 1.08 inWC for a 50-cfm airflow. The test would be accurate if this small nozzle chamber is free of air leaks.

A third method uses a Bureau of Standards box for testing vacuum cleaners. This method is fast, simple, and accurate. Figure 19.3 shows a schematic sketch of this test box. The top plate of the box has a rectangular opening 10 × 14 in. The vacuum cleaner will exhaust the air through this opening. For testing the exhaust fan only, a

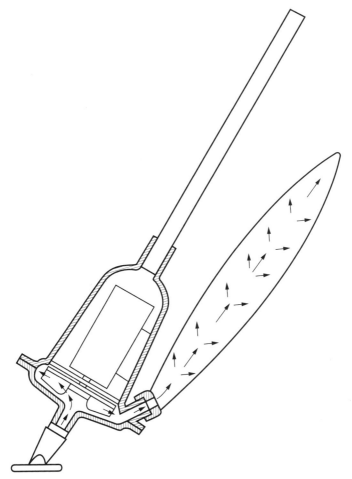

FIGURE 19.2 Schematic sketch of upright vacuum cleaner with single-stage turbo exhauster and outside filter bag.

separate cover plate will be bolted to the top plate of the box. This cover plate will have a circular hole about 2½ in i.d. to fit the inlet diameter of the exhaust fan. For testing the complete vacuum cleaner, another cover plate will be bolted to the top plate of the box. This cover plate will have a rectangular hole, about 12 × 1½ in, to fit the opening of the floor nozzle.

Figure 19.3 also shows a short pipe, 3½ in o.d., welded to the outside of one of the side plates of the box. Through this pipe, an air stream will enter the box, to replace the air that has been exhausted by the vacuum cleaner. The pipe has a flange that will accept various orifice rings, used as a throttling device, to obtain the test points for plotting the performance curves. Thirteen sharp-edged orifice rings should be available for the tests.

Finally, Fig. 19.3 shows a small pipe, ¼ in o.d., welded to the outside of another side plate of the box. This pipe is used for connection by rubber tubing to a manome-

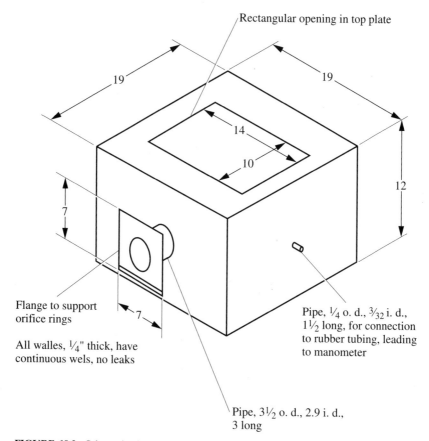

Rectangular opening in top plate

19

19

14

10

12

7

Flange to support orifice rings

7

All walles, $\frac{1}{4}''$ thick, have continuous wels, no leaks

Pipe, $\frac{1}{4}$ o. d., $\frac{3}{32}$ i. d., $1\frac{1}{2}$ long, for connection to rubber tubing, leading to manometer

Pipe, $3\frac{1}{2}$ o. d., 2.9 i. d., 3 long

FIGURE 19.3 Schematic sketch of box for testing vacuum cleaners, recommended by the Bureau of Standards.

ter that measures the suction pressure in the box. No other pressure needs to be measured. For comparison, in a chamber test (as you will recall), two manometers are needed to measure two pressures (a static pressure and a pressure drop across the nozzles). In a duct test, again, two manometers are needed to measure a static pressure and several velocity pressures. In this box test, only one manometer is needed. It measures the negative suction pressure S inside the box for the sharp-edged orifice on the flange. A box test, therefore, takes only about 20 minutes.

From the suction S, the air volume can be calculated as follows: If the sharp-edged orifice has an inside diameter d (in inches), the area A (in square feet) will be

$$A = \frac{d^2\pi}{4 \times 144} = 0.005454d^2 \tag{19.1}$$

For a sharp-edged orifice, the coefficient of discharge is 0.611, and the velocity through the orifice will be

$$V = 0.611 \times 4005 \times \sqrt{S} \tag{19.2}$$

TABLE 19.1 Constant K versus Orifice Inside Diameter d, According to Eq. (19.5), Used in Calculating the Air Volume When Testing Vacuum Cleaners on a Bureau of Standards Box

Orifice i.d. (in)	2	1¾	1½	1¼	1⅛	1	⅞
Constant K	53.5	40.9	30.1	20.9	16.9	13.35	10.2
Orifice i.d. (in)	¾	⅝	½	⅜	¼	⅛	
Constant K	7.51	5.21	3.34	1.88	0.835	0.209	

where the suction pressure S is in inches of water column and the velocity V is in feet per minute. The air volume, therefore, can be calculated as

$$\text{cfm} = A \times V = 13.35 d^2 \sqrt{S} \qquad (19.3)$$

or

$$\text{cfm} = K\sqrt{S} \qquad (19.4)$$

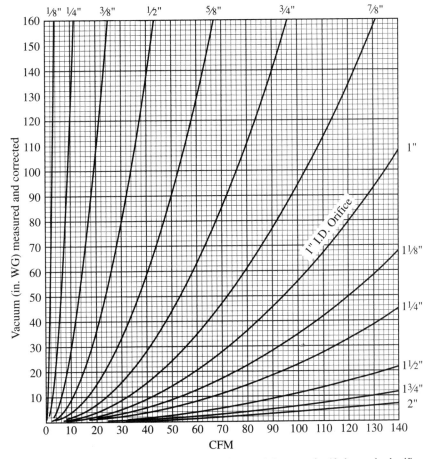

FIGURE 19.4 Graph showing parabolic system characteristic curves for 13 sharp-edged orifice rings used on Bureau of Standards test box for vacuum cleaners.

TABLE 19.2 Vacuum Cleaner Test Form (*Bureau of Standards Box*)

Fan wheel: o.d., number of blades   , blade width   , number of stages

Motor: hp,   rpm at full load

Tested by: Date: Calculated by: Date:

Model: Tank or upright.   Fan outlet area OA =

Remarks:

Orifice ring d (in) Barometric pressure BP	2	1¾	1½	1¼	1⅛	1	⅞	¾	⅝	½	⅜	¼
DB WB												
AR (from graph) Volts												
Amps Watts												
rpm Suction s												
$W = AR \times w$ $S = AR \times s$												
K cfm $= K\sqrt{S}$	53.5	40.9	30.1	20.9	16.9	13.35	10.2	7.51	5.21	3.34	1.18	.835
ahp $=$ cfm $\times S/6356$ EE (from graph)												
bhp $= EE \times W/746$ $ME =$ ahp/bhp												

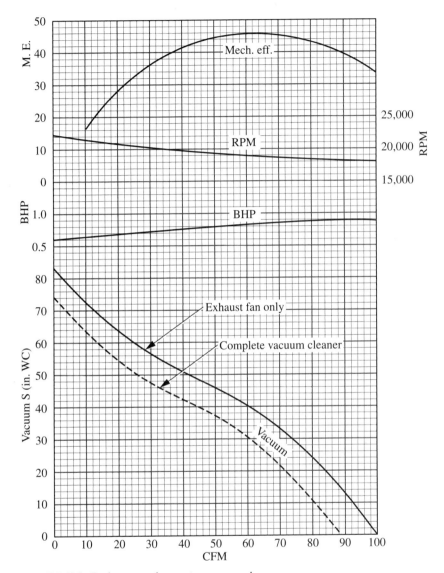

FIGURE 19.5 Performance of a two-stage vacuum cleaner.

where the constant K is

$$K = 13.35d^2 \tag{19.5}$$

K is a constant for each orifice ring. Table 19.1 shows the K values for 13 d values.

Figure 19.4 is a graph showing 13 parabolic system characteristic curves per Eq. (19.3). For example, if a ¾-in i.d. orifice ring is used and a suction pressure of 50 inWC is observed on the manometer, the air volume will be 53 cfm. This graph is a time saver. When calculating a test, we do not even have to use Eq. (19.3) to calculate the air volume. We can just read the suction pressure s on the manometer, correct it to S for standard air density, and read the corresponding air volume (cfm) in Fig. 19.4.

Table 19.2 shows the test sheet for the Bureau of Standards box. If we compare it with Table 18.5, for an inlet chamber test, we find the following two differences:

1. In Table 18.5, the pressure drop PD across the nozzles was measured, and the air volume was calculated from the pressure drop. In Table 19.2, the air volume is calculated from the negative suction pressure s (Eq. 19.4), which can be considered the pressure drop across the sharp-edged orifice ring.

2. In Table 18.5, the fan outlet velocity was calculated, and from it, the velocity pressure, the total pressure, and the fan efficiency were calculated. In Table 19.2, the suction pressure S is so large that the velocity pressure would be insignificant in comparison. The suction pressure S, therefore, can be used instead of the total pressure for calculating the fan efficiency.

As a result of these two differences, Table 19.2 is shorter and simpler.

PERFORMANCE

Figure 19.5 shows the performance for a typical two-stage vacuum cleaner per Fig. 19.1. You will note the following:

1. The suction produced by the complete vacuum cleaner is considerably lower than the suction produced by the exhaust fan only. This loss is the result of the flow resistances from the floor nozzle, the pickup pipe, the flexible hose, and the filter bag.

2. The motor speed varies by about 20 percent from free delivery to no delivery. This considerable variation is caused by the single-phase, high-speed universal motor, which is more sensitive to a change in load (bhp) than a three-phase induction motor, with which the variation would be less than 5 percent.

3. The maximum efficiency of the exhaust fan only is shown as 46 percent, but it is often lower. For the complete vacuum cleaner, the maximum efficiency (not shown in Fig. 19.5) might be 25 to 32 percent.

CHAPTER 20
FAN PERFORMANCE AS SHOWN IN CATALOGS

THREE WAYS TO PRESENT FAN PERFORMANCE

There are three ways how a catalog can present the performance of a fan or of a line of fans: as performance curves, as rating tables for direct drive, and as rating tables for belt drive.

Performance Curves

Figure 4.43 is an example: It shows seven static pressure curves for a 36-in vaneaxial fan at 1750 rpm, with the tip angles varying from 13° to 33°. These curves are the results of laboratory tests. Curves for other sizes and speeds may be derived by using the fan laws.

Rating Tables for Direct Drive

The top part of Table 4.5 is an example: This table was derived from several test curves by simple conversions for size and speed in accordance with the fan laws. The procedure was described in detail in Chap. 5 (Tables 5.1 and 5.2 and Fig. 5.1). After each conversion, new curves are plotted, and the new air volumes, for the static pressures shown at the top of the table, are taken from the new curves. As mentioned previously, the efficiencies will increase slightly for larger sizes (size effect), even if the fans are in geometric proportion. Conversions, therefore, are only permissible to larger sizes but not to smaller sizes, because they would not quite come up to the efficiencies of the larger sizes. Several sizes of a line, therefore, have to be tested, and the performances of the in-between sizes are calculated.

Rating Tables for Belt Drive

An example is Table 4.3: It is for a 30-in belt-driven vaneaxial fan. This table was derived from the test performance shown in Fig. 20.1, again by means of conversions in accordance with the fan laws, but these computations are more complex. They will be explained now.

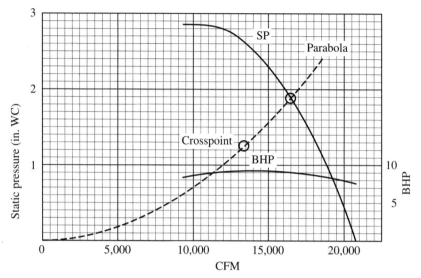

FIGURE 20.1   Test performance of a 30-in vaneaxial fan, 1760 rpm, showing three curves: static pressure, brake horsepower, and parabolic curve through a cross-point from Table 4.3.

COMPUTATION OF RATING TABLES
FOR BELT DRIVE, DERIVED FROM
TEST PERFORMANCE CURVES

The graph in Fig. 20.1 shows the test performance that was obtained for the 30-in vaneaxial fan running at 1760 rpm. This graph shows only the operating range of the fan. The stalling range (static pressure dip, etc.) was omitted because it will not be used in calculating the rating tables. The rating table was derived from the performance curves shown in Fig. 20.1. Table 4.3 has the customary format, with the static pressures shown on top and the outlet velocities and the corresponding air volumes shown in the first two columns on the left side. The air volumes are the product of the outlet velocities times the outlet area of the fan. We now have to calculate the speed and brake horsepower for each cross-point of air volume and static pressure. As an example, let's calculate the speed and the brake horsepower for 13,409 cfm at a static pressure of 1¼ inWC, figures taken from Table 4.3. We have marked this point on the graph Fig. 20.1 and have drawn a parabolic curve through this point and let it intersect with the static pressure curve. This is similar to the parabolas in Fig. 5.1. As explained in Chap. 5, this parabola indicates how the performance point will move if the speed is increased gradually. There are various ways to determine the points of the parabola. For example, one way is to use the good, old-fashioned slide rule with a square scale.

Table 20.1 shows a form that can be used in making these computations. We find that the parabola through the point 13,409 cfm at 1¼ inWC of static pressure will intersect the static pressure curve in Fig. 20.1 at the point 16,440 cfm at 1.88 inWC of static pressure. We can now enter these two figures in Table 20.1. We know that for that point the fan will run at 1760 rpm. For our catalog cross-point (13,409 cfm at 1¼

TABLE 20.1 Form for Computing the Various Speeds and Brake Horsepowers for Performance Table 4.3 for a 30^{7}/$_{16}$-in Vaneaxial Fan, Belt Drive, $OA = 5.16$ ft^2, $OV = 2600$ fpm, 13,409 cfm

| Catalog SP (inWC) | Point of intersection with SP curve in Fig. 20.1 at 1760 rpm | | Catalog rpm | Fig. 20.1, 1760 rpm bhp$_x$ | Catalog bhp |
	cfm$_x$	SP_x			
¼					
½					
¾					
1					
1¼	16,440	1.88	1436	9.02	4.89
1½					
1¾					
2					
2½					
3					
3½					
4					
4½					
5					

Note: Data derived from test performance shown in Fig. 20.1, 30 in, 1760 rpm.

inWC of static pressure), the fan speed then will be $13{,}409/16{,}440 \times 1760 = 1436$ rpm. We now can enter 1436 rpm in Table 20.1 and at our cross-point in Table 4.3.

To determine the brake horsepower for our cross-point in Table 4.3, we first look at Fig. 20.1 and find that at 1760 rpm and 16,440 cfm, the brake horsepower is 9.02. We now enter this figure in Table 20.1. For our catalog cross-point, then, the brake horsepower will be $(13{,}409/16{,}440)^3 \times 9.02 = 4.89$ bhp. We now can enter 4.89 bhp in Table 20.1 and at our cross-point in Table 4.3.

This time-consuming procedure to determine the speed and brake horsepower for each cross-point in Table 4.3 will have to be repeated many times. A separate computation form per Table 20.1 will be needed for each outlet velocity and corresponding air volume.

COMPUTATION OF RATING TABLES FOR BELT DRIVE FROM ANOTHER RATING TABLE

This computation can only be done to a larger size, but it is simpler and quicker. Let's demonstrate it on Table 4.3 (30-in size) to be converted to Table 4.4 (36-in size). The figures shown in Table 4.4 were calculated directly from the corresponding figures in Table 4.3 using the exact wheel diameters. The figures for corresponding cross-points (same OV and SP) are obtained as follows: Each speed is reduced in the wheel diameter ratio 30^{7}/$_{16}$/36^{3}/$_{8}$ = 0.8368, and each brake horsepower is increased in the ratio (36^{3}/$_{8}$/30^{7}/$_{16}$)^2 = 1.4282. For example, for 2600 fpm and 1¼ inWC of static pressure, we get $0.8368 \times 1436 = 1202$ rpm and $1.4282 \times 4.89 = 6.99$ bhp.

ANALYZING THE RATING TABLES PUBLISHED
BY VARIOUS FAN MANUFACTURERS

Before placing an order for a fan, the customer should study the catalogs of various fan manufacturers and try to analyze the rating tables. We have seen that the rating tables disclose five parameters: air volume, static pressure, output velocity, speed, and brake horsepower. Sometimes they also disclose the fan efficiency for each cross-point (see Table 7.1), but in most rating tables, efficiencies are not shown. It is useful for the customer, however, to know the fan efficiencies for the following two reasons:

1. A high efficiency is an indication of a good fan and of operating economy.

2. An overly high efficiency is an indication of exaggerated ratings. (I once analyzed a rating table and calculated from it an 85 percent maximum efficiency, but when I ran a test on the unit, I found it to be only 37 percent.)

In Chap. 7 we showed how the fan efficiencies can be calculated from the five quantities given for each cross-point. To analyze a rating table, we select a few cross-points in the range where we can expect the best efficiencies. For these points, we calculate the efficiencies and compare them with the efficiencies that can be expected for this type of fan, as shown in Table 4.2, in Fig. 7.1, and in Table 7.4. We also might plot efficiency versus air volume, to check whether this is a smooth curve, as it is supposed to be. This analysis will give you an idea of whether or not the fan will fulfill your requirements.

CHAPTER 21
AIR CURTAINS

FLOW PATTERN AND FUNCTION

An *air curtain* is a sheet of moving air that is blown down an open doorway or across some other open area. This air stream is produced by one or several fans that usually are mounted above a doorway. Occasionally, the fans are mounted on one side or on both sides of a doorway or of some other open area. Figure 21.1 is a schematic sketch showing a fan box mounted above a doorway and blowing an air curtain down the doorway. This configuration will minimize the loss of heated or cooled air from the building, almost like a real door. It also prevents insects, rain, and dust from the outside entering the building, even if wind tries to penetrate the open doorway. Outside winds up to 30 mi/h can be stopped by an air curtain. At the same time, air curtains will permit trucks to enter the door openings of factories or warehouses for loading and unloading. Air curtains are also used at the entrances to office buildings, department stores, public buildings, and hospital emergency sections for people to pass through. Some air curtains can handle heated air during the winter months.

OUTSIDE WIND

In some buildings, there is a negative pressure as a result of exhaust ventilation by propeller fans or roof ventilators. This negative pressure invites the outside wind to penetrate the air curtain and enter the building. The negative pressure, therefore, should be relieved by a supply roof ventilator, providing makeup air, before an air curtain is installed.

After the negative pressure has been relieved, another step is taken to keep the outside wind from penetrating the air curtain: The outlet slot of the air curtain unit is somewhat tilted. As a result, the air stream leaving the slot is not exactly vertical (as shown in Fig. 21.1), but it starts with a flow slanted toward the outside and then becomes curved inward due to the wind velocity, as shown in Fig. 21.2.

INDUCED AIRFLOW

In Chap. 17 on fanless air movers, we explained how the air jets induce some additional air to be drawn in. A similar phenomenon occurs in air curtains, as shown in Fig. 21.3. Here, the primary airflow induces secondary airflows on each side of the primary airflow. However, while in the fanless air mover the additional air was desired, in the air curtain the secondary airflow is an undesirable loss of energy.

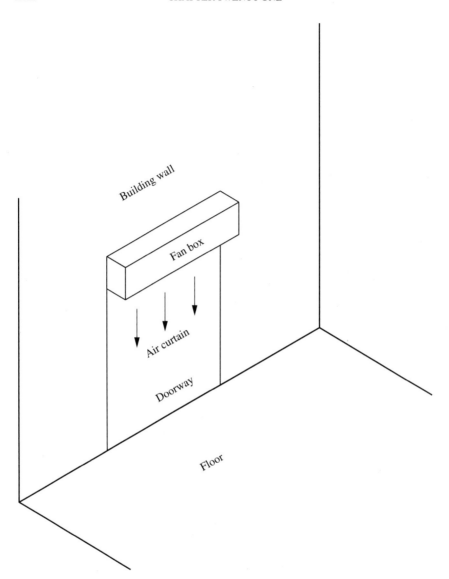

FIGURE 21.1   Schematic sketch of a typical air curtain.

REQUIREMENTS

The volume needed for the primary airflow depends on the width and height of the doorway. It can vary from 1000 to 100,000 cfm. This air volume will leave the fan box through a slot with a 3- to 8-in width and a length about equal to the width of the doorway. For industrial applications, such as factory doorways for trucks, the initial

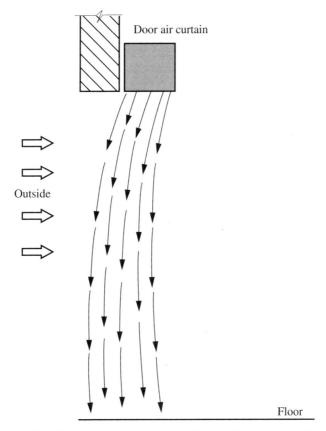

FIGURE 21.2   Schematic sketch showing the curved air stream past the fan box.

air velocity past the slot is anywhere between 2000 and 5000 fpm. Since the airflow spreads out while moving down, the air velocity will decrease gradually. At 3 ft above the floor, the primary air velocity should still be 1600 fpm. For commercial applications, such as office buildings, etc., the air velocities are lower so that the people will not be inconvenienced.

Three types of fans are used to produce air curtains: vaneaxial fans, FC centrifugal fans, and cross-flow blowers. Each type has some advantages and some disadvantages, as will be seen.

Vaneaxial Fans

The vaneaxial fan first blows the air into a horizontal round duct, as shown in Fig. 21.4, mounted above the doorway. The duct is closed at the other end, but at the bottom it has a longitudinal slot 2 to 8 in wide. Sometimes, the slot width is adjustable. The axial length of the slot corresponds to the width of the doorway. The air stream

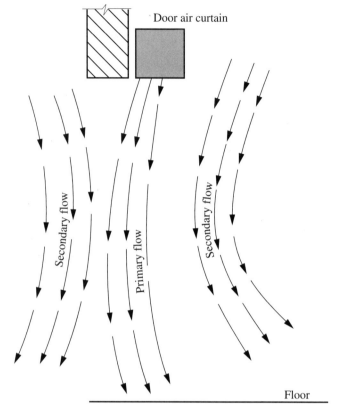

FIGURE 21.3   Schematic sketch showing a secondary airflow on each side of the primary airflow.

has to make a 90° turn from the axial flow in the duct to the downward flow through the slot. This turn will result in a minor loss of pressure and efficiency. Nevertheless, the flow distribution over the length of the slot is fairly even.

The use of vaneaxial fans in air curtains has two advantages: Only one fan is needed, and a high fan efficiency (about 80 percent) can be obtained, provided that the slot is wide enough for the vaneaxial fan to operate in the good performance range. If the slot is too narrow, the vaneaxial fan will operate in the stalling range and will be inefficient and noisy. Caution has to be used to avoid this. Wheel diameters range from 12 to 24 in.

FC Centrifugal Fans

FC centrifugal fans have larger air volumes and higher outlet velocities than other types of centrifugal fans of the same size and speed. For this reason, FC centrifugal fans are best suited for air curtains. The wheel diameters range from 4 to 18 in. The doorways usually range from 3 to 20 ft but sometimes are even larger. To obtain air

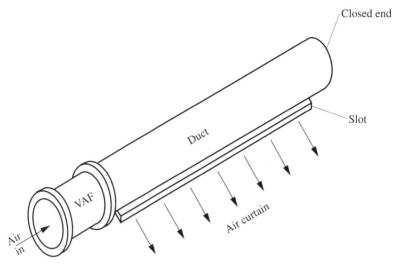

FIGURE 21.4 Schematic sketch of a vaneaxial fan blowing air into a round duct and through a slot at the bottom, producing an air curtain.

curtains this wide, several FC centrifugal fans, DIDW, are used, operating in parallel, mounted inside a rectangular box, as shown in Fig. 21.5. This may result in a slightly uneven flow distribution over the width of the doorway. The maximum fan efficiency here is 50 percent in small sizes and up to 60 percent in the larger sizes. The cost is moderate, even though several fans are used.

FIGURE 21.5 Air curtain unit for a 38-in-wide doorway. The unit contains several FC centrifugal fans operating in parallel and blowing down the doorway to protect the inside of the building from wind, rain, and insects. *(Courtesy of Leading Edge, Inc., Miami, Fla.)*

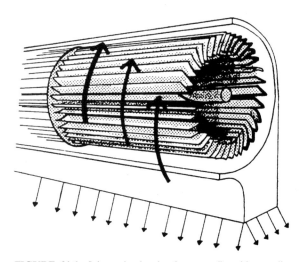

FIGURE 21.6 Schematic sketch of a cross-flow blower discharging air down across a doorway.

Cross-Flow Blowers

Figure 21.6 shows an 8-in cross-flow blower drawing air in radially along its entire axial length and discharging it through a 3-in-wide slot at the bottom. In cross-flow blowers for air curtains, the wheel diameters range from 5 to 12 in, and the motor sizes range from 3 to 60 hp. Air volumes up to 100,000 cfm have been obtained. The use of cross-flow blowers in air curtains has two advantages and two disadvantages compared with vaneaxial fans and FC centrifugal fans. Here are the two advantages:

1. As explained in Chap. 13, cross-flow blower wheels can be built to any axial width desired. One wheel will do it. Figure 21.7 shows a 10-in cross-flow blower wheel 6 ft long.

2. Due to the one-wheel construction, the flow distribution is 100 percent even.

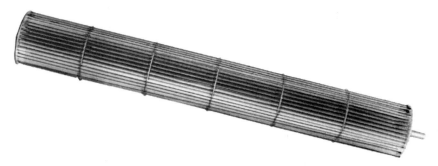

FIGURE 21.7 A 10-in cross-flow blower wheel 6 ft long.

Here are the two disadvantages:

1. The cross-flow blower has a lower efficiency, only about 40 percent maximum, compared with 50 to 60 percent for the FC centrifugal fan and even higher for the vaneaxial fan. Figure 13.4 showed a comparison of performance between a cross-flow blower and an FC centrifugal fan.

2. The cross-flow blower wheel is more expensive to build and to balance.

THREE EXAMPLES

Example 1. The main industrial application of the air curtain is in protecting the inside of a building, such as a factory or a warehouse, at the doorways where trucks are loaded and unloaded. Suppose we use an initial velocity past the slot of 3000 fpm and a slot area of 5 in × 20 ft. The outlet area then will be $OA = 8.33$ ft^2, and the required air volume will be $8.33 \times 3000 = 25,000$ cfm.

Example 2. For commercial applications, such as the entrance to an office building or the entrance to a department store, a lower initial outlet velocity of 1900 fpm will be needed to prevent excessive air currents that would disturb people passing through. In this case, a wider and thicker air current will be required. If the slot outlet will be 8 in × 30 ft, the outlet area will be $OA = 20$ ft^2, and the required air volume will be $20 \times 1900 = 38,000$ cfm.

Example 3. In Examples 1 and 2, the air curtain was vertical and the fans were installed above the doorway. Figure 21.8 shows an example where the air curtain is horizontal, providing a roof over a crusher pit, to contain the dust and at the same time push it over to the exhaust opening on the other side.

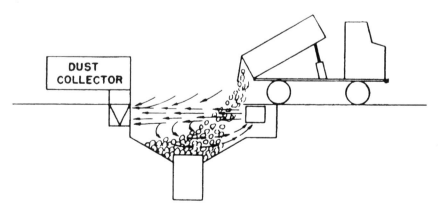

FIGURE 21.8 Horizontal air curtain blowing across a dust-laden space to contain the dust and to push it over to an exhaust opening on the other side.

PERFORMANCE TESTING

Performance tests on air curtain units are described in the AMCA booklet AMCA 220-91. Here the unit is connected to an inlet chamber, such as shown in Fig. 18.3, where the air volume and static pressure are measured, but in addition, some measurements are taken on the discharge side, where the air curtain actually spreads out. The test unit may be mounted in horizontal or in vertical position, and the following three performance features are tested:

1. The air volume will be measured for free delivery and for negative pressures of 0.1 and 0.2 inWC in the inlet chamber in the same way as would be done for any fan exhausting from a chamber.
2. The throw or reach of the air curtain unit is determined. To be more precise, the discharge velocity at various distances from the nozzle is measured.
3. The uniformity of the discharge velocity is checked at various locations along the width of the air curtain.

SUMMARY

The air curtain is an inexpensive method to protect the inside of a building from the cold or hot air on the outside and from insects, rain, and dust on the outside while still keeping the doorway open for traffic by trucks and people. Most air curtains are vertical, either blowing down or blowing across doorways, to replace a door. Some air curtains are horizontal and provide a roof over a dust-laden space and carry the dust across to a dust collector. Vaneaxial fans, FC centrifugal fans, and cross-flow blowers can be used to produce air curtains. Vaneaxial fans have the best fan efficiencies but carry a risk of operation in the stalling range. Several FC centrifugal fans, operating in parallel and blowing down the doorway, are needed to obtain the desired width of the air curtain. Cross-flow blowers have lower efficiencies and higher cost but an even flow distribution.

CHAPTER 22
CEILING FANS

DESCRIPTION

Ceiling fans are large, lightweight propeller fan wheels suspended from the ceiling and blowing a gentle breeze down to the floor. They are made in sizes from 36 to 72 in, handling from 6000 to 52,000 cfm. The air velocities are low, due to small blade angles and low speeds. The maximum air velocity, just below the fan, is about 600 fpm. At floor level, it is about 100 fpm.

Ceiling fans use direct drive from single-phase, low-speed motors. Most of them have 18 poles, but some have 16, 20, or 22 poles. They run at 270 to 400 rpm, but even lower speeds can be obtained with two- and three-speed motors and with variable-speed controls. The motors consume between 60 and 200 W of power. They are inside-out motors, with the inside stationary and the outside rotating. Three to six blades are attached to the rotating outside. The hub diameter is small, about 15 percent of the wheel diameter. This is adequate because these fans operate at or near free delivery. The blade angle (pitch) usually is between 10° and 15° and constant from hub to tip. A larger pitch, of course, would result in more air volume but also in an increased power consumption. Despite the low-speed motors, these are inexpensive fans, since they are only fan wheels without a fan housing. Figure 22.1 shows a 60-in ceiling fan with three blades attached to the rotating part of the motor.

FUNCTION DURING THE HEATING SEASON

The main purpose of a ceiling fan is destratification during the heating season, i.e., bringing down the warm air near the ceiling and mixing it with the cool air at floor level. The temperature difference depends on the height of the building. For a 20-ft ceiling height, there often is a 15°F temperature difference. By bringing down the warm air, the temperature difference can be reduced to 2°F. This will minimize the heating requirements and result in a heating cost reduction of at least 5 percent.

FUNCTION DURING THE SUMMER

A second function of a ceiling fan is the cooling effect it produces during the summer months as the gentle breeze causes some evaporation of the moisture on the skin. Furthermore, some models have a reversible motor so that the fan can blow upward during the summer for general air circulation in the space.

FIGURE 22.1   A 60-in industrial ceiling fan with three blades.

PERFORMANCE

Figure 22.2 shows the flow pattern for a typical ceiling fan mounted 20 ft above the floor. As the air stream moves down, it spreads out to ever-increasing areas, resulting in decreasing air velocities. The largest floor area that can be covered is a circle of 80 ft diameter. This can be accomplished with a large fan, 60- to 72-in diameter, mounted 20 to 25 ft above the floor. This is the optimal mounting height. For higher or lower mounting heights, the covered area becomes smaller.

Figure 22.3 is a graph showing how the air velocity decreases as the air stream moves down and diffuses radially outward. The rate of decrease is fast at first and becomes slower as the floor level is approached.

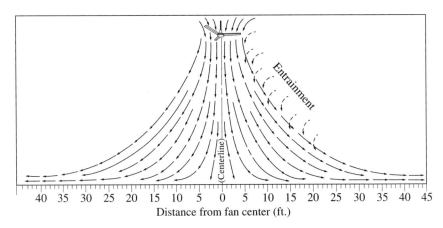

FIGURE 22.2   Flow pattern of a typical 60-in ceiling fan mounted 20 ft above the floor. *(Courtesy of Leading Edge, Inc., Miami, Fla.)*

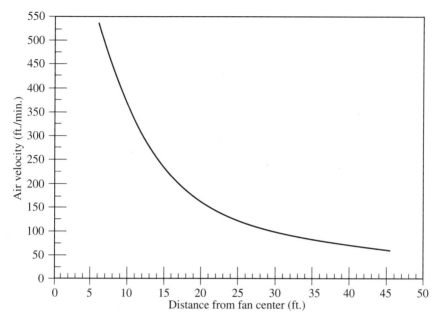

FIGURE 22.3 Graph showing how the air velocity decreases as the air stream moves down toward the floor and diffuses radially outward. *(Courtesy of Leading Edge, Inc., Miami, Fla.)*

APPLICATIONS

Ceiling fans are used in residential, commercial, agricultural, and industrial applications. Here are some examples.

Residential Applications

The ceiling fan can be used in any room but is most often found in exercise rooms and finished basements. Here the fan should be positioned 1 to 2 ft from the ceiling and not less than 8 to 9 ft from the floor. Occasionally, a ceiling fan is used in outdoor patios.

Commercial Applications

Ceiling fans are used in retail stores, office buildings, public buildings, schools, gymnasiums, libraries, churches, indoor swimming pools, and indoor tennis courts.

Agricultural Applications

Ceiling fans are used in animal shelters, poultry houses, stables, and dairy barns.

Industrial Applications

Ceiling fans are used in factories, foundries, and warehouses. Here, the fan is mounted 20 to 25 ft above the floor for maximum area coverage.

SUMMARY

Ceiling fans are not an absolute necessity, but they are beneficial because they economize regarding the heating cost in winter and they improve the comfort of the occupants in summer. They are used in many applications because they are simple to install and are inexpensive, both in first cost and in operation.

CHAPTER 23
AMCA STANDARDS

STANDARDS HANDBOOK

The Air Movement and Control Association, Inc. (AMCA), Arlington Heights, Illinois, has established some standards for axial fans, centrifugal fans, and tubular centrifugal fans (also called *mixed-flow fans* or *axial-centrifugal fans*), as well as for some accessories, such as inlet boxes. AMCA's *Standards Handbook,* Publication 99-86, starts with definitions for 27 products, such as the several fan types, air-handling units, heaters, roof ventilators, air curtains, makeup air units, louvers, penthouses, and dampers. This chapter will present and discuss the most important portions of these AMCA standards.

CENTRIFUGAL FANS

There are two groups, called *centrifugal fans* and *industrial centrifugal fans.* Both these groups were discussed in Chap. 7. Figures 23.1 and 23.2 show the AMCA recommendations for these two groups of fans. You will note that there are 25 standard sizes of centrifugal fans and 16 standard sizes of industrial centrifugal fans. Reviewing some of the data presented in Chap. 7, the two groups have the following differences in their designs: The centrifugal fans with AF and BI blades usually have d_1/d_2 ratios of about 75 percent, maximum blade widths equal to 46 percent of d_1 and mostly blade angles of 15° to 20° at the leading edge and about 45° at the blade tip. The industrial centrifugal fans with AH wheels have smaller d_1/d_2 ratios (about 58 percent), narrower blades (46 percent of a smaller d_1), and steeper blade angles (about 23° at the leading edge and 62° at the blade tip). These differences will result in less air volume and more static pressure for the industrial centrifugal fans, as was indicated in Fig. 7.20, which showed a comparison of the performances.

DRIVE ARRANGEMENTS
FOR CENTRIFUGAL FANS

Figures 23.3a and 23.3b present 16 different drive arrangements for centrifugal fans, some of them for direct drive, some of them for belt drive, and some of them for either direct or belt drive. Twelve arrangements are for SWSI, and four are for DWDI. Four of the arrangements are with inlet boxes. In most of the belt-drive arrangements, the bearings are supported by the housing structure, but in four of the arrangements, separate bearing pedestals are used.

Figure 23.4 shows how the airflow through the inlet box of a centrifugal fan can approach the fan inlet from above, from below (with possible interference with the floor structure), from a side, or from an angular direction.

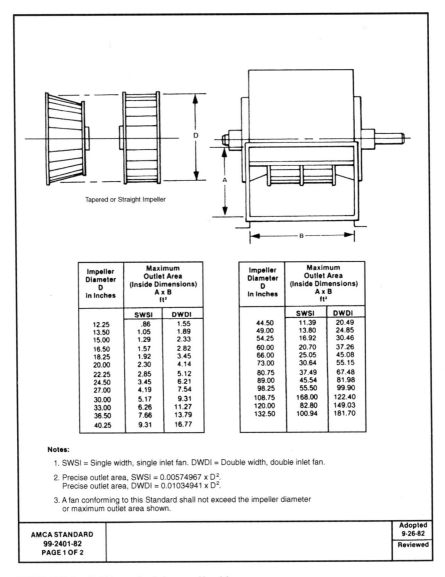

Tapered or Straight Impeller

Impeller Diameter D In Inches	Maximum Outlet Area (Inside Dimensions) A x B ft²		Impeller Diameter D In Inches	Maximum Outlet Area (Inside Dimensions) A x B ft²	
	SWSI	DWDI		SWSI	DWDI
12.25	.86	1.55	44.50	11.39	20.49
13.50	1.05	1.89	49.00	13.80	24.85
15.00	1.29	2.33	54.25	16.92	30.46
16.50	1.57	2.82	60.00	20.70	37.26
18.25	1.92	3.45	66.00	25.05	45.08
20.00	2.30	4.14	73.00	30.64	55.15
22.25	2.85	5.12	80.75	37.49	67.48
24.50	3.45	6.21	89.00	45.54	81.98
27.00	4.19	7.54	98.25	55.50	99.90
30.00	5.17	9.31	108.75	168.00	122.40
33.00	6.26	11.27	120.00	82.80	149.03
36.50	7.66	13.79	132.50	100.94	181.70
40.25	9.31	16.77			

Notes:

1. SWSI = Single width, single inlet fan. DWDI = Double width, double inlet fan.

2. Precise outlet area, SWSI = $0.00574967 \times D^2$.
   Precise outlet area, DWDI = $0.01034941 \times D^2$.

3. A fan conforming to this Standard shall not exceed the impeller diameter or maximum outlet area shown.

AMCA STANDARD 99-2401-82 PAGE 1 OF 2		Adopted 9-26-82
		Reviewed

FIGURE 23.1   AMCA standards for centrifugal fans.

Figure 23.5 shows 16 possible designations for the rotation and discharge of centrifugal fans. While top horizontal and bottom horizontal may be the most commonly used arrangements, upblast, downblast, and angular discharge are used quite often, too.

Figure 23.6 shows four motor positions for belt or chain drive centrifugal fans. Here the motors are mounted separately, as is sometimes done in large fan sizes. In

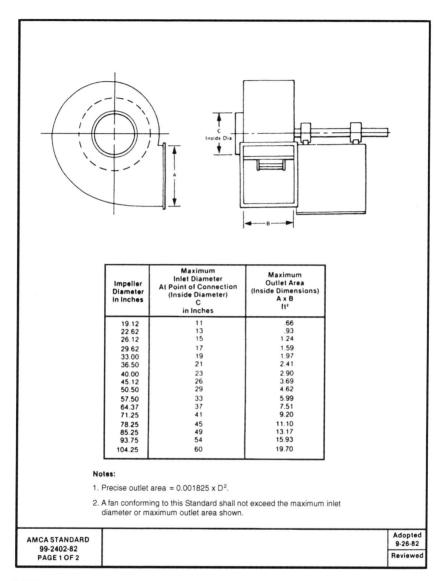

Impeller Diameter In Inches	Maximum Inlet Diameter At Point of Connection (Inside Diameter) C in Inches	Maximum Outlet Area (Inside Dimensions) A x B ft²
19.12	11	.66
22.62	13	.93
26.12	15	1.24
29.62	17	1.59
33.00	19	1.97
36.50	21	2.41
40.00	23	2.90
45.12	26	3.69
50.50	29	4.62
57.50	33	5.99
64.37	37	7.51
71.25	41	9.20
78.25	45	11.10
85.25	49	13.17
93.75	54	15.93
104.25	60	19.70

Notes:

1. Precise outlet area = 0.001825 x D^2.

2. A fan conforming to this Standard shall not exceed the maximum inlet diameter or maximum outlet area shown.

AMCA STANDARD 99-2402-82 PAGE 1 OF 2		Adopted 9-26-82
		Reviewed

FIGURE 23.2   AMCA standards for industrial centrifugal fans.

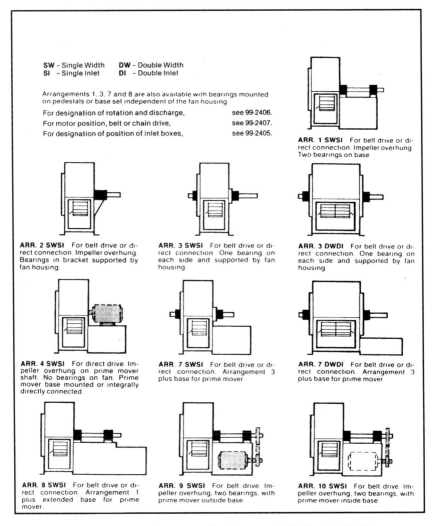

SW – Single Width **DW** – Double Width
SI – Single Inlet **DI** – Double Inlet

Arrangements 1, 3, 7 and 8 are also available with bearings mounted on pedestals or base set independent of the fan housing

For designation of rotation and discharge,	see 99-2406.
For motor position, belt or chain drive,	see 99-2407.
For designation of position of inlet boxes,	see 99-2405.

ARR. 1 SWSI For belt drive or direct connection. Impeller overhung Two bearings on base

ARR. 2 SWSI For belt drive or direct connection. Impeller overhung. Bearings in bracket supported by fan housing.

ARR. 3 SWSI For belt drive or direct connection. One bearing on each side and supported by fan housing

ARR. 3 DWDI For belt drive or direct connection. One bearing on each side and supported by fan housing

ARR. 4 SWSI For direct drive. Impeller overhung on prime mover shaft. No bearings on fan. Prime mover base mounted or integrally directly connected.

ARR. 7 SWSI For belt drive or direct connection. Arrangement 3 plus base for prime mover

ARR. 7 DWDI For belt drive or direct connection. Arrangement 3 plus base for prime mover

ARR. 8 SWSI For belt drive or direct connection. Arrangement 1 plus extended base for prime mover.

ARR. 9 SWSI For belt drive. Impeller overhung, two bearings, with prime mover outside base

ARR. 10 SWSI For belt drive. Impeller overhung, two bearings, with prime mover inside base

FIGURE 23.3a Drive arrangements for centrifugal fans per AMCA Standard 99-86.

smaller sizes, the motor usually is mounted on some part of the housing structure, as shown in Fig. 23.3.

OPERATING LIMITS FOR CENTRIFUGAL FANS

Figure 23.7 shows the operating limits for single-width centrifugal fans with airfoil or backwardly inclined blades, dividing them into classes I, II, and III according to their

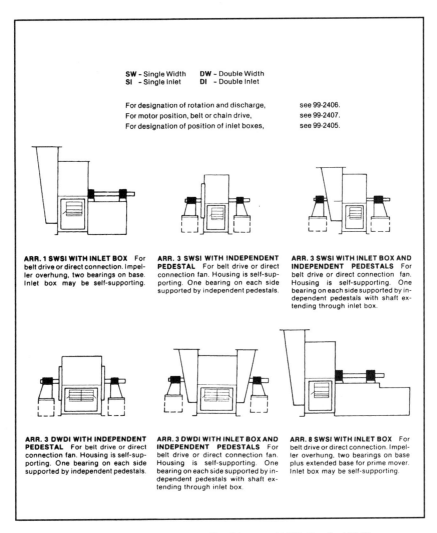

SW - Single Width **DW** - Double Width
SI - Single Inlet **DI** - Double Inlet

For designation of rotation and discharge, see 99-2406.
For motor position, belt or chain drive, see 99-2407.
For designation of position of inlet boxes, see 99-2405.

ARR. 1 SWSI WITH INLET BOX For belt drive or direct connection. Impeller overhung, two bearings on base. Inlet box may be self-supporting.

ARR. 3 SWSI WITH INDEPENDENT PEDESTAL For belt drive or direct connection fan. Housing is self-supporting. One bearing on each side supported by independent pedestals.

ARR. 3 SWSI WITH INLET BOX AND INDEPENDENT PEDESTALS For belt drive or direct connection fan. Housing is self-supporting. One bearing on each side supported by independent pedestals with shaft extending through inlet box.

ARR. 3 DWDI WITH INDEPENDENT PEDESTAL For belt drive or direct connection fan. Housing is self-supporting. One bearing on each side supported by independent pedestals.

ARR. 3 DWDI WITH INLET BOX AND INDEPENDENT PEDESTALS For belt drive or direct connection fan. Housing is self-supporting. One bearing on each side supported by independent pedestals with shaft extending through inlet box.

ARR. 8 SWSI WITH INLET BOX For belt drive or direct connection. Impeller overhung, two bearings on base plus extended base for prime mover. Inlet box may be self-supporting.

FIGURE 23.3b Drive arrangements for centrifugal fans per AMCA Standard 99-86.

outlet velocities and static pressures. The fans must be structurally strong enough to produce the outlet velocities (fpm) and static pressures (inWC) indicated for each class.

Figure 23.8 shows the operating limits for double-width centrifugal fans with air-foil or backwardly inclined blades. Figure 23.9 shows the operating limits for single-width FC centrifugal fans. Figure 23.10 shows the operating limits for double-width FC centrifugal fans. Figure 23.11 shows the operating limits for tubular centrifugal fans.

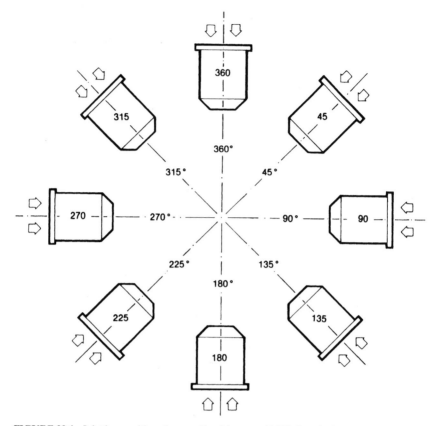

FIGURE 23.4   Inlet box positions for centrifugal fans per AMCA Standard 99-86.

Figure 23.12 shows various drive arrangements for tubular centrifugal fans. Figure 23.13 shows various drive arrangements for axial-flow fans with or without evasé and with or without an inlet box. However, inlet boxes are not often used in combination with axial-flow fans, because inlet boxes produce uneven flow conditions and axial-flow fans are more sensitive to inlet turbulence and to an uneven distribution of the inlet velocity.

SPARK-RESISTANT CONSTRUCTION

This section of the standard deals with fans handling potentially explosive or flammable gases or vapors where sparks must be avoided. It applies to centrifugal fans, axial-flow fans, propeller fans, and power roof ventilators. It discusses permissible materials, spark-preventing constructions, and various arrangements to improve

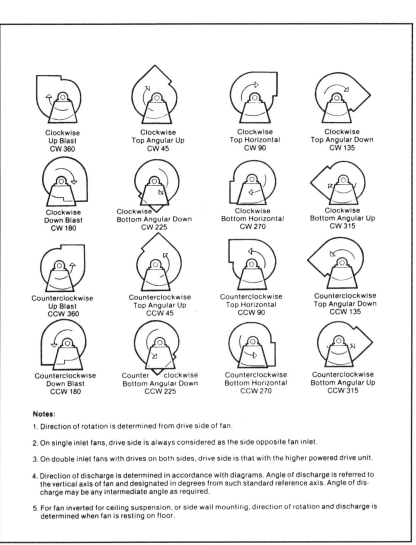

Clockwise
Up Blast
CW 360

Clockwise
Top Angular Up
CW 45

Clockwise
Top Horizontal
CW 90

Clockwise
Top Angular Down
CW 135

Clockwise
Down Blast
CW 180

Clockwise
Bottom Angular Down
CW 225

Clockwise
Bottom Horizontal
CW 270

Clockwise
Bottom Angular Up
CW 315

Counterclockwise
Up Blast
CCW 360

Counterclockwise
Top Angular Up
CCW 45

Counterclockwise
Top Horizontal
CCW 90

Counterclockwise
Top Angular Down
CCW 135

Counterclockwise
Down Blast
CCW 180

Counter clockwise
Bottom Angular Down
CCW 225

Counterclockwise
Bottom Horizontal
CCW 270

Counterclockwise
Bottom Angular Up
CCW 315

Notes:

1. Direction of rotation is determined from drive side of fan.

2. On single inlet fans, drive side is always considered as the side opposite fan inlet.

3. On double inlet fans with drives on both sides, drive side is that with the higher powered drive unit.

4. Direction of discharge is determined in accordance with diagrams. Angle of discharge is referred to the vertical axis of fan and designated in degrees from such standard reference axis. Angle of discharge may be any intermediate angle as required.

5. For fan inverted for ceiling suspension, or side wall mounting, direction of rotation and discharge is determined when fan is resting on floor.

FIGURE 23.5 Designations for rotation and discharge of centrifugal fans per AMCA Standard 99-86.

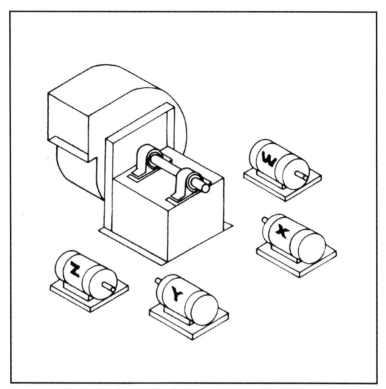

Location of motor is determined by facing the drive side of fan and designating the motor positions by letters W, X, Y, or Z as the case may be.

FIGURE 23.6   Motor positions for belt or chain drive centrifugal fans per AMCA Standard 99-86.

the safety of fans and fan systems, including bearings, drive components, and electrical devices.

SUMMARY

The standards established by AMCA are valuable in several ways. For example, the fan industry needed definitions to avoid misunderstandings. It also needed fan classifications with regard to outlet velocities and static pressures and other guidelines for uniform provisions, such as drive arrangements, rotations, discharge positions, and inlet boxes. Another value of the standards is the promotion of safety regarding structural strength and spark resistance. Finally, the standards will permit customers to make easier comparisons of fans made by different manufacturers, if they conform with the AMCA standards. This means that customers can study catalogs on competitive fans and decide in favor of one not only by comparing prices but also by comparing design and performance features.

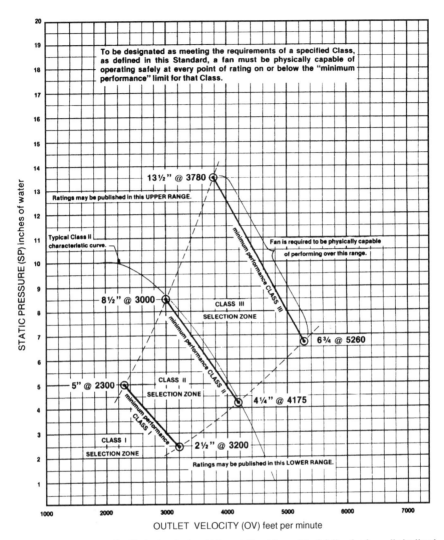

FIGURE 23.7   Operating limits for single-width centrifugal fans with airfoil or backwardly inclined blades per AMCA Standard 99-86.

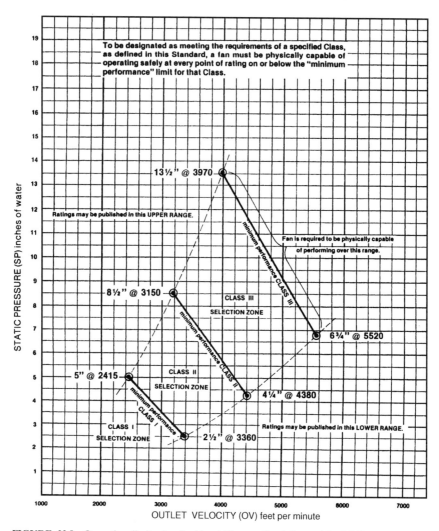

FIGURE 23.8 Operating limits for double-width centrifugal fans with airfoil or backwardly inclined blades per AMCA Standard 99-86.

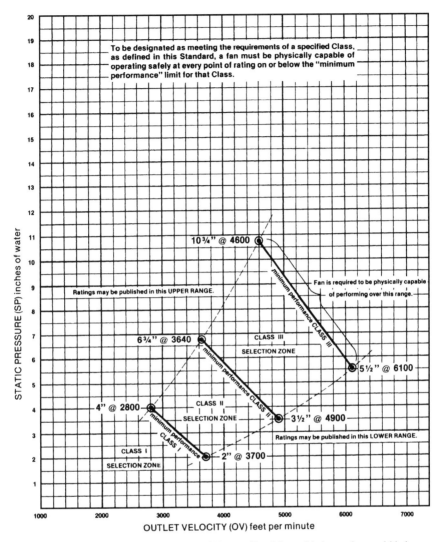

FIGURE 23.9 Operating limits for single-width centrifugal fans with forward-curved blades per AMCA Standard 99-86.

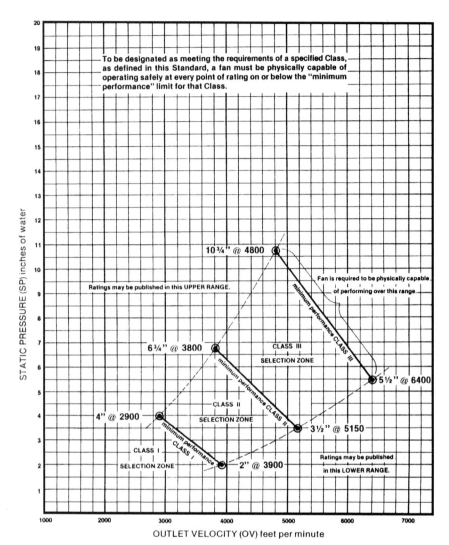

FIGURE 23.10 Operating limits for double-width centrifugal fans with forward-curved blades per AMCA Standard 99-86.

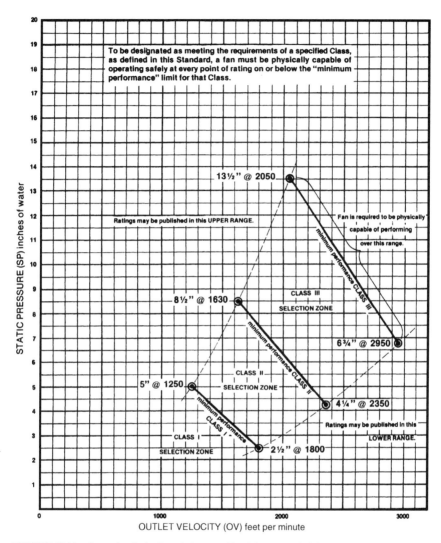

FIGURE 23.11 Operating limits for tubular centrifugal fans per AMCA Standard 99-86.

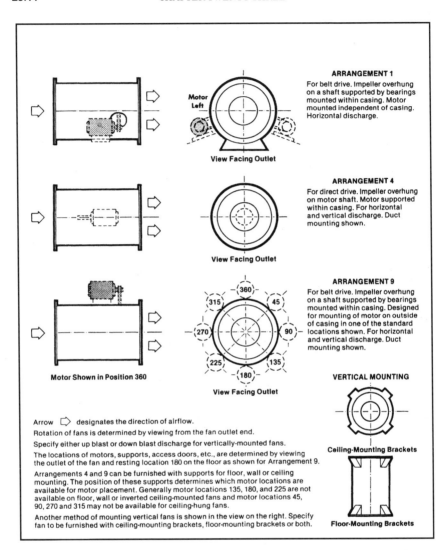

ARRANGEMENT 1
For belt drive. Impeller overhung on a shaft supported by bearings mounted within casing. Motor mounted independent of casing. Horizontal discharge.

View Facing Outlet

ARRANGEMENT 4
For direct drive. Impeller overhung on motor shaft. Motor supported within casing. For horizontal and vertical discharge. Duct mounting shown.

View Facing Outlet

ARRANGEMENT 9
For belt drive. Impeller overhung on a shaft supported by bearings mounted within casing. Designed for mounting of motor on outside of casing in one of the standard locations shown. For horizontal and vertical discharge. Duct mounting shown.

Motor Shown in Position 360

View Facing Outlet

VERTICAL MOUNTING

Ceiling-Mounting Brackets

Floor-Mounting Brackets

Arrow ▷ designates the direction of airflow.

Rotation of fans is determined by viewing from the fan outlet end.

Specify either up blast or down blast discharge for vertically-mounted fans.

The locations of motors, supports, access doors, etc., are determined by viewing the outlet of the fan and resting location 180 on the floor as shown for Arrangement 9.

Arrangements 4 and 9 can be furnished with supports for floor, wall or ceiling mounting. The position of these supports determines which motor locations are available for motor placement. Generally motor locations 135, 180, and 225 are not available on floor, wall or inverted ceiling-mounted fans and motor locations 45, 90, 270 and 315 may not be available for ceiling-hung fans.

Another method of mounting vertical fans is shown in the view on the right. Specify fan to be furnished with ceiling-mounting brackets, floor-mounting brackets or both.

FIGURE 23.12 Drive arrangements for tubular centrifugal fans per AMCA Standard 99-86.

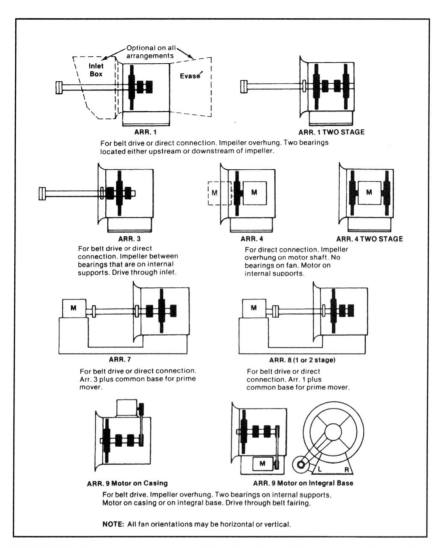

FIGURE 23.13 Drive arrangements for axial-flow fans with or without evasé and with or without an inlet box per AMCA Standard 99-86.

CHAPTER 24
MECHANICAL STRENGTH

CENTRIFUGAL FORCE

As a fan wheel rotates, the blades are subject to centrifugal force, pulling the blades radially outward. In order to check whether this force might break the fan blade from the hub or from the back plate, we want to calculate this centrifugal force. In order to do this, we need to know only three quantities: the weight W of one blade, the distance R of the blade center from the center of rotation, and the speed of the rotating fan wheel. If we know these three quantities, we can calculate the centrifugal force F using the following formula:

$$F = 3.409 \times 10^{-4} \times W \times R \times \text{rpm}^2 \qquad (24.1)$$

where F = force in pounds
  W = weight in pounds
  R = distance in feet

Example: Suppose the blade weighs 5.5 lb, its center is 2 ft from the center of rotation, and the fan wheel runs at 1200 rpm. The centrifugal force then will be

$$F = 3.409 \times 10^{-4} \times 5.5 \times 2 \times 1200^2 = 5400 \text{ lb}$$

Figure 24.1 is a graph from which the centrifugal force F for $W = 1$ lb can be read directly. We first find 1200 rpm on the abscissa and a 2-ft radius on the ordinate. Where these two meet, we find a centrifugal force of 980 lb if the blade weighs only 1 lb. If we multiply this number by the actual weight of 5.5 lb, we get a centrifugal force of 5400 lb, the same as from Eq. (24.1).

TENSILE STRESS

If the cross section of the blade is 3.4 in^2, the simple tensile stress will be $5400/3.4 = 1588$ lb/in^2 (psi). With a safety factor of 3, we get 4765 psi. If our material has a tensile strength of 60,000 psi (structural steel), we are on the safe side. However, we could not use aluminum of 20,000 psi tensile strength. For shear stress, the calculation would be similar.

If a safety factor of 3 seems higher than necessary, we have to consider three side effects that require this high safety factor:

1. The tensile strength is the strength of actual fracture. The yield strength may be only 65 percent of the tensile strength.
2. In centrifugal fans with wide blades and small blade angles, the blades may buckle out before they break.

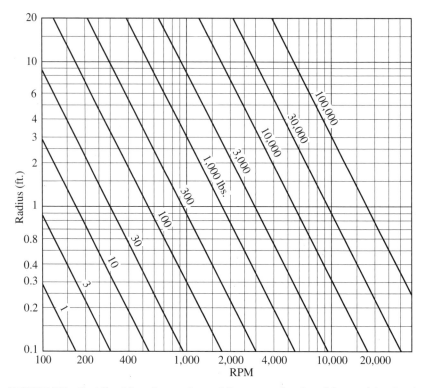

FIGURE 24.1 Centrifugal force in pounds on a 1-lb mass as a function of the speed (rpm) and the radius R.

3. Some slight blade vibration may result in fatigue failure even before the yield strength is reached. Blade vibration is a particular danger if the speed produces a natural blade frequency.

For these three reasons, a high safety factor is advisable in axial-flow fans as well as in centrifugal fans.

AXIAL REACTION FORCE

In axial-flow fans, the shaft is subject to an axial reaction force opposite the direction of airflow. Suppose the fan produces a static pressure of 5 inWC. One inch of water column equals 0.036127 psi (see Conversion Factors before Chap. 1). The 5 inWC then will be equal to $5 \times 0.036127 = 0.1806$ psi. If this is a 20-in axial-flow fan, the housing area will be 2.18 ft^2 or 314 in^2. The axial force, therefore, will be $314 \times 0.1806 = 57$ lb. Note that the axial reaction force is small and will not be a problem.

SHAFT TORQUE

Finally, we should check the torque T transmitted by the shaft driving the fan wheel by belt drive. This torque can be calculated from the following formula:

$$T = 63030 \times \text{bhp/rpm} \qquad (24.2)$$

where T is in pound-inches.

Suppose that the fan consumes 100 bhp while running at 1750 rpm. The torque then will be $T = 63030 \times 100/1750 = 3600$ lb·in. Again, an adequate safety factor will be needed to take care of the starting torque, which may be considerably larger. The shaft has to be strong enough to transmit this torque.

TEST PIT

Fan wheels can be tested for mechanical strength in a test pit with heavy steel walls as a protection in case of fractures with flying blades. The speed is increased gradually, either by rheostat speed control or simply by changing pulleys. After the actual operating speed has been reached without any deformation, the speed often is further increased for overspeed testing to ensure a safety margin. This increased speed may require an excessive motor brake horsepower. If the test pit is built airtight, the required motor brake horsepower can be greatly reduced by rarefying the air pressure inside the test pit with a vacuum pump. Equation (5.23) shows that the brake horsepower is proportional to the air density, which in turn is proportional to the atmospheric pressure. Therefore, if the pressure inside the pit is reduced to one-hundredth (from 29.92 to 0.30 inHg), the brake horsepower is reduced to one-hundredth too (say from 100 to 1 bhp). While the overspeed testing could be continued to actual fracture and destruction, most of the time this is not done, but the test is continued only until a satisfactory safety margin has been reached.

CHAPTER 25
TROUBLE SHOOTING AND PROBLEM SOLVING

GUIDELINES

Here is a list of troubles and problems frequently encountered in the operation of fans.*

Symptom	Possible cause
1. Fan will not start	Blown fuses Broken belts Loose pulleys Impeller touching housing Wrong voltage
2. Excessive noise level	Impeller hitting inlet cone Loose motor bolts Wrong pulley size Defective bearing Bent or undersized shaft Poor wheel balance
3. Air volume too small	Wrong fan rotation Fan speed too slow Dampers closed too much Coils and filters dirty Inlet or outlet obstructions Fan too small for application
4. Air volume too large	Wrong fan rotation Fan speed too high Dampers not installed Access door open Fan too large for application
5. Horsepower too high	Wrong fan rotation Fan speed too high Fan type or size not best for application Incorrect motor selection Gas density too heavy

In addition to these simple events, here are a few examples of problems that I have encountered. They might be of help in avoiding similar problems.

---

* Some of this information has been taken from Bleier, F. P., *Fan Design and Application Handbook* (Hopkin, Minn.: Ammerman Company).

CONVERGING CONE

In Fig. 1.8 we reported about the excessive air spin developed as the airflow passed through a converging cone. *Conclusion:* A converging cone sometimes should be equipped with some longitudinal vanes to prevent excessive air spin.

WRONG ROTATION

A 22-in pressure blower was designed by me. It was directly driven by a 5-hp, 3500-rpm motor. It had narrow blades that were flat and backward inclined by 15° from a radial line. After the shop had built a test sample, the factory manager phoned. He was perplexed because the motor was badly overloaded and the fan was quite noisy. I traveled to the plant, inspected the unit, and found that the blades were welded to the back plate at the 15° angle but inclined to the wrong side. In other words, these blades were not backward inclined but were forward inclined and therefore overloaded the motor. A new fan wheel was built. It performed quietly, in accordance with the requirements and without motor overload. This example belongs to the preceding group 5 (brake horsepower too high due to wrong fan rotation). *Conclusions:*

1. Check the blade angles in relation to the fan rotation.
2. Forward-inclined, flat blades are not customary in centrifugal fans. Forward-curved blades, radial blades, and backward-inclined, flat blades are customary, but not forward-inclined, flat blades.

WRONG INLET SPIN

An 18-in vaneaxial fan with inlet vanes was designed by me. After the shop had built a test sample, the president phoned. He pointed out that the air stream past the unit spread outward in a conical pattern, leaving the inner portion with hardly any airflow. I inspected the unit and found that the inlet vanes were curved the wrong way, giving the air stream a spin *in* the direction of fan rotation instead of *against* it. In other words, the inlet vanes reinforced the air spin instead of counteracting it. This resulted in an even stronger air spin past the unit that—by centrifugal force—caused the air stream to spread outward. After new inlet vanes were installed, the air stream left the unit axially and evenly distributed from hub to tip. At the same time, the air volume was increased by 16 percent on average. *Conclusion:* Check the direction of the air spin past inlet vanes. It should be opposite the fan rotation so that the subsequent fan wheel will produce an approximately axial airflow.

WRONG UNITS (METRIC)

A coworker who had been running and calculating fan tests for years and was quite good at it studied a book on fan design. This book was an English translation of a

German book, with the formulas in metric units. The translator did not bother to convert the formulas to American units. My coworker put actual values into the formulas, using American units, and was disappointed that the results did not make sense. After I went over his calculations with him, the problem was cleared up and he continued with his studies. *Conclusion:* Before using a formula, check whether it is for American or for metric units. (American units are used throughout this book.)

CHAPTER 26
INSTALLATION, SAFETY, AND MAINTENANCE*

For successful operation of a fan, three requisites are necessary: satisfactory performance, installation, and maintenance.

1. The fan selected must be adequate for the required duty with regard to air volume, static pressure, horsepower, and noise.
2. The fan must be installed properly for ensured safety, for a minimum of inlet turbulence and vibration, and for optimal line-up of bearings, couplings, and sheaves.
3. The fan must be inspected periodically and maintained properly with regard to belt tension, lubrication, and clean surfaces. Preventive maintenance is better than subsequent expensive repairs.

Item 1 was discussed in Chaps. 8 and 11. Items 2 and 3 will be discussed now.

SAFETY PRECAUTIONS

Guards

Fans contain various rotating components, such as fan wheels, shafts, couplings, sheaves, and shaft cooling wheels. All these rotating components are potential hazards. Guards, therefore, should be provided at exposed fan inlets and outlets as well as over the couplings, V-belt drives, and so on.

Maximum Speed and Air Temperature

These limits for each fan size can be found in the rating tables published by the manufacturer. These limits *must* be observed.

Foundations

Roof ventilators are mounted on top of roof curbs that are factory-built for proper support of the weight. Larger fans require more substantial foundations. These foundations should be level, rigid, and of sufficient mass for the equipment. Concrete is preferable. The mass should be equal to three or four times the fan weight. If a structural-steel foundation is used, it must be rigid enough to ensure permanent alignment and to prevent excessive vibrations. The minimum natural frequency of

---

* Some of the information presented in this chapter has been taken from various bulletins published by the Chicago Blower Corporation, Glendale Heights, Ill., and by FloAire, Inc., Bensalem, Pa.

any foundation part should be 25 to 50 percent higher than the fan speed. When the fan is mounted on the foundation, the fan shaft should be level, and shims should be used at the support points before the foundations bolts are tightened. This will prevent any distortion or twisting of the equipment and any possible rubbing of the rotating parts.

AIRFLOW AT FAN INLET AND OUTLET

For a fan to perform in accordance with the rating tables, the airflow ahead and past the fan must be smooth and evenly distributed. This requires either no duct connection or a straight duct at least three duct diameters long before an elbow is placed. If this is not possible because of space limitations, the elbow should be equipped with turning vanes to prevent a loss in performance. Duct connections at the fan inlet or outlet should be flexible to isolate the fan from an expanded duct diameter, from vibration, and from noise. The duct should be separately anchored and not be supported by the fan.

HIGH-TEMPERATURE FANS

These fans require special provisions in order to prevent rubbing of the fan wheel against the inlet cone of centrifugal fans. This is so because the housing expands up from the foundation while the fan wheel expands concentrically from the shaft centerline. In order to obtain an even radial clearance all around when the fan handles hot air, the cold fan should have a clearance that is twice as much at the top as at the bottom. The allowance for clearance, therefore, must be larger to begin with.

V-BELT DRIVE

In axial-flow fans, the air volume can be varied with adjustable-pitch blades. In centrifugal fans, the air volume can be varied by using different blade widths. The widest flexibility, however, is obtained with belt drive. Here, the speed can be varied and thereby the air volume and also the static pressure. In Chap. 16 we mentioned that belt drive has three disadvantages: the expense of bearings, sheaves, etc., the belt drive losses, and the extra maintenance. Nevertheless, belt drive is prevalent in large sizes because it results in the widest flexibility and it avoids expensive low-speed motors.

V-belt drives require careful alignment of the sheaves and adjustment of the belt tension, both to be done after the fan has been mounted on its foundation because the tightening of the mounting bolts may distort some of the previously aligned surfaces. Figure 26.1 shows four possible misalignments to be avoided. The following steps should be taken:

1. Check that fan and motor shafts are parallel; adjust and shim the motor as required.

2. Move fan and motor sheaves axially so that their faces are not only parallel but also aligned. This can be checked with a cord, as shown in Fig. 26.2.

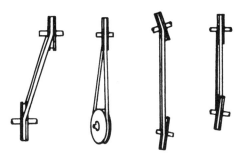

FIGURE 26.1 Fan belt misalignments to be avoided. *(Courtesy of Chicago Blower Corporation, Glendale Heights, Ill.)*

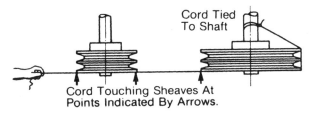

FIGURE 26.2   Use of a cord to check the alignment of two sheaves. *(Courtesy of Chicago Blower Corporation, Glendale Heights, Ill.)*

3. Check fan and motor sheaves for balance whenever there is too much vibration. Use balanced sheaves.

4. It is normal on V-belt drives handling more than 20 hp to squeal on start-up. Do not tighten belts too much.

Figure 26.3 shows three possible shapes a fan belt will assume, depending on its tightness. Belts either too loose or too tight may cause vibration and excessive wear. The following steps should be taken to obtain the proper belt tension:

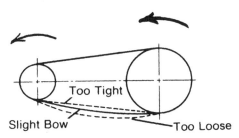

FIGURE 26.3   The shape of a fan belt on the slack side is an indication of the belt tension. *(Courtesy of Chicago Blower Corporation, Glendale Heights, Ill.)*

1. With all belts in their proper grooves, adjust the motor position to take up all the slack until the belts are fairly taut.

2. Run the fan and observe the belt shape. Continue to adjust until the belts have only a slight bow on the slack side of the drive while operating under load.

3. After 2 or 3 days of operation, the belts will seat themselves in the sheave grooves. Furthermore, a new belt may stretch 10 percent during the first month of operation. It therefore is necessary to readjust the belt tension until the drive again shows only a slight bow on the slack side. Once this has been done, the fan should operate satisfactorily with only an occasional recheck every few months.

LUBRICATION OF BEARINGS AND COUPLINGS

After the bearings and couplings have been aligned properly and the set screws have been tightened, their lubrication should be checked. (The motor bearings are prelubricated for 5 to 10 years.) The grease fittings should be wiped clean. The fan either should be run or slowly rotated by hand while the lubricating is done. The grease should be pumped in very slowly until a slight bead of grease forms around the bearing seal. Overgreasing, however, should be avoided, because it may cause the bearings to heat up. After proper lubrication, the fan is ready for operation. For a continuously running fan, the lubrication should be rechecked every 2 months. Bearings and couplings may heat up somewhat, but as long as the bare hand can be held on them briefly, the heat-up is acceptable.

VIBRATION

Vibration amplitudes are measured by electronic instruments in mils (1 mil = 0.001 in). They are checked on the outside of the bearings, in vertical, horizontal, and axial directions. A small vibration amplitude is unavoidable, but a larger amplitude would reduce bearing life and should be corrected. The acceptable amplitude depends on the fan speed. It becomes smaller as the speed increases. The following three formulas indicate the vibration amplitude V in mils for three conditions:

$$V = \frac{2865}{\text{rpm}} \quad \text{normal and acceptable} \tag{26.1}$$

$$V = \frac{4200}{\text{rpm}} \quad \text{alarm, potential hazard} \tag{26.2}$$

$$V = \frac{9550}{\text{rpm}} \quad \text{immediate shutdown} \tag{26.3}$$

Table 26.1 shows the vibration amplitudes calculated from the preceding formulas. Should excessive vibration develop, the following possible causes should be investigated:

1. Buildup of dirt or other foreign matter on the fan wheel might cause an imbalance.

2. The bolts on the housing, the bearings, or the motor might not be tight enough.

TABLE 26.1   Vibration Severity Chart

Maximum or design speed (rpm)	Vibration in mils (peak to peak)		
	Normal = $\dfrac{2865}{\text{rpm}}$	Alarm = $\dfrac{4200}{\text{rpm}}$	Shutdown = $\dfrac{9550}{\text{rpm}}$
400	7.1	10.5	23.9
600	4.8	7.0	15.9
800	3.6	5.3	11.9
1000	2.9	4.2	9.6
1200	2.4	3.5	7.8
1400	2.0	3.0	6.8
1600	1.8	2.6	6.0
1800	1.6	2.3	5.3
2000	1.4	2.1	4.8
2200	1.3	1.9	4.3
2400	1.2	1.8	4.0
2600	1.1	1.6	3.7
2800	1.0	1.5	3.4
3000	0.9	1.4	3.2
3200	0.9	1.3	3.0
3400	0.8	1.2	2.8
3600	0.8	1.2	2.7
3800	0.7	1.1	2.5
4000	0.7	1.1	2.4

Source: Chicago Blower Corporation, Glendale Heights, Ill.

3. The V-belt drive's alignment, the belt tension, or the balance of the sheaves may be inadequate.
4. The bearing locking collars may not be tight enough.
5. Check the set screws of the fan wheel.
6. Check for foreign matter that may have hit and damaged the fan wheel, the shaft, or the bearings.
7. Check whether the vibration may come from a source other than the fan. To do this, stop the fan and determine if the vibration still exists. Disconnect the motor from the fan and operate it by itself to determine if it produces any vibration.
8. Check whether there is sufficient clearance between the fan wheel and the inlet cone.

PROTECTION OF FAN WHILE NOT IN USE

If the fan is to remain idle and will be stored for an extended period, the exposed surfaces as well as the bearings and couplings should be protected. The shaft should be rotated periodically to prevent corrosion. The fan wheel should be blocked to prevent windmilling. Upon removal from storage, bearings should be regreased with an ample supply of fresh grease to purge and replace the old grease.

INDEX

ABOUT THE AUTHOR

Frank P. Bleier is a consulting engineer who specializes in the aerodynamic design of air moving equipment. He has done work for more than 130 manufacturers in the United States, Canada, and Europe, and has designed and tested over 800 fans of various types. Mr. Bleier was formerly director of research and development for the IIg Electric Ventilating Company in Chicago.